The Institute of British Geographers
Special Publications Series

27   Impacts of Sea-level Rise on European Coastal Lowlands

INTERNATIONAL UNION FOR QUATERNARY RESEARCH
UNION INTERNATIONAL POUR L'ETUDE DU QUATERNAIRE
INTERNATIONALE QUARTÄRVEREINIGUNG
МЕЖДУНАРОДНЫЙ СОЮЗ ПО ИЗУЧЕНИЮ ЧЕТВЕРТИЧНОГО ПЕРИОДА
UNION INTERNACIONAL PARA EL ESTUDIO DEL CUATERNARIO

Commission on Quaternary Shorelines
国际第四纪研究联合会   第四纪海岸线委员会

The Institute of British Geographers
Special Publications Series

EDITOR: N. J. Thrift
University of Bristol

18 Rivers: Environment, Process and Form
    Edited by Keith Richards
19 Technical Change and Industrial Policy
    Edited by Keith Chapman and Graham Humphrys
20 Sea-level changes
    Edited by Michael J. Tooley and Ian Shennan
21 The Changing Face of Cities: A Study of Development Cycles and Urban Form
    J. W. R. Whitehand
22 Population and Disaster
    Edited by John I. Clarke, Peter Curson, S. L. Kayasha and Prithvish Nag
23 Rural Change in Tropical Africa
    David Siddle and Kenneth Swindell
24 Teaching Geography in Higher Education
    John R. Gold, Alan Jenkins, Roger Lee, Janice Monk, Judith Riley, Ifan Shepherd and David Unwin
25 Wetlands: A Threatened Landscape
    Edited by Michael Williams
26 The Making of the Urban Landscape
    J. W. R. Whitehand
27 Impacts of Sea-level Rise on European Coastal Lowlands
    Edited by M. J. Tooley and S. Jelgersma

In preparation
   Salt Marshes and Coastal Wetlands
   Edited by D. R. Stoddart
   Demographic Patterns in the Past
   Edited by R. Smith
   The Geography of the Retailing Industry
   John A. Dawson and Leigh Sparks
   Asian-Pacific Urbanization
   Dean Forbes

   For a complete list see pp. 266–267

Impacts of Sea-level Rise on European Coastal Lowlands

Edited by
M. J. Tooley and S. Jelgersma

BLACKWELL
Oxford UK & Cambridge USA

Copyright © The Institute of British Geographers 1992

First published 1992
First published in USA 1992

Blackwell Publishers
108 Cowley Road
Oxford OX4 1JF
UK

238 Main Street, Suite 501
Cambridge, Massachusetts 02142
USA

All rights reserved. Except for the quotation of short passages for the purposes of criticism and review, no part may be reproduced, stored in a retrieval system, or transmitted, in any form or by any means, electronic, mechanical, photocopying, recording or otherwise, without the prior permission of the publisher.

Except in the United States of America, this book is sold subject to the condition that it shall not, by way of trade or otherwise, be lent, resold, hired out, or otherwise circulated without the publisher's prior consent in any form of binding or cover other than that in which it is published and without a similar condition including this condition being imposed on the subsequent purchaser.

British Library Cataloguing in Publication Data

A CIP catalogue record for this book is available from the British Library.

Library of Congress Cataloging in Publication Data

Impacts of sea-level rise on European coastal lowlands / edited by
  M. J. Tooley and S. Jelgersma.
      p. cm. — (The Institute of British Geographers special
  publication series ; 27)
    Includes bibliographical references and indexes.
    ISBN 0–631–18183–0 (alk. paper)
    1. Sea level—Europe. 2. Coast changes—Europe. I. Tooley, M.
  J. II. Jelgersma, Saskia, 1929– . III. Series: Special
  publications series (Institute of British Geographers) ; 26.
  GC89.I46 1992
  551.4′57′094—dc20   91–32557
                                                                CIP

Typeset in 11/13pt Plantin by TecSet Ltd, Wallington, Surrey.
Printed in Great Britain by T.J. Press Ltd, Padstow, Cornwall.

This book is printed on acid-free paper.

Contents

Foreword	vii
Preface	viii
Acknowledgements	xiii
List of Plates	xiv

1. Impacts of a Future Sea-level Rise on European Coastal Lowlands — 1
   S. Jelgersma and M. J. Tooley

2. Relative Rise of Mean Sea-levels in the Netherlands in Recent Times — 36
   J. van Malde

3. Vulnerability of the Belgian Coastal Lowlands to Future Sea-level Rise. — 56
   C. Baeteman, W. de Lannoy, R. Paepe and C. van Cauwenberghe

4. Impacts of Sea-level Rise on The Wash, United Kingdom — 72
   I. Shennan

5. Vulnerability of the Coastal Lowlands of the Netherlands to a Future Sea-level Rise — 94
   S. Jelgersma

6. Carbon Dioxide Increase, Sea-level Rise and Impacts on the Western Mediterranean: The Ebro Delta Case — 124
   M. G. Marino

7. Sea-level Changes and Impacts on the Rhône Delta Coastal Lowlands — 136
   A. L'Homer

8	The Coastline of the Gulf of Lions: Impact of a Warming of the Atmosphere in the Next Few Decades *J.-J. Corre*	153
9	The Impact of Climatic Changes and Sea-level Rise on Two Deltaic Lowlands of the Eastern Mediterranean *G. Sestini*	170
10	The Evolution of the Coastal Lowlands of Huelva and Cadiz (South-west Spain) during the Holocene *C. Zazo, C. J. Dabrio and J. L. Goy*	204
11	The Future of the European Coastal Lowlands *M. J. Tooley and S. Jelgersma*	218

The Contributors	251
Author Index	253
Subject and Place Index	257
Related Titles: List of IBG Special Publications	266

Foreword

Global climatic change and sea-level rise are the most striking manifestations of the impact of man on the ecological balance of our planet. Fortunately, the public and governments have realized that preventative action is the first priority. Actions to reduce greenhouse gas emissions have been initiated at all levels of society. But it will take time – maybe 100 years – to stabilize the concentration of these gases in the atmosphere. Meanwhile, a considerable change in global climates and sea-level rise seems inevitable.

Planning for change is the message. However, such planning requires detailed knowledge of past and continuing geological processes. This book describes these processes for the European coastal area. The two editors are world-renowned authorities in the field of sea-level changes and coastal zone responses over the last 10,000 years. In this book, they bring together the contributions of many experts on sea-level rise and its impacts on the European coastal lowlands.

The work has already proved to be invaluable for the formulation of the response strategies presented by the United Nations Intergovernmental Panel on Climate Change (IPCC) in its Coastal Zone Management Report of 1990.

I share the editors' conclusion that vulnerability assessment and coastal zone planning must be a priority issue for all coastal regions of the world, not only for reasons of climatic change but also in view of the increasing ecological stress and land use conflicts in the world's coastal areas.

It is my hope that this book will help us to reverse the trend of environmental degradation and assist us in laying a sound basis for a well-informed planning process.

Pier Vellinga
Co-ordinator, National Climate Programme
Ministry of Housing, Physical Planning and the Environment
The Hague, Netherlands

Preface

At the end of the International Conference on the assessment of the role of carbon dioxide and of other greenhouse gases in climate variations and associated impacts sponsored jointly by ICSU (International Council of Scientific Unions), UNEP (United Nations Environment Programme) and WMO (World Meteorological Organization) at Villach, Austria from 9 to 15 October 1985 several conclusions and recommendations were made. These can be summarized in the following ways:

1 Many long-term capital projects are arising from economic and social decisions being taken today. These include major resource management activities, such as irrigation and hydro-electric generation; agriculture; coastal engineering projects; energy resource planning. All of these are based on the assumption that past climatic data are a reliable guide to the future. This is no longer a sound assumption, because the increasing concentration of greenhouse gases is expected to cause significant warming of the global climate in the next century. It is a matter of urgency to refine estimates of future climatic conditions to improve decision-making.

2 Climatic change and sea-level rise due to greenhouse gases are closely linked to other major environmental issues, such as acid rain and threats to the earth's ozone shield, mostly the consequence of man's activities. Reduction of coal and oil use and energy conservation undertaken to reduce acid rain will also reduce the emission of greenhouse gases: a reduction in the release of the synthetic CFCs (chlorofluorocarbons) will help protect the ozone layer, and will also slow the rate of climatic change.

3 While some warming of climate now appears inevitable due to past actions, the rate and degree of future warming could be profoundly affected by governmental policies on energy conservation, use of fossil fuels and the emission of some greenhouse gases.

4 Whilst uncertainties remain, government and funding agencies should increase research support and focus efforts on crucial unsolved problems related to greenhouse gases and climate change. Priority should be given to national and international scientific programme initiatives, e.g. the World Climate Research Programme, biogeochemical cycling and tropospheric chemistry in the framework of the Global Change Programme and national climatic research programmes. Special emphasis should be placed on improved modelling of the ocean, cloud-radiation interactions and land surface processes.

5 Support for the analysis of policy and economic options should be increased by governments and funding agencies. The assessments should provide an analysis and evaluation of the widest possible range of social responses aimed at preventing, or adapting, to climatic change. Some analyses should be undertaken in a regional context to link available knowledge with economic decision-making and to characterize regional variability and adaptability to climatic change. Candidate regions may include the Amazon basin, the Indian subcontinent, Europe, the Arctic, the Zambezi basin and the Great Lakes of North America.

In order to implement one of these recommendations (5), the government of the Netherlands offered to act as host to the European Workshop on Interrelated Bioclimatic and Land-use Change in December 1986. The Dutch Minister for Housing, Physical Planning and Environment, Mr E. H. T. M. Nijpels, invited governments and related international organizations in the European region to be represented at the Workshop to discuss the consequences that climatic change and sea-level rise may have for societies in Europe and the research directions emerging from these changes.

The postponed meeting was held at the Leeuwenhorst Congress Centre, Noordwijkerhout, the Netherlands, from 17 to 21 October 1987 and was attended by eighty-one government representatives and scientists. The Workshop was structured in two plenary sessions (opening and closing) and two days of parallel sessions. Discussion papers were printed and available in ten volumes for the parallel sessions. These covered the following topics:

Volume A: Climate scenarios
Volume B: Palaeoclimatic data
Volume C: Impact of climatic change and CO_2 enrichment on exogenic processes and the biosphere
Volume D: Impact analysis of climatic change in the Fennoscandian part of the boreal and sub-arctic zone

Volume E: Impact analysis of climatic change in the central European lowlands
Volume F: Impact analysis of climatic change in the Mediterranean region
Volume G: Impact analysis of climatic change in the central European mountain ranges
Volume H: The impacts of a future rise in sea-level on the European coastal lowlands
Volume I: Sensitivity of natural ecosystems in Europe to climatic change
Volume J: Society's reaction to scientific information about greenhouse effects

Within each of the parallel sessions several papers were presented, as an example of which the session on 19 October 1987 on the impacts of a future rise in sea-level on the European coastal lowlands had the following papers:

Session A *Impacts of sea-level change on the southern North sea basin*

J. van Malde	Tide gauge measurements
C. Baeteman	The Belgian coast
I. Shennan and M. J. Tooley	The east coast of England
J. de Ronde	The Dutch coast
H. Streif	The coast of the German Bight

Session B *Impacts of sea-level changes on the Mediterranean coast*

M. Marino	The Ebro delta
A. L'Homer	The Rhône delta
G. Sestini	The eastern Mediterranean
C. Perissoratis and V. Zimianitis	Land-sea relationships on the northern Aegean coast: a case study and impacts of a 3 m rise of sea-level
N. El-Fishawi	The Nile delta

Session C *Impacts of sea-level changes on the Atlantic coast*

C. Zazo	The Guadalquivir delta

Report: conclusions and recommendation M. J. Tooley (Chairman)
and S. Jelgersma
(rapporteur)

Forty-two participants attended the three sessions altogether, more than any other parallel session, bearing witness to the importance attributed to the impacts of sea-level on the European coastal lowlands (Tooley and Jelgersma 1989).

It was the intention that all the papers, together with the reports, conclusions and recommendations, would be published shortly after the Workshop in 1989. In the event, only the *Final Report* was published (Kwadijk and de Boois 1989). At this late stage, the chairman and rapporteur sent a copy of the papers to Mr John Davey of Basil Blackwell, and, after an assessment, it was accepted for publication. However, as time had elapsed, the authors were invited to revise their contributions and take this opportunity to add new material. The Commission of the European Communities had approved the European Programme on Climatology and Natural Hazards (EPOCH I) which ran from 1986 until 1990 and included many projects on sea-level changes and impacts of sea-level rise. This was succeeded in 1991 by a two-year programme (EPOCH II) on climatic change, sea-level rise and associated impacts in Europe in which twenty-two institutes are collaborating within the European Communities. At the same time two international scientific groups were active in the European area – the International Union for Quaternary Research (INQUA) Commission on Quaternary Shorelines, and International Geological Correlation Programmes 200 and 274. Of the former, there are two relevant subcommissions – North-western Europe with Professor D. E. Smith as president and Dr C. Baeteman as secretary, and the Mediterranean and Black Seas with Dr C. Zazo as president and Dr A. Ulzega as secretary. One of the priorities of the INQUA Commission was the compilation of maps of selected populous coastal lowlands showing the impact of projected sea-level change to AD 2100 (Jelgersma and Tooley 1988), and this book is a contribution to the work of the Commission on Quaternary Shorelines during the inter-congress period.

It was gratifying that all but three of the contributors were able to take advantage of such work that was under way. In addition, the editors have delayed final submission until the publication of *Climate Change: The Intergovernmental Panel on Climate Change Scientific Assessment* (Houghton et al. 1990) in September 1990, in order to incorporate some of the conclusions into this text. Essentially, the views expressed are those current at the workshop in 1987, with some revisions in 1990.

Whilst there appears to be a consensus amongst informed scientific and political opinion of the risks involved in coastal residence, and the need to reduce the risk by restraining and redirecting development, the accelerating development of coastal lowlands of not only Europe but also the rest of the world portends a greater disruption to human society than would otherwise have occurred if these areas had been reserved for agriculture and wildlife. Clearly some overall plan for the rational development of coastal lowlands is required.

We should like to thank Mr Alex de Jong for his editorial assistance, and Mrs Lesley Yeung and Mrs Elizabeth Pearson for typing assistance. Saskia Jelgersma thanks the Director, Geological Survey of the Netherlands; and Michael Tooley, the Chairman, Board of Studies in Geography, University of Durham, for making available technical facilities to us. Assistance with proof reading was given by Mr David Bedlington, Dr J.B. Innes, Dr Antony Long and Mr Ian Sproxton.

REFERENCES

Houghton, J. T., Jenkins, G. J. and Ephraums, J. J. 1990: *Climate Change: The IPCC Scientific Assessment*. Cambridge, Cambridge University Press

Jelgersma, S. and Tooley, M. J. 1988: The INQUA Commission on Quaternary Shorelines. *Journal of Coastal Research* 4(2), 507–10

Kwadijk, J. and de Boois, H. 1989: *Final Report. European Workshop on Interrelated Bioclimatic and Land-use Changes*. Bilthoven, National Institute of Public Health and Environmental Protection

Tooley, M. J. and Jelgersma, S. 1989: The INQUA Commission on Quaternary Shorelines. *Journal of Coastal Research* 5(1), 166–9

Michael Tooley, Durham
Saskia Jelgersma, Haarlem

Acknowledgements

We acknowledge with gratitude permission to reproduce the following figures:

figures 1.3, 11.8, 11.9 and 11.10	Director, Mediterranean Blue Plan Regional Activity Centre
figure 4.6	Dr I. Shennan and Kluwer Academic Publishers, Dordrecht
figure 5.4	Professor R. G. West and Longmans, London
figure 11.1	Nature Conservancy Council, Peterborough
figure 11.2	Joint Nature Conservation Committee and the European Union for Coastal Conservation
figure 11.3	Dr H. Rohde and the German Committee for Coastal Engineering Research
figure 11.4	Professor C. Verlaque and Pergamon Press, Oxford

List of Colour Plates

Plate 1 The Tees coastal lowlands, North-east England, showing Seaton Carew nuclear power station, the initial capital investment of which was £91 million. Seal sands, between the inset dock for Phillips Petroleum and the power station, is a Site of Special Scientific Interest and is of international importance for its populations of breeding and over-wintering birds. The reclaimed tidal flats range in altitude from 2 to 5m above mean sea level.
(Photograph: Tees and Hartlepool Port Authority)

Plate 2 Dungeness nuclear power station in Kent, South-east England, after the storms of 27 February 1990. The initial capital investment for the two reactors was £165 million. The shingle foreland here is younger than AD 1596. The flinty shingle is eroded from the south shore of Dungeness and deposited on the east shore, requiring beach nourishment of 26,000m^3 and more in some years. The shingle ridges at 6m above mean sea level protect reclaimed tidal flats at 2–3m above mean sea level. High tides reach 3.5m and wave crests 6.8m above mean sea level.
(Photograph: Michael Tooley)

Plate 3 Bergen aan Zee, North Holland on 2 March 1990 after the storms in the North Sea of 25 January and 27 February 1990, when extreme still water levels reached c.+1.8m and +2.5m NAP and wave heights up to +9m NAP. The protecting dune ridge has been destroyed and peats formed in a freshwater slack in the dunes and dated to 600 BC are exposed in the intertidal zone. The dunes are over 30m high and about 5km wide here and protect the North Holland Canal and the reclaimed polderlands that lie below mean sea level.
(Photograph: Rijkswaterstaat: Public Works Department: North Sea Directorate)

Plate 4 The Hondsbossche Sea dyke beyond the high dunes north of Bergen aan Zee, North Holland, 2 March 1990. The reclaimed lagoonal deposits are now rich farmland and the small settlements of Groet and Petten are protected from inundation by the dunes and the sea embankment. At Petten on 27 February 1990, the still water level rose to +2.6m NAP and wave heights to + 8m NAP.
(Photograph: Rijkswaterstaat: Public Works Department: North Sea Directorate)

Plate 5 Callantsoog, North Holland on 2 March 1990 is located at mean sea level and has been relocated four times further landward as shoreline and dune erosion has occurred. The church dates from AD 1582. The town is protected by a single dune ridge, a narrow intertidal zone and a series of groynes. Erosion continues at 1.0 to 1.5m/year.
(Photograph: Rijkswaterstaat: Public Works Department: North Sea Directorate)

Plate 6 Nieuwpoort, Belgium, described in Baedeker's guide at the turn of the century as 'prettily situated and the most fashionable of the small Belgian sea-bathing resorts' has now become part of the 'Atlantic Wall' along the Belgian coast. The coast is subject to erosion from North sea storm surges and waves, and, although formerly possessing a natural wide dune sea defence rising to 30 metres, is now fronted by dykes protecting reclaimed farmland, close to or below mean sea level.
(Photograph: C. Baeteman)

Plate 7 Aigues-Mortes on the Rhône Delta in France is a medieval town of historical and architectural importance, lying a few metres above present-day sea level. It is a defensive town on a grid iron plan laid out by Louis IX in AD 1240. The Tower of Constance, in the foreground, overlooking the Canal Maritime, dates from AD 1241–1250, and the town walls from AD 1270. Beyond the town, towards the coast, is reclaimed farmland and the *salinas* in the Etang de la Ville.
(Photograph: Daniel Philippe)

Plate 8 The Po Delta, Italy, near the village of La Pila. The lobate delta of the River Po began to grow at an accelerated rate from the sixteenth century onwards, and the Pila branch has extended at a rate of 65m/year. The reclaimed delta is used for agriculture and industry.
(Photograph: Tappeiner Werbefoto)

1

Impacts of a Future Sea-level Rise on European Coastal Lowlands

S. Jelgersma and M. J. Tooley

1.1 INTRODUCTION

There is a consensus that the measured and projected increase of carbon dioxide and other radiatively active gases in the troposphere, the so-called 'greenhouse gases', will affect the radiation balance of the earth, causing climatic changes. The two main consequences will be a rise in the average global temperature, due to the trapping of an increasing proportion of longwave radiation leaving the earth, and a rise in sea-level, unevenly distributed, due to an increase in temperature of surface ocean water masses leading to expansion and some melting of valley glaciers and high latitude ice masses.

There is no debate among scientists about the existence of the 'greenhouse effect' and the enhancement resulting from the industrial and agricultural activity of man. There is, however, controversy about the following questions:

1. If the carbon dioxide (CO_2) content or its equivalent in the lower atmosphere is doubled, what will be the rise in temperature?
2. What will be the rate of temperature change and what will be the time lag between the increase in all the radiatively active gases (water vapour, carbon dioxide, ozone, nitrous oxide, methane and chlorofluorocarbons) and global temperature?

3 What feedbacks will there be to global climates of rising temperature and concomitant changes in precipitation, cloudiness, especially cirrus clouds, and cyclogenesis?
4 What impacts will there be on oceanic circulations, and what are the time lags?
5 At what rate will sea-level rise, and what is the range of variation of the impacts?
6 What are the time lags and relationships between changes in the volume of radiatively active gases, temperature and sea-level?

During the 1980s many workshops and conferences addressed these and other questions. At the Villach conference (Bolin et al. 1986), it was agreed that a doubling of CO_2 from pre-industrial levels would occur as early as AD 2030. The associated increase of global mean equilibrium surface temperature would be between 1.5°C and 4.5°C. At the 1987 First North American Conference on Preparing for Climate Change Hansen et al. (1988) argued that a doubled CO_2 level of forcing from 315 to 630 ppmv yielded an increase in global mean temperature of 1.25°C by AD 2030, assuming no climatic feedbacks. Schneider (1989) indicated that the doubling of CO_2 should occur between AD 2030 and AD 2080 with a concomitant increase of temperature of 3.0°C to 5.5°C. Working Group I of the Intergovernmental Panel on Climate Change (Houghton et al. 1990) estimated that by AD 2030 global mean temperature will rise 1.1°C above 1990 temperatures, and by AD 2090 by 3.3°C as CO_2 rises above its present-day level of 353 ppmv and other greenhouse gases increase by 0.2 and 0.5 per cent per annum.

There are many estimates of sea-level rise. Hoffman (1984) estimated that the range of sea-level rise scenarios by AD 2075 would be 38 to 212 cm. This was revised down to 36 to 191 cm by AD 2075 (Hoffman et al. 1986 in National Research Council (US) 1987). Robin (1986) calculated a rise of 25 to 165 cm by AD 2080, whereas Wigley and Raper's (1992) best guess is between 19 and 35 cm by AD 2030. For the IPCC 'Business-as-Usual' scenario for AD 2030 the estimated range of global mean sea-level rise is 8 to 29 cm, and for AD 2070, 21 to 71 cm (figure 1.1) (Warrick and Oerlemans 1990). The rates of sea-level rise for the Hoffman (1984) and IPCC (1990) scenarios are given in tables 1.1 and 1.2 respectively.

Variations in the sea-level rise estimates are due to the fact that the inputs to the models are uncertain and difficult to quantify. Robin (1986) expressed this problem elegantly: 'With our lack of knowledge of the hydrological cycle and the dynamics of the oceans and the polar ice sheets,

Impacts of Future Sea-level Rise

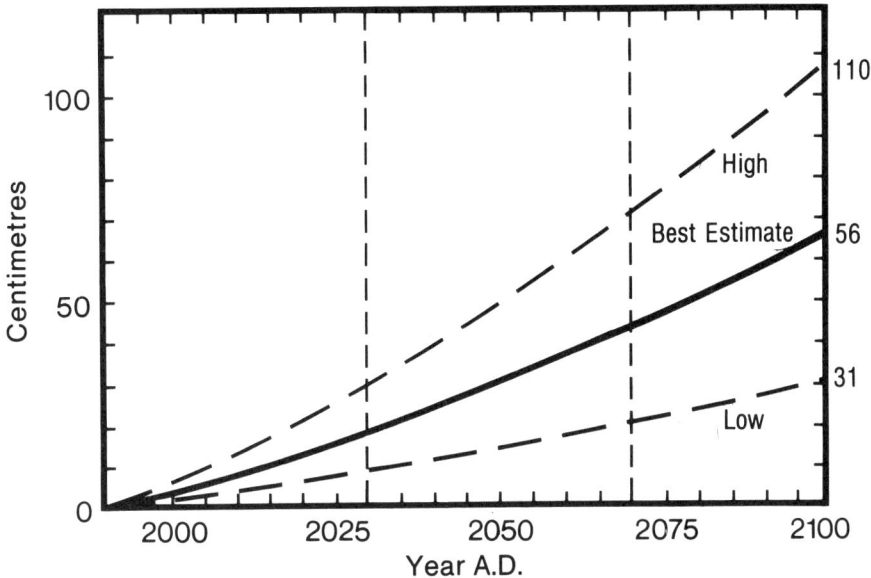

Figure 1.1 Predicted sea-level rise AD 1990–2100, showing the best estimate and the range for Scenario A, 'Business-as-Usual'. The 'B-a-U' scenario assumes that the energy supply is coal-intensive, modest increases in energy efficiency occur, carbon monoxide controls are modest, deforestation continues until all tropical forests are depleted, and agricultural emissions of methane and nitrous oxide are uncontrolled. World population was assumed to have risen to 10.5 billion by AD 2050. Economic growth was assumed to be between 2 and 5 per cent for at least the next ten years.
Source: (After Houghton et al. 1990).

Table 1.1 Estimated sea-level rise (in cm), AD 2000–2100 (after Hoffman 1984) (rate of rise of sea-level in mm/year in brackets)

Year AD	Conservative estimate	Mid-range low	Mid-range high	High estimate
2000	4.8 (3.0)	8.8 (5.5)	13.2 (8.2)	17.1 (10.6)
2025	13.0 (3.1)	26.2 (6.3)	39.3 (9.5)	54.9 (13.3)
2050	23.8 (3.6)	52.3 (7.9)	78.6 (11.9)	116.7 (17.6)
2075	38.0 (4.1)	91.2 (10.0)	136.8 (15.0)	212.7 (23.3)
2100	56.2 (4.8)	144.4 (12.4)	216.6 (18.6)	345.0 (29.7)

Table 1.2 Estimated sea-level rise (in cm) for policy scenario 'Business-as-Usual' to AD 2100 (after Warrick and Oerlemans 1990) (rate of rise of sea-level in mm/year in brackets)

Year AD	Low estimate	Best estimate	High estimate
2030	8 (2.0)	18 (4.5)	29 (7.2)
2070	21 (2.6)	44 (5.5)	71 (8.8)
2100	31 (2.8)	66 (6.0)	110 (10.0)

forecasting of global changes of sea level involves considerable extrapolation and speculation.'

There are also different and diverging estimates of the contribution of different factors to sea-level rise. These are summarized in the IPCC report (Warrick and Oerlemans 1990); and for the best estimate, the 'Business-as-Usual' scenario for AD 2030, of the 18.3 cm total sea-level rise estimated, 10.1 cm is accounted for by thermal expansion, 7.0 cm by the partial melting of mountain glaciers, 1.8 cm from the Greenland ice cap and −0.6 cm from Antarctica due to increased precipitation and ice accumulation there. No consideration has been given to variable rates of sea-level rise, for which there is abundant evidence during the Holocene and Eemian (Tooley 1989), or to the implications of the addition of 8 to 29 cm to the topography of the sea surface, for which there are many models (e.g. Clark and Primus 1987).

Whilst the amount and rate of future sea-level rise may be uncertain, it has been concluded (Warrick and Oerlemans 1990) that it will be greater than the rise of sea-level recorded from tide gauge data during the last 100 years by a factor of 3 to 6.

Estimates of global sea-level rise during the past 100 years range from 0.5 to 3.0 mm/year, with most estimates within the range 1.0 to 2.0 mm/year (Warrick and Oerlemans 1990). Within this envelope lies the estimate of Gornitz et al. (1982) of 1.2 mm/year, based on an analysis of 193 tide gauge stations using data assembled by the Permanent Service for Mean Sea Level at the Proudman Oceanographic Laboratory (formerly Institute of Oceanographic Sciences), Bidston, United Kingdom. Pirazzoli (1989) and Pirazzoli et al. (1989) have cautioned against acceptance of all the published rates of sea-level over the past 100 years and argue that if there is a global (eustatic) sea-level signal, it is weak and is obscured by tectonic, glacio-isostatic, oceanographic, anthropogenic and atmospheric effects. Removing land movements on the Atlantic coasts of Europe, sea-level rise has been calculated over the past 100 years by Pirazzoli (1989) to 0.4 to 0.6 mm/year.

However, taking the IPCC estimates and assuming a three- to six-fold increase in the next 100 years in the mean rate of sea-level rise (1.5 mm/year) over the last 100 years, yields rates of 4.5 to 9.0 mm/year (cf. tables 1 and 2). Woodworth (1990) has analysed the European tide gauge records and concluded that since AD 1870 there is no evidence of an acceleration of Mean Sea Level. However, an analysis of the four tide gauge stations with the longest records (Brest 1807, Sheerness 1834, Amsterdam 1700 and Stockholm 1774) showed a positive acceleration of Mean Sea Level of 0.4 mm/year/century. Using similar increases in Mean Sea Level as the IPCC 'Business-as-Usual' scenario during the twenty-first century, Woodworth (1990) concluded that accelerations at the 95 per cent confidence level would be observable by AD 2010.

These prognostications bring little comfort to about half the world's population, living as it does in subsiding deltaic areas, coastal lowlands and river flood plains. These extensive areas are already zones of intense economic, industrial and agricultural activity. This can be iterated from UNEP's (1990) report on *The State of the Marine Environment*: 'Half the world's population dwells in coastal regions which are already under great demographic pressure, and exposed to pollution, flooding, land subsidence and compaction, and to the effects of upland water diversion. A rise in sea level would have its most severe effects in low-lying coastal regions, beaches and wetlands.' Most of the world's coastlines are already subject to erosion, not only due to the measured increase in Mean Sea Level and associated High Water Mark, but also an increase in storminess in middle latitudes. Bird (1985), however, concluded that more than 70 per cent of sandy coastlines (which constitute 20 per cent of the world's coastlines) displayed erosion over the past few years, and put forward fourteen factors favouring an acceleration of beach erosion, of which relative sea-level rise was only one.

A complex of factors affects the balance of erosion and deposition along a coastline, and interference by man embanking, reclaiming and armouring the coastline for more than two millennia in Europe has disturbed this balance.

Natural processes of subsidence in deltaic areas and within sedimentary basins of continental shelf seas and adjacent land areas are exaggerated by land drainage, the extraction of ground water and the creation of polder land by reclamation of tidal flats. At Holme Fen in the English Fenlands, Hutchinson (1980) has demonstrated that the installation of more effective pumps between 1851 and 1962 has resulted in a reduction in ground altitudes from +1.6 m Ordnance Datum (O.D.) in 1848 to −2.3 m O.D. in 1978 due to dewatering of the peats, consolidation and peat wastage following oxidation. The rate of lowering was 30.5 mm/year, but the rate in the first five years of drainage was 220 mm/year. Holme Fen is some 45 km from the present embanked coastline of The Wash where Mean Water of High Spring Tides touches the coast at +3.9 m O.D. (Comparable data for the Netherlands can be found in Jelgersma, this volume). Hence, sea-level rise in the future will have an impact on coasts and coastal lowlands profoundly affected by man. The scale of impact is directly related to the extent and density of man's activities in these areas.

A context is provided for an assessment of the impact of future sea-level rise on coasts and coastal lowlands of Europe by considering first the impact of sea-level changes in the past on the environments of these areas.

1.2 THE ENVIRONMENTS OF EUROPEAN COASTAL LOWLANDS

Coastal lowlands are those areas where sediments are laid down both directly under the influence of the sea with its semi-diurnal tides and tidal currents, and indirectly where brackish water and freshwater backs up in lagoons and swamps. The characteristic sediments are peats, clays, silts and sands. The peats can be both autochthonous and allochthonous and can be further subdivided into detrital, limnic and *turfa* types. In general, these sediments occur below the spring tide levels, but are found at slightly higher altitudes around the margins of the coastal lowlands and along the river valleys that feed them. Locally, where ombrogenous peats developed following the climatic deterioration of the sub-Atlantic some 3,000 years ago, raised bogs developed, carrying ground surfaces 3 to 4 m higher, before drainage and reclamation. Within the oxidized remains of the ombrogenous peats and other organic deposits in the coastal lowlands dating from 4,000 years ago, coarse sand is often encountered within distinct strata (Tooley 1978a, b; 1990), bearing witness to episodes of sand blowing and dune accumulation on the coast. The origin, lithology, vegetational history and chronology of the coastal dunes in the Netherlands have been elaborated by Jelgersma et al. (1970), and subsequent studies (for example, reviewed in Bakker et al. 1990) support their conclusions for a cyclical pattern of dune building and erosion in Europe related to climatic and sea-level change.

The coastal dunes of western Europe (see figure 11.2) appear to have begun to accumulate in late neolithic times and to be anchored on coastal barriers. There are successions of dune ridges and high parabolic dunes that have developed along prograding coasts. Sand thicknesses 40 m and more have accumulated above high tide marks and provide massive though finite natural defences against storm surges and sea-level rise.

Within the sand thickness are organic beds and lenses, bearing witness to periods of dune stability, and in the Netherlands permitting a subdivision into an Old Dune and Young Dune chronostratigraphy (Jelgersma et al. 1970, 1979).

A model for the palaeogeography and environments of the Netherlands has been proposed by Hageman (1969) and revised by Jelgersma et al. (1979) and Zagwijn (1986). It can be applied with modifications to all the European coastal lowlands.

In addition to the diverse lithology of unconsolidated sediments in the coastal lowlands and dune fields along the coast, there is a great variety of micro-morphological features. Tidal, saltmarsh creeks with micro-levees of laminated silt and clay now choked with coarser clastic deposits and eroded organic deposits mark in their meandering courses former tidal flats and saltmarshes. Several generations, associated with specific sea level stands, overlie each other like a palimpsest in the coastal lowlands. In the Fenlands of East Anglia in England they have been described and mapped as roddens (Fowler 1934; Godwin 1938; Godwin 1978). They were the locus of prehistoric settlements and salt-making works throughout coastal Europe. In Germany, the relationship of human settlements to sea-level has been summarized by Behre et al. (1979) and Rohde (1978) and in the Netherlands have been described by Louwe Kooijmans (1974). In the lower river valleys, estuaries and distributaries several generations of river dunes or *donken* have formed, and provide an uncompressible basal sandy deposit against which buried topography sea-level changes during the Holocene can be measured (for example, Jelgersma 1961).

The maximum mean rate of rise of sea-level during the Holocene has been given as 22 mm/year (Kwadijk and de Boois 1989), but this conceals periods when sea-level rise accelerated, and was associated with catastrophic melting of high and middle latitude ice caps. In Denmark, Petersen (1981) has described the rate of rise of sea-level between 8,260 and 7,680 radiocarbon years ago to be 46 mm/year. In north-west England, rates of rise between 7820 and 7605 radiocarbon years ago ranged from 34 to 44 mm/year (Tooley 1974, 1978a, 1989, 1992). Similar rates have been recorded from the last interglacial (Eemian) in Europe (in Tooley 1989, Zagwijn 1983). In Germany, Streif (1990) has reported rates of sea-level rise of 40 mm/year for the early Eemian and 21 mm/year for the early Holocene between 8,600 and 7,100 years ago.

The consequences of these rapid rates of sea-level rise were extensive marine transgression of the landward margin of the continental shelf and the coastal lowlands, at rates ranging from 15 to 60 km/1,000 years or up to 60 m/year (Evans 1979). In Germany, the coastline advanced 250 km landward between 8,600 and 7,100 radiocarbon years ago as a consequence of a rise of sea-level of 30 m from 46 to 15 m below present sea-level (Streif 1989); that is more than 160 m/year.

Although none of the rates for the sea-level rise scenarios given in tables 1.1 and 1.2 approach the rates calculated from the recent geologic record, the implication of these lower rates derived from the sea-level rise scenarios is a landward migration of all the genetic zones of the coastal lowlands. In the case of the Fenlands and Wash in the United Kingdom, Shennan (1987 and chapter 4) has shown that a sea-level rise in the past of 5 mm/year is the critical rate tipping the balance from accretion to erosion and initiating shoreline retreat. Each coastal lowland will have a different threshold rate and this requires examination.

The shorelines and coastal lowlands of Europe (figure 1.2) can be divided into three geographical zones which will be treated separately: the Atlantic, North Sea and Irish Sea coasts; the Baltic Sea coasts; the Mediterranean Sea coasts.

The Atlantic, North Sea and Irish Sea coasts

The Quaternary history of these coasts has been reviewed by Oele et al. (1979), Kidson and Tooley (1977) and Romariz (1989). In this zone, there are no important deltas. Most rivers discharge into the sea via an estuarine tract, such as the Elbe, Weser and Ems estuaries in Germany, the combined Rhine–Meuse–Scheldt estuary in south-west Netherlands, the Severn and Thames estuaries in England, the Shannon estuary in Ireland, the Tajo estuary in Portugal and the Guadalquivir estuary in Spain. On open low-lying coasts with an adequate sediment supply, fresh, brackish and saline lagoons have developed behind beach barriers and dune systems. Whilst many of these have been reclaimed for agriculture, and residential and industrial development around the shores of the southern North Sea and locally around the Irish Sea, the Atlantic seaboard retains many examples, a few of which have been investigated. In Ireland, Ballycotton Bay and Lady's Island Lake are examples on the south coast (Carter et al. 1989). On the west coast of Portugal many unreclaimed lagoons exist, for example the Albufeira, Obidos, St Andre, Melides and Aveiro lagoons. Of these, the Aveiro lagoon occupies some 4700 ha at high tide (Gomes 1989). On the Algarve coast, many lagoons such as those at Faro (Andrade 1989) exist.

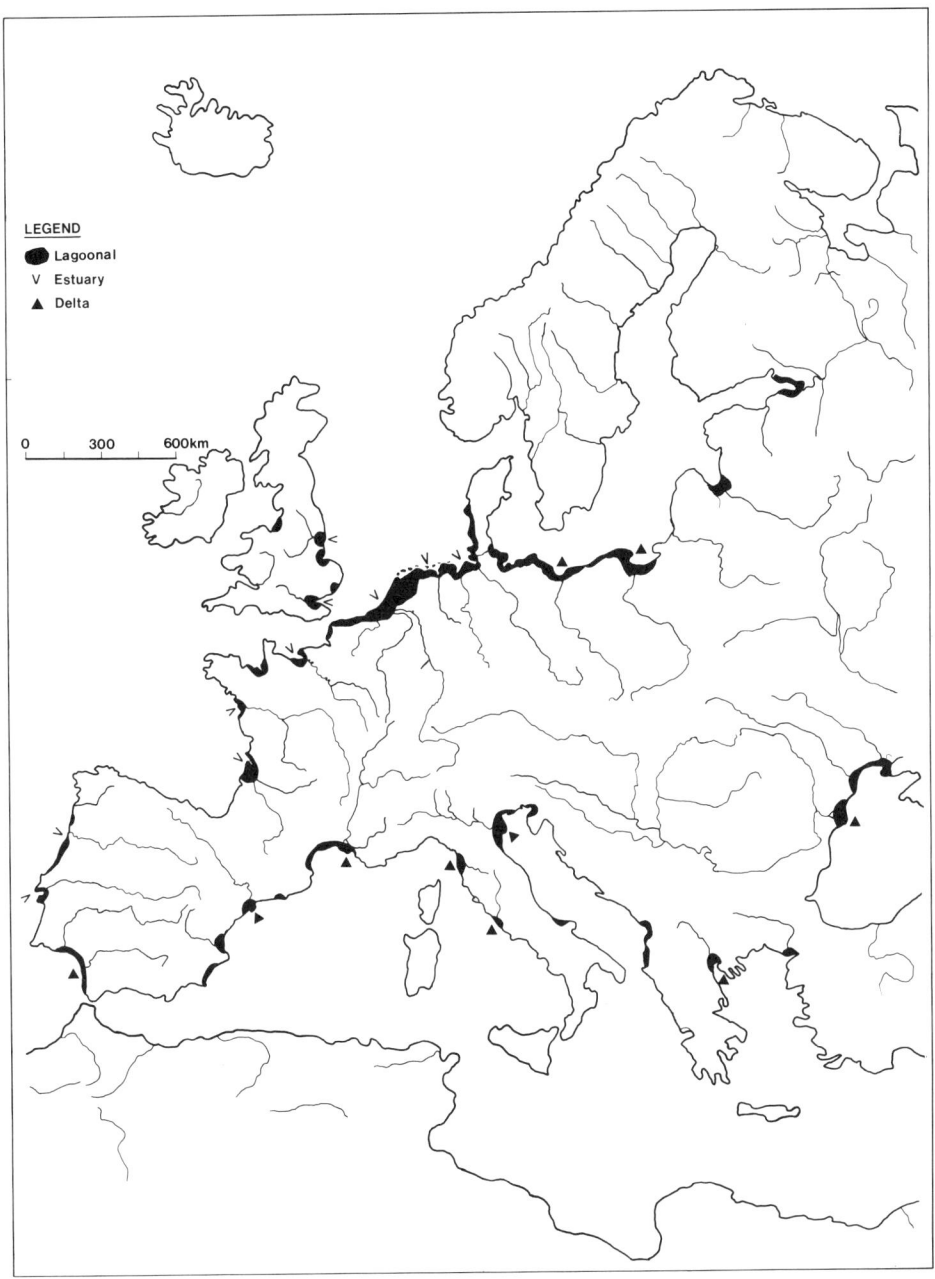

Figure 1.2 Map of Europe, showing diagrammatically the distribution of the main coastal lowlands with tidal flat and lagoonal sediment infills, estuaries and deltas

The prevailing wind direction is from the west, south-west and north-west and the Atlantic seaboard is exposed directly to a storm wave environment. Due to the configuration of the epicontinental seas off the European mainland, the North Sea, Irish Sea, Celtic Sea and English Channel are affected by storm surges and extreme water levels are an aperiodic phenomenon. During the infamous 1953 storm surge in the North Sea water levels were elevated up to 3 m above predicted levels. In the Irish Sea in 1977, the storm surge added up to 2.5 m to predicted levels and the extreme water level intersected the embanked coast at Preston, on the Ribble estuary at +7.5 m OD (Graff 1978); on the south side of Morecambe Bay some 2875 ha of agricultural land were flooded or waterlogged up to +5.5 m OD. A storm surge on the North Wales coast in February 1990 carried water levels to +5.4 m OD and 2 km landward and wave heights over 6 m were recorded (Englefield et al. 1990). In the storms of January and February 1990, the Dutch coast was eroded by waves 8 to 9 metres above NAP (see plates 3 and 4). Such extreme water levels are not unique, and appear to occur in temporal clusters. The historic record of storm surges has been summarized for the North Sea by Gottschalk (1971, 1975, 1977, for the Netherlands. The consequence of the storms was shoreline retreat and the relocation of settlements. See plate 5.) and by Lamb (1980) and for north-west Europe by Lamb (1991). On the German coast, Rohde (1978) has tabulated the storm surges that have affected this coast from the twelfth century onwards, and shown that from the sixteenth century onwards there has been a rise in mean water levels of c. 25 cm/century due in part to embanking. Consequently, extreme water levels during storm surges have risen progressively. Hofstede (1990) has summarized a more complex pattern of fluctuations of Mean High Water in the inner German Bight and correlated them with the Medieval Warm Period and the Little Ice Age. Streif (1989) also notes a lowering of Mean High Water between AD 1125/1395 and AD 1650 which overlaps with the Little Ice Age. Embankments on the German coast have been increased in altitude from 2 to 9 m and cross-sectional widths from 4 to 70 m (Kramer 1978).

Although Davies (1980) has categorized the coasts of western Europe as macrotidal with a storm wave environment, there are significant geographical variations (Kidson 1982). Macrotidal conditions (>4 m range) prevail in the English Channel and extreme tidal ranges are found in Mount St Michael's Bay (12 m) and in the Bristol Channel (9 m). The range of astronomical tides is much greater: at Avonmouth in the Severn estuary it is 13.8 m whereas at Lisbon it is 4.2 m.

The coasts of the British Isles, France, Belgium, the Netherlands, Germany, Denmark and Norway have been affected both directly and indirectly by vertical land movements associated with glaciation and deglaciation. Using a sidereal timescale, Shennan (1987) has computed for the past 8,000 years rates of uplift in formerly glaciated areas and of subsidence in marginally glaciated or unglaciated coasts around the North Sea. In Germany, subsidence rates of 0.6 mm/yr have been calculated; in the Netherlands 0.4 to 0.6 mm/yr; in the Thames estuary 1 mm/yr. Uplift rates range from 1.9 mm/yr in Scotland to 1.7 mm/yr in Norway. In Germany, Streif (1990) attributes 95 per cent of the Holocene sea-level rise of about 100 m to eustatic factors and 5 per cent to tectonic subsidence, isostatic movements, *inter alia*.

Although uplift and subsidence rates are portrayed as smooth and continuous, there is some evidence (for example from the Forth valley of Scotland, Sissons, 1972, 1983; from Sweden, Mörner 1979) of the dislocation and non-uniform uplift or subsidence of raised and buried beaches following their formation. The processes include not only the vertical movements involved, but also water and sediment loading. In addition seismic activity may have resulted in the reactivation of fault lines.

Seismicity is not a process usually associated with the passive plate margin of north-west Europe, and yet there are well-documented events such as the London earthquake of 6 April 1580 (Neilson et al. 1984) and the Lisbon earthquake of 1 November 1755 (Reid 1914). The former is thought to have been caused by a movement along a fault of the Variscan trend at a depth of *c*. 33 m with an epicentre offshore in the Strait of Dover, and damage points to an intensity of between 6.2 and 6.9 on the Richter Scale. Other marked seismic events in south-east England have been identified in AD 1133, 1247, 1382, 1449, 1692 and 1938 and the epicentres of the last two events were located in the Low Countries (Nielson et al. 1984). Seismic risk in the North Sea has been reviewed by Ritsema and Grupiner (1983).

The consequence of seismic activity is the generation of tsunamis. The primary wave 13 m high struck Lisbon and secondary waves 2 to 3 m high affected the English Channel coasts after the Lisbon earthquake of 1755. The earthquake-induced Storrega slide off the Norwegian coast may have resulted in a seismic sea wave which impacted on the coast of eastern Scotland 7,000 years ago, laying down an extensive grey micaceous, silty fine sand layer (Long et al. 1989, 1990).

The Baltic Sea coasts

Tidal variations are slight and measure only a few centimetres. The greatest variations are in the large gulfs feeding into the Baltic. In the Gulf of Finland and the Bay of Kronstadt further east the range is from 8 to 18 cm (Lisitzin 1974). Extreme water levels are associated with storms. In the

Baltic, there appears to be a marked cyclicity of storms, during which water levels can be raised by up to 2 m. Along the Polish coast storm surges occur about fifteen times a century, raising water levels by 2.0 to 2.5 m and leading to cliff recession of 0.6 to 1.0 m/year (Borowkwa 1985). During the winter months, the presence of pack ice can have two effects: firstly water levels are not raised as the result of wind stress, and secondly as the ice breaks up it is driven several tens of metres landward (Alestalo 1985) eroding saltmarshes and moving large boulders. Much erosion of till cliffs and dune coasts is recorded under these conditions that occur every eleven to twenty-five years.

The whole area has been affected by glaciation, the stages of which have been well reviewed (Gudelis and Königsson 1980; Eronen 1983; Donner and Raukas 1988). The consequences are still manifest at present, with uplift in the northern part of the Bothnian Gulf and subsidence along the German and Polish coasts. In the past 9,000 years the maximum amount of uplift has been about 280 m (Eronen 1983). Uplift has not been uniform, and irregularities in Sweden (Mörner 1979) and throughout Fenno-Scandinavia (Kakkuri 1987) have been noted. Indeed, Kakkuri (1987) concludes that there is a tectonic component in the uplift phenomena, and notes the record of seismic events in the area over the past 500 years. At the head of the Bothnian Gulf rates of uplift are 9 mm/year and shorelines are advancing at rates exceeding 1 m/year (Normann 1985). The consequences in both Sweden and Finland are the emergence of new islands, the rapid emergence of new land, land ownership problems and migrating harbours. Large islands in the Gulf of Bothnia, such as Bergö, Replot and Björköby have increased in size by one third since the mid-eighteenth century. Pori in Finland was founded as a port in AD 1000 but land uplift and rapid sedimentation have left it more than 40 km from the sea (Palomäki 1987). At present, the zero isobase separating rising land to the north and subsiding land to the south (see map in Jelgersma 1979) bisects Scania, South Sweden and the Nemunas delta in Lithuania, passing then through Riga (Kakkuri 1987). Further south, rates of subsidence exceed 1 mm/year. In the Koszalin Gulf in Poland subsidence rates are 1.8 mm/yr, compared with 0.7 mm/year in the Pomeranian Gulf (Borowkwa 1985). The multi-barred and dune coastline of Poland, backed by extensive lagoon and marshes is a manifestation of these long-term movements together with sediment supplied by the River Vistula discharging into the Gulf of Gdansk and sediment released during storm surges in this part of the Baltic Sea.

The Mediterranean coast

Tides and tidal currents play a more significant part in coastal processes in the Mediterranean than in the Baltic Sea. They attain their greatest amplitudes (170 cm) at the heads of gulfs and seas, but are usually only 20 to 30 cm. However, during winter storms, extreme water levels and storm waves are recorded: for example, Corre (chapter 8) notes extreme water levels of 1.80 m and waves with crest heights of 5 to 10 m with onshore winds on the Rhône delta.

Coasts and coastal processes are profoundly affected by tectonic activity, and the Mediterranean is characterized by earthquakes, volcanoes and earth movements, which have resulted in both uplift and subsidence of coastal landforms and sediments and of archaeological sites (Flemming 1978). Pirazzoli (1986) has drawn attention to the abruptness of these movements and described the 'Early Byzantine Tectonic Paroxysm' that occurred about 1530 BP and affected localities 1,200 km apart. Locally, the western part of Crete was elevated 10 m by this event, probably attributable to collision processes between the African and Eurasian plates.

The Mediterranean Sea comprises a series of basins with water depths exceeding 3,000 m. Subsiding sedimentary basins are filled with terrigenous sediments 10 km or so thick that have accumulated over the past 5my at a rate of *c.* 0.3 mm/year (Pirazzoli 1987). The Mediterranean coast is also characterized by large deltas (figure 1.3) such as the Ebro delta in Spain, the Rhône delta in France, the Po delta in Italy and the Nile delta in Egypt (see chapters 6 to 9). These great deltas lie on top of the sedimentary basins and are subsiding as the result of lithospheric processes, sediment and water loading.

Sea-level changes in the Mediterranean are manifest by marine eroded features, such as rock platforms, notches and sea-caves, which are well developed in the widely distributed limestone. Beachrock also occurs and serves as a sea-level indicator, as do subfossil marine organisms such as barnacles and Vermetids (Pirazzoli 1987).

The Mediterranean was, until the early 1960s, the type area for sea-level stands during the interglacial ages of the Quaternary. The extreme tectonic activity and block faulting over extensive areas led Hey (1978) to conclude that all the Mediterranean shorelines had been affected, and Pirazzoli (1987) argued that without more unequivocal dating, only the last interglacial (Tyrrhenian) and Holocene beaches could be discussed regionally.

Rates of sedimentation in coastal lagoons have increased from prehistoric times onwards as the result of forest clearance by man. Eisma (1978) has described the seaward advance of the Küçük Menderes delta, and the

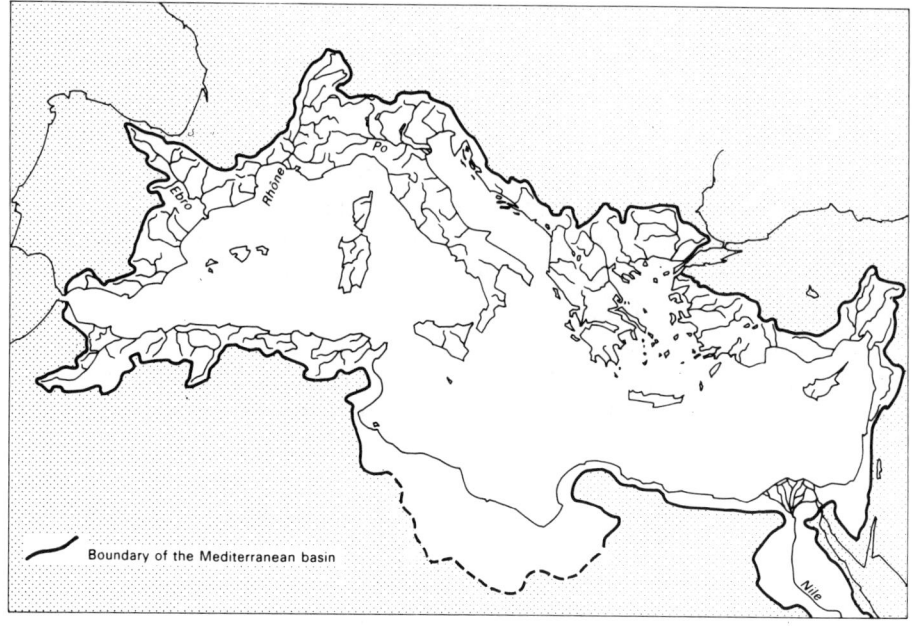

Figure 1.3 The Mediterranean region, showing the catchment of the basin (Grenon and Batisse 1989), but excluding the Nile and Black Sea catchments. All the river deltas will become increasingly at risk from erosion and inundation as sea-level rises. (Details of the Ebro, Po and Nile deltas are given in chapters 6 and 9. See also figure 9.1.)
Source: Grenon and Batisse (1989). (Reproduced by permission of the Director, Mediterranean Blue Plan Regional Activity Centre)

silting up the inner and outer part of Ephesus between 750 BC and AD 726 and explained it as a result of soil erosion in the catchment. Although Tziavos and Kraft (1985) regard the Marathon Bay in Greece to be an enigma with its strong littoral drift and beach accretion plain forming-barrier between the Aegean Sea and the swamps and marshes of the Plain of Marathon, it is not an uncommon association in the Mediterranean (for example, the Ebro delta, described in chapter 6 and the Po delta described in chapter 9), or, for that matter throughout coastal Europe. Baeteman (1985) has recorded a 7 m thick accumulation of lagoonal muds, carbonate muds, marine sands and intercalated peats, of which over 4 m has accumulated in the past 5,000 years. Brückner (1986) attributes coastal sedimentation in the Mediterranean in part to climatic and sea-level changes, and in part to man's impact clearing forests and agriculture.

1.3 LAND USE IN THE COASTAL LOWLANDS

In Europe, most of the coastal lowlands are used or are strongly influenced by man. The only extensive natural areas left are the tidal flats of the Dutch, German and Danish Wadden Sea, The Wash in England, and some of the lagoons and deltas of the Mediterranean. Since at least Roman times, and subsequently on an increasing scale, 'dredge, drain and reclaim' has been the slogan. European coastal lowlands have been reclaimed for agriculture, industry and housing. The drainage of these coastal wetlands has caused land subsidence due to compaction of unconsolidated sediments and oxidation of peat soils. In some areas compaction has been so great that ground levels are lying now well below mean sea-level. Accordingly they have to be drained by pumping and many areas are protected by dikes against storm surges and high tides (see Jelgersma, chapter 5). Another side effect of this intense drainage is the intrusion of salt water due to pumping in areas now below sea-level. In a number of estuaries salt water moves inwards due to dredging in connection with port development (Jelgersma, chapter 5). The rivers of the coastal lowland are dammed, embankments are constructed and dredging of sand takes place either for the purpose of sand mining or for shipping. The construction of dams and reservoirs in the river valleys to generate electricity and water for agriculture has strongly reduced the sediment load of rivers. Accordingly less sediment is supplied to the mouth of the rivers to build up shorelines and wetlands. As a consequence serious coastal erosion occurs in the Ebro, Rhône and Po delta regions (Marino, Corre and Sestini, chapters 6 to 9).

In the coastal area itself barrier islands and dune areas are extensively used for recreation and for pumping groundwater for drinking water, especially in the Netherlands. Many of the European coastal lowlands are in critical balance with the present sea-level and are in great danger of flooding if storm surges occur. Important parts of the shorelines are affected by erosion especially during storm surges. At places where shorelines are eroding, stone jetties and concrete or wooden groynes are built to lessen sand drift. For example, on the Lincolnshire coast of Great Britain, Shennan (1988) has noted an increase in the number of groynes between 1953 and 1981 from 82 to 263, and some 22.4 km of coast thus protected. This method of shoreline protection is debatable as it seems to be that as a reaction to these artificial works erosion on the rest of the unprotected shoreline is increasing. Beach nourishment (using sand dredged from the continental shelf) seems to be a more successful method. The erosion observed on the shorelines is for a major part thought to be caused by human activities, the observed relative sea-level rise by means of tide gauge readings is less likely to be the major erosion agent although it is a contributory factor.

On the map of Europe (figure 1.4) the concentrations of population and industrial activities in the coastal lowlands and their hinterlands are clearly shown. In Spain, Cendrero and Charlier (1989) have described the decrease in population of the interior provinces and an increase in population in the coastal provinces. Of the four million people living in the Cantabrian region, three million live in a 15 km wide coastal strip. Here the coastal landscape has been transformed: only 12 per cent of the natural vegetation remains and 50 per cent of the estuaries, intertidal flats and coastal wetlands have been reclaimed in the past.

More details about human activities, land use and vulnerability of flooding especially arising from a rise in sea level can be given on smaller-scale maps from the individual European countries. An example is the maps (figures 1.5a,b,c) of the southern North Sea Basin. In this region about 20 million people live below the high tide level. The economic value of these low-lying areas is enormous. They are also a gateway to industrial areas in the hinterland. Figure 1.5a gives the distribution of the coastal lowlands, the coastal dunes, estuaries, deltas and wadden (tidal flat) areas.

Figure 1.5b indicates the concentration of population in the low areas. Nuclear power stations lying near the high water mark are also shown. Plates 1 and 2 show, respectively, the nuclear power stations of Seaton Carew in the Tees coastal lowlands and of Dungeness on the south coast of England. Figure 1.5c indicates important industrial areas in the lowlands, and also the oil and gas fields of the southern North Sea Basin are shown.

Pipelines of oil and gas to the shore and oil harbours and refineries are indicated. It may be concluded that the concentration of all kinds of industries and the dense population in the coastal lowlands surrounding the southern North Sea Basin make these areas very vulnerable to a rise of sea-level.

1.4 EFFECTS OF A FUTURE SEA-LEVEL RISE

As mentioned in the introduction, estimates of the rise in sea-level in the next 100 years show a range from 0.48 to 3.45 m. In our study of the European coastal lowlands we assume a rise in sea-level of 1 m in the next

Figure 1.4 Map of Europe, showing diagrammatically industrial and population concentrations on the coast and the hinterlands

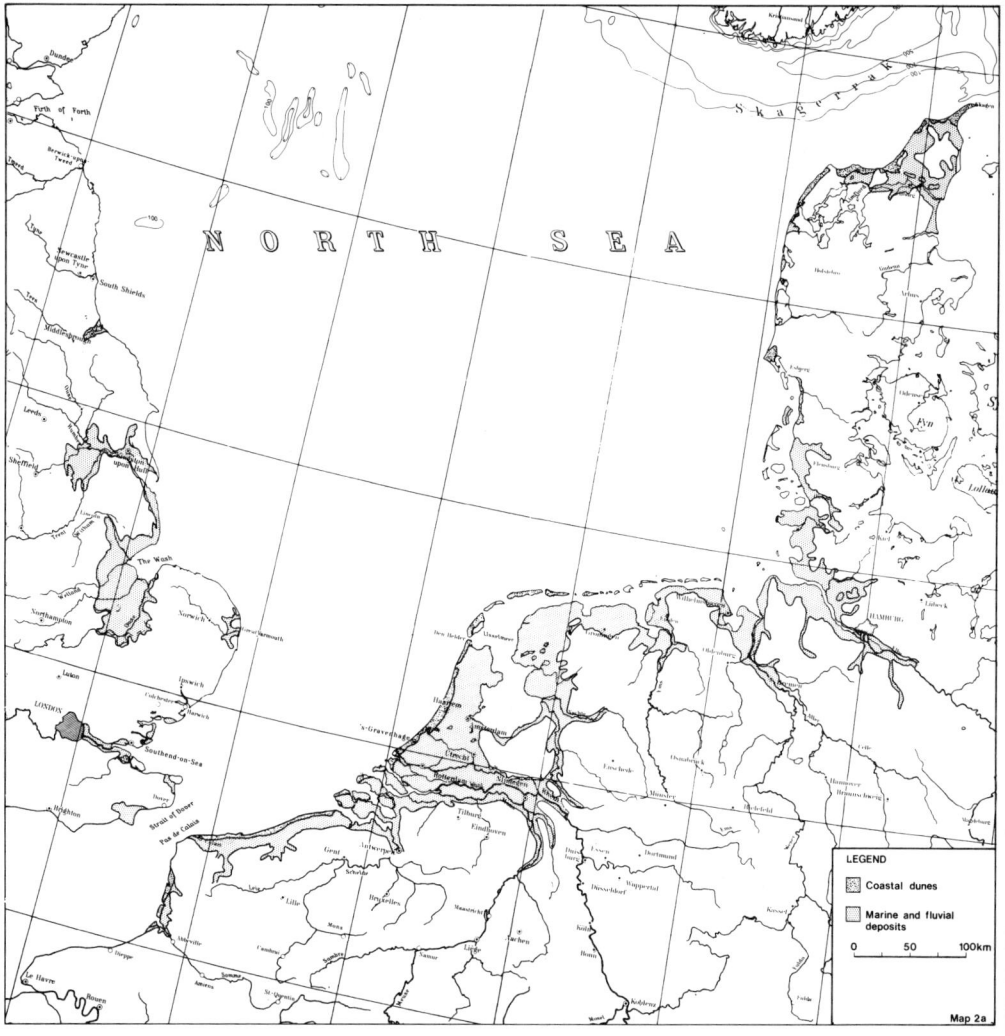

Figure 1.5a The southern North Sea: distribution of coastal dunes, marine and fluvial deposits, including salt marshes

100 years. General remarks of the effect of such a future rise in sea-level are given below. The predicted sea-level rise will:

- increase the risk of inundating reclaimed lowlands;
- accelerate the rate at which coastal erosion is already occurring;
- increase the risk of flood disasters from an increased incidence of storm surges;
- create problems with respect to drainage systems;

Impacts of Future Sea-level Rise 19

Figure 1.5b The southern North Sea: distribution of population centres and location of nuclear power stations

- increase salt water intrusion into ground water, rivers, bays and farmland;
- damage port facilities and coastal protection structures and interfere with navigation;
- destroy high quality farmland;
- disrupt fisheries and natural habitats and disrupt the ecological balance in coastal systems;
- result in the loss of recreational beaches;

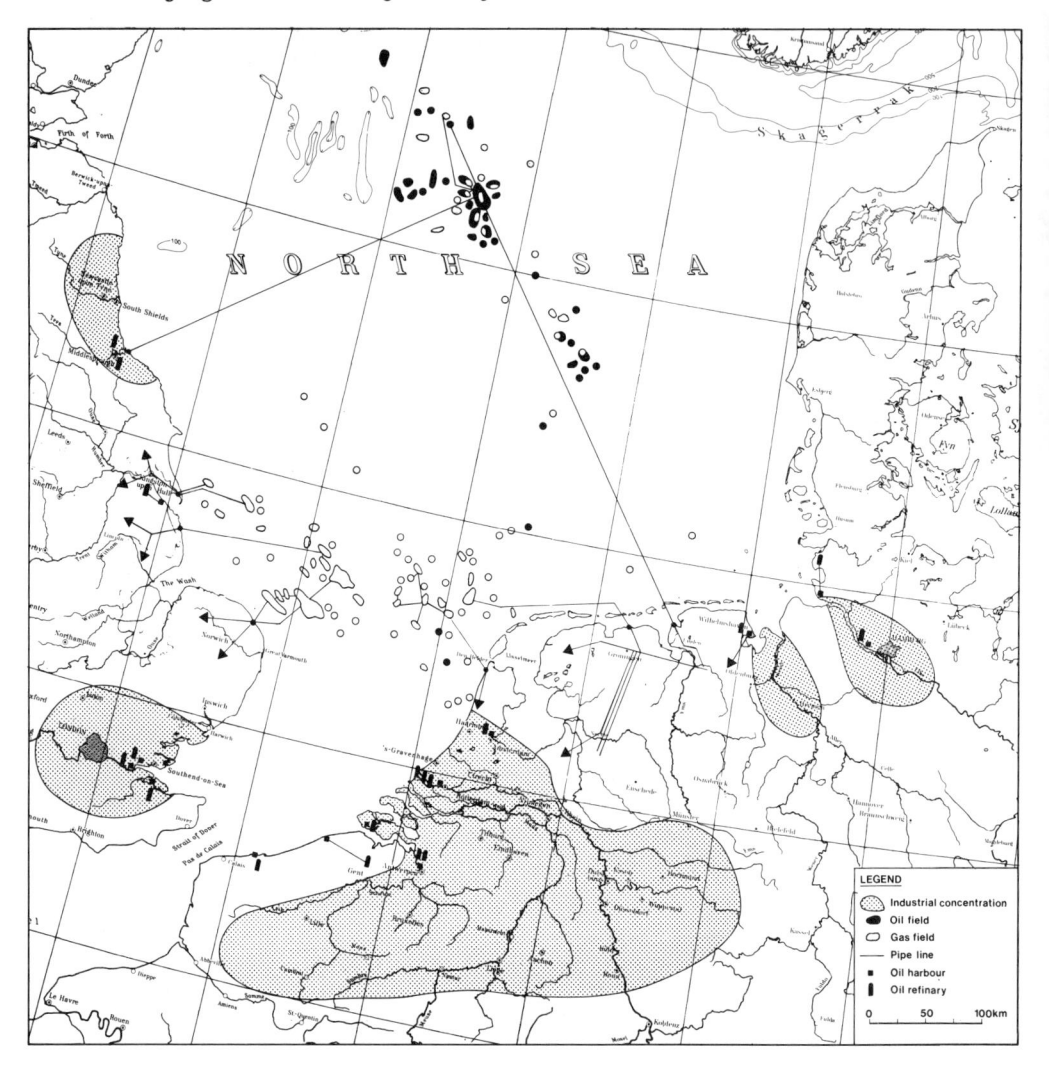

Figure 1.5c The southern North Sea: distribution of heavily industrialized areas, oil and gas fields and associated services

- shift sedimentation in rivers more up stream;
- remobilize toxic waste in landfill sites in coastal lowlands;
- remobilize radionuclides trapped in saltmarsh and fine-grained intertidal sediments.

Another important factor that can influence the coastal areas is how the climate will change due to the predicted rise in temperature. If the weather pattern and climate change, will the run-off from rivers increase in winter and decrease in summer? The increased run-off can give problems for the embankments. A decreased run-off can cause important salt water intrusion in the upstream area and reduce sediment loads and affect coastal sediment budgets. In the Netherlands (Jongman 1987), drinking water is in part extracted from the groundwater resources, which in the coastal provinces is brackish. Only in the dune area, an extensive freshwater lense was available, floating on salt water, but this resource has been strongly reduced due to overexploitation in the last decades. The majority of the drinking water, industrial and irrigation water is, however, derived from Lake Ijssel. Depleted groundwater is replenished from pretreated Rhine water infiltrated in the dunes south of Haarlem. Rotterdam's drinking water is purified Rhine/Meuse water. In 1984, 1334 million m^3 were extracted of which 434 million m^3 were surface water, and in 1995, the estimated extraction values are 1536 million and 465 million respectively. As mean sea-level has risen on the Dutch coast (see chapter 2), so salt water intrusion has increased to the extent that the intake of fresh water supply in Rotterdam Waterway has had to be moved gradually 20 km further upstream. The situation has been aggravated by the construction of deeper entrance channels, larger locks and bigger harbours (Colenbrander 1986). With climatic change, and sea-level rise consequences of the greenhouse effect, it is expected that river discharges will decline, the saline effect will increase and there will be shortages of potable water for drinking and irrigation.

Another unknown aspect is that the tidal range along the coast of Europe can change. Tides are influenced by dynamic factors involved in atmospheric–ocean interactions that are related to climatic change. The coastal morphology is in certain ways related to the tidal range. Changes in this range can have important impacts on the coastlines. Another aspect of a change in climate can be an increase in storminess along certain parts of the coasts of Europe. It is well known that the most important damage on coastlines occurs during storm surges at the time of high tide. If this increase in storminess should occur along the coasts of the Southern Bight of the North Sea the results in the coastal lowlands, with a high concentration of population and industry, would be disastrous. Rossiter (1962) indicated that a rise of sea-level as little as 15cm would double the probability of storm surges exceeding danger level on the North Sea coast of England and treble the probability on the Irish Sea coast.

1.5   RESPONSE TO SEA-LEVEL RISE

It must be stressed that Europe, a developed part of the world, will have the organization, the technology and resources to counter the effects of a future sea-level rise. Less developed countries lack these resources. The question arises: how will the predicted rise in sea-level be managed? Either the lowlands can be defended or present activities and developments can be checked and moved landwards. The protection can be done by dikes, sea walls, beach nourishment and other engineering solutions. It must be realized that economic and environmental impacts can make such a protection strategy unacceptable. All coastal cities and the sites of strategic industries will require enhanced protection. An example can be given of Hamburg in Germany, which, although 120 km from the North Sea is located at low altitude and is at risk from rising sea-levels, rising tidal levels and extreme water levels during storm surges. Ascher (1989) has described the breaching of sea defences in 1962 when 16 per cent of the city was flooded, 300 people died and damage to property amounted to DM100 million. Between 1962 and 1989, DM1,700 million have been expended on improving sea defences up to +9 m Mean Sea Level. Nevertheless, 180,000 residents and 140,000 employees are at risk, and the value of property likely to be affected by sea flooding amounts to DM1,6000 million. At least DM3,000–4,000 million will have to be expended raising sea embankments and other protection over the next twenty-five years, and this is for one coastal city alone. Abandoning land in coastal lowlands and moving present activities landwards to sites 5 m above Mean Sea Level will also have a serious economic and social effects. Nuclear power station sites cannot be abandoned and the sites require enhanced protection. The proposal that decommissioning could involve pumping sand over the plant and landscaping it (Passant 1990) is not viable given the instability of coastal environments.

Sea-level rise and the implementation of response strategies will have serious effects on the individual, the regional, the national and international economic levels. Impacts on real income include the loss of production from land and seas as well as the effects of employment changes from reconstruction. Migration of people and enterprises will disrupt the existing economic and social structures.

In the Netherlands some calculations have been made about the costs involved to defend the country against a future sea-level rise. These calculations made by Goemans (1986) and the Ministry of Transport and Public Works (discussion paper for the government) are given by Jelgersma (chapter 5 and below, p. 118).

1.6 THE ANALYSES OF RISK FACTORS

One of the recommendations of the European Workshop on Interrelated Bioclimatic and Land-use Changes, 1987, at Noordwijkerhout in the session on the impact of a future rise in sea level on the European coastal lowlands (Kwadijk and de Boois 1989 and chapter 11) was the development of objective criteria to be applied to landuse planning in the coastal lowlands. The production of maps of coastal lowlands showing high, intermediate and low risk zones, as recommended twenty years ago (Tooley 1971, and, again, 1979), but now in relation to the expected sea-level rise in the next hundred years was also recommended.

An inescapable fact about the coastal lowlands of Europe is that they are already at risk, given that their topography is low, invariably below +5 m Mean Sea Level and often well below mean sea-level itself. The consequences of land drainage and reclamation are sediment consolidation and a man-induced reduction of ground altitudes enlarging the areas at risk, and funded by grants from national governments and the European Commission. Given the present extreme water levels during storm surges and extreme wave heights, without the natural and man-made sea defences, the sea water depths could exceed 5 m to 8 m in places some 45 km and 100 km landward of the present coastlines of The Wash, England and the Netherlands respectively.

A rise in sea-level due to the enhanced greenhouse effect and regional subsidence in all the deltaic areas of Europe will increase the risks through an increase in the number and magnitude of storm surges and long-period inundation.

The first requirements before any restrictive or permissive planning decisions are taken at local, regional, national or supranational (European Parliament) level is an inventory of the European coastal lowlands. Where are they, what do they contain, how many people live there, what percentage of gross national product is generated there and what value are they ecologically and economically to Europe? The inventory would require the following data:-

1 the recent geological history, the palaeogeography and the thickness of the unconsolidated sediments forming the coastal lowland; the rates and directions of coastline movement;
2 the nature of the present coast and intertidal zone; the direction of longshore drift, its persistence and strength and the volume of sediment moved; the distribution of natural coastal landforms – sand dunes and saltmarshes, and the volume of sediment locked

into them; the extent, nature and condition of sea defences, and their history;
3 the rate and direction of land movements, both geological and man-induced;
4 the rate and direction of sea-level movements;
5 the record and magnitude of storm surges and their penetration points on the coast, and the record of extreme water levels;
6 groundwater reserves, water quality and depth;
7 river discharge and sediment load variations;
8 present and past landuse, rates of changes in the patterns of landuse; the ratio of industrial, agricultural, residential and conserved land;
9 the topography of the coastal lowlands with a contour interval of 1 m up to 5 m above Mean Tide Level;
10 population density, structure and change;
11 the pattern of land values, of capital investment and of insured risk.

Other inventories have been proposed by Cendrero and Charlier (1989). All these data require acquisition, storage, retrieval and display in an information system that can be continually updated. It must be accessible and at a scale that permits realistic management and planning at the regional or national level, and ultimately the European level, given social and economic constraints.

A simple geographic information system has been developed (Shennan and Tooley 1987; Shennan and Sproxton 1989, 1990) and applied to the Teesside coastal lowlands in north-east England. Within the geographic information system, tidal data, current flood defence data, sea-level scenarios, topographic map data, land use data and census data are stored and integrated with outputs in the form of maps, tables, diagrams and reports. By overlaying population and altitudinal data and census data, a measure of impact can be gained for each of the sea-level rise scenarios and the extreme water levels associated with them. Urban areas, industrial areas, transport networks, strategic industries, landfill sites, conservation sites can all be shown in relationship to a specific contour, and for a specified extreme water level with a recurrence interval, the likely impact can be evaluated (See also Shennan, chapter 4).

Zoning coastal lowlands into areas of high, intermediate and low risk requires a consideration of all the items in the inventory given above. High risk areas in coastal lowlands are those where ground altitudes are below present spring tide altitudes (on macro- and meso-tidal coasts), and where marine inundation has been frequent. The risk is increased in those areas where tectonic subsidence and reduction of ground altitudes due to drainage amplifies the rise in sea-level due to thermal expansion and melting of small glaciers. High risk areas are also associated with high tidal range, persistent onshore winds and the regular occurrence of storms and surge phenomena. Risk is increased where natural sea defences are narrow (e.g. a single low dune ridge) or poorly maintained sea embankments (e.g. earth embankments into which moles, badgers or rabbits have burrowed). These shorelines can be found around the southern North Sea, the Irish Sea and the English Channel, the Atlantic coast of the Iberian peninsula, and locally around the Mediterranean, for example the head of the Adriatic Sea.

In the Netherlands, Germany and parts of south-east England the high risk status has already been recognized and heavy capital investment has been approved by national governments to protect valuable property in coastal lowlands. Examples are the Deltaplan in the Netherlands, the Thames barrage in Great Britain, the city of Hamburg in Germany and the 'Progetto Insulae' for the city of Venice in Italy. Open coastlines with a meso-tidal range, backed by lowlands, can also be considered as high risk, where the dune belt is less than 100 m wide and shoreline erosion is occurring. In addition, cyclonic conditions and surge phenomena increase the risk. Such areas occur in parts of eastern England and Scotland and in some western parts of western Holland.

Low risk areas are those approaching altitudes of +5 metres Mean Tide Level, where the coastline is fronted by sand dunes several kilometres wide and the back dune area has been elevated by sand blown from the main dune area. Shoreline erosion of some tens of metres due to an increased rate of sea-level rise will not destroy the bulk of the dune system, the primary function of which is to protect the lowlands to landward from flooding by the sea. Such situations exist in the Netherlands, in France, in Portugal and in western Britain.

There is much debate at present about the impact of sea-level rise on the Dutch, German and Danish Wadden Sea area: these tidal flats are the last remaining extensive natural area in north-west Europe. Any change in the morphology and tidal flat extent will have a profound impact on the coastal ecology, and, in particular, the fisheries and avifauna. In general, the Wadden Sea is considered to be a sink for the silt brought into the North Sea by the Rivers Rhine and Meuse and the sand fraction eroded from the dunes of north Holland. Misdorp et al. (1990) have concluded that with a 50 cm rise in sea level the tidal flats of the tidal basins, such as Texelstroom, will disappear, and it is debatable whether future sedimentation in larger basins, such as Eierlandsegat and Vliestroom, will keep pace with a

50 to 100 cm sea-level rise. If sea-level rise attains 200 cm, the tidal flats in these large basins will disappear. It has also been posited that the consequences of sea-level rise are increased shoreline erosion, the liberation of sediments, accelerated sedimentation rates in the Wadden Sea, maintaining or extinguishing the tidal flats.

The responses to the risk, both real and perceived, have been many. Newman and Fairbridge (1986) argued that there were three alternative strategies for the coastal lowlands: evacuate them, defend them or arrest sea-level rise by interrupting the hydrological cycle by damming rivers and diverting fresh water into reservoirs, ground water storage and inland basins. Such diversions could result in regional-scale disruptions, including the initiation of earth movements in inland basins. These topics were considered at a Nato Advanced Research workshop (Paepe et al. 1990), and one of the main conclusions was that the control of sea-level rise by current and foreseeable technologies for storing water underground was not economically feasible. It was calculated (Roebart 1990) that the impact on sea-level of storing 1 per cent ($c.$ 4,250 km^3) per year of the world's precipitation as groundwater recharge in semi-arid areas would be a lowering of less than 15 mm.

Doornkamp (1990) has noted that the response to natural hazards in industrialized countries is confrontational, and that engineering solutions and projects requiring high capital investment are usual. Hence, the response to a rise in sea level in Europe is likely to involve engineering solutions such as armouring coasts, raising sea embankments, water management of river estuaries and ports, and restoration of dune systems. In the Netherlands, Goemans (1986) has estimated that a rise of sea level of 2 m over the next 100 years would cost 20 billion guilders or 0.1 per cent of Dutch gross national product – a figure that could readily be accommodated in a national budget.

Alternatively, a planning strategy could be adopted embracing the high, intermediate and low risk categories of land in coastal lowlands by equating them with a zone in which development was prohibited, one in which development was restricted and one in which warnings of inundation would be given. Doornkamp (1990) tabulated a series of managerial choices that can be made in the case of large-scale floods in the United States of America, which could be transferred to a coastal lowland context. The five action strategies comprise:

Pre-emptive action	All land liable to marine inundation is transferred to public ownership.
Mandatory action	Appropriate local building codes are introduced.

Influencing action	Subsidized flood damage insurance.
Informative action	Local awareness increased by the provision of information.
Incidental action	Income tax reductions for flood losses.

In theory, the existing national planning legislation in Europe is sufficient to constrain development, but in fact high investment in inappropriate structures in high risk areas has occurred and is continuing at an accelerating rate.

In the United States of America, Flood Insurance Rate Maps exist for coastal lowlands. The maps have been produced since 1968 as part of the National Flood Insurance Program of the Federal Emergency Management Agency. There are five zone designations – A, B, C, D and V. Zones A and V are zones affected by the 100–year flood, and for some areas expected water depths are given. Particular attention is given to methods for the determination of wave heights, wave crest elevations, wave run-up and coastal erosion, but no allowance is made for sea-level rise (FEMA 1984, 1988, Flood Insurance Rate Maps).

1.7 CONCLUSIONS

The coastal lowlands of Europe present a great variety of natural environments, upon which the rising sea-level will work. There will be a range of responses and impacts. Sea-level rise will not be uniform because of the operation of geoidal forces. The dilation of tidal inlets and estuaries will affect the tidal range. The nature of the shoreline, the sediment budget and the rates of erosion will respond to the sea-level and tidal changes, but changes in the catchments of the large European rivers due to temperature and precipitation changes will also be manifest in the coastal lowlands, particularly ground water storage and quality and sediment load. Carter (1987) has drawn attention to the concern about coastal hazards, as relatively infrequent, temporary rises in sea-level associated with cyclones, hurricanes and typhoons causing inundation, whereas the sea-level rises predicted will lead to long-term inundation. Evidence from the Holocene indicates that long-term may be from 200 to 500 years.

Human response to short, aperiodic inundation has been preventive, reactive or mitigating. The national or local response depends on economic ability and political will, and once measures have been agreed, enforcement is required (Schröder 1988). Examples are essential and it is for this reason that the inventory (p. 23) includes the natural responses to sea-level changes as well as the human responses. Schröder (1988) draws attention to the need for 'laboratory' cases, contrasting a well-protected coastal state, such as the Netherlands, with the case of the Caspian Sea which is unprotected and is suffering from rapidly rising water levels. To these could be added the subsiding deltaic shorelines, largely unprotected, of the Mediterranean Sea, and the isostatically rising shorelines of the Gulf of Bothnia, both of which present different problems to coastal managers against a background of sea-level rise. The legal implications of boundary changes due to sea-level rise and coastline retreat have not been addressed in any detail: an exception is the impact of coastline changes on baseline construction by Prescott and Bird (1990). In the Gulf of Bothnia, a legal framework exists for apportioning land gained from the sea due to isostatic uplift and a relative fall of sea-level (Palomäki 1987), but environmental lawyers need to consider the implications of coastline retreat and the liabilities that will arise from the reactivation and discharge of toxic materials from landfill sites in the coastal lowlands and on former salt-marshes.

Whatever the local or regional responses, note should be taken of Schneider's (1989) statement that, 'sea-level rise is undoubtedly the most dramatic and visible effect of global warming into the greenhouse century'. Whilst all European coastal lowlands are already at risk from extreme water levels, exacerbated by subsidence and the lowering of ground surfaces by water extraction and drainage, this risk will increase during the next century as sea-level rises and these areas become industrialized and more densely populated.

REFERENCES

Alestato, J. 1985: Finland. In, Bird, E. C. F. and Schwartz, M. L. (eds) *The World's Coastline*. New York, van Nostrand Reinhold, 295–302

Andrade, C. F. de 1989: Litoral Sul, Guia de Excursoes. VI Simpósio de Geologia Aplicada e do Ambiente – Ambientes Geológicos Litorais. Lisbon, Departamento de Geologia, Universidade de Lisboa

Ascher, G. 1989: The Hamburg Case. In, Frankenfield Zanin, J. (ed.) *Impact of Sea Level Rise on Cities and Regions* (summaries of presentations). Venice, Marsilio Editori, 25

Baeteman, C. 1985: Late Holocene geology of the Marathon Plain (Greece). *Journal of Coastal Research* 1(2), 173–85

Bakker, Th. W. M., Jungerius, P. D. and Klijn, J. A. (eds) 1990: *Dunes of the European Coasts: Geomorphology – Hydrology – Soils Catena*, Supplement 18

Behre, K.-E., Menke, B. and Streif, H. 1979: The Quaternary geological development of the German part of the North Sea. In, Oele E. et al. (eds), *op.cit.* 85–113

Bird, E. C. F. 1985: The study of coastline changes. *Zeitsch. für Geomorph. Suppl. Bd.* 57, 1–9

Bolin, B., Döös, B., Jäger, J. and Warrick, R. A. (eds) 1986: *The Greenhouse Effect, Climatic Change and Ecosystems* (SCOPE 29). Chichester, John Wiley and Sons

Borowka, R. K. 1985: Poland. In, Bird, E. C. F. and Schwartz, M. L. (eds) *The World's Coastline*. New York, van Nostrand Reinhold, 311–14

Brückner, H. 1986: Man's impact on the evolution of the physical environment in the Mediterranean region in historical times. *Geojournal* 13(1), 7–17

Carter, R. W. G. 1987: Man's response to sea-level change. In, Devoy, R. J. N. (ed.) *Sea Surface Studies: A Global View*. London, Croom Helm, 464–98

Carter, R. W. G., Devoy, R. J. N. and Shaw, J. 1989: Late Holocene sea levels in Ireland. *Jl. Quaternary Science* 4(1), 7–24

Cendrero, A. and Charlier, R. M. 1989: Resources, landuse and management of the coastal fringe. *Geolis* 3(1–2), 40–60

Clark, J. A. and Primus, J. A. 1987: Sea-level changes resulting from future retreat of ice sheets: an effect of CO_2 warming of climate. In, Tooley, M. J. and Shennan, I. (eds) *Sea-level Changes*. Oxford, Basil Blackwell, 356–70

Colenbrander, H. J. (compiler) 1986: *Water in the Netherlands*. The Hague, TNO Committee on Hydrological Research

Davies, J. L. 1980: *Geographical Variation in Coastal Development*. London, Longman

Donner, J. and Raukas, A. (eds) 1988: Problems of the Baltic Sea history. *Ann. Acad. Sci. Fennicae* A.III 148, 1–35

Doornkamp, J. C. 1990: Policy, planning and engineering reactions to sea level rise. In, Doornkamp, J. C. (ed.) *The Greenhouse Effect and Rising Sea Levels in the UK*. Nottingham, MI Press, 135–45

Eisma, D. 1978: Stream deposition and erosion by the eastern shore of the Aegean. In, Brice, W. C. (ed.) *The Environmental History of the Near and Middle East since the Last Ice Age*. London, Academic Press, 67–81

Englefield, G. J. H., Tooley, M. J. and Zong, Y. 1990: *An Assessment of the Clwyd Coastal Lowlands after the Floods of February 1990*. Durham, Environmental Research Centre

Eronen, M. 1983: Late Weichselian and Holocene shore displacement in Finland. In, Smith, D. E. and Dawson, A. G. (eds) *Shorelines and Isostasy*: Institute of British Geographers Special Publication No. 16. London, Academic Press, 183–207

Evans, G. 1979: Quaternary transgressions and regressions. *Jl. Geol. Soc. Lond.* 136, 125–32

Federal Emergency Management Agency (FEMA) 1984: *Design and Construction Manual for Residential Buildings in Coastal High Hazard Areas*. Washington, D.C., United States Department of Housing and Urban Development

Federal Emergency Management Agency (FEMA) 1988: *Guidelines and Specifications for Wave Elevation Determination and V Zone Mapping*. Washington, D.C., FEMA (2nd Draft, July 1988)

Flemming, N. C. 1978: Holocene eustatic changes and coastal tectonics in the north east Mediterranean: implications for models of crustal consumption. *Phil. Trans. R. Soc.* A. 289(1362), 405–58

Fowler, G. 1934: The extinct waterways of the Fens. *Geogr. J.* 83, 30–9

Godwin, H. 1938: The origin of roddens. *Geogr. J.* 91, 241–50

Godwin, H. 1978: *Fenland: Its Ancient Past and Uncertain Future.* Cambridge, Cambridge University Press

Goemans, T. 1986: The sea also rises. In, *Effects of Changes in Stratospheric Ozone and Global Climate.* UNEP/EPA International Conference on the Health and Environmental Effects of Ozone Modification and Climate Change, Crystal City, Va., USA. Washington, D.C., Environmental Protection Agency, 29–38

Gomes, C. 1989: Ria de Aveiro. In, Romariz, C. (ed.) *Excursion C. Littoral North of Peniche.* VI Simpósio de Geologia Aplicada e do Ambiente: Ambientes Geológicos Litorais, Guia do Simpósio. Lisbon, Departamento de Geologia, Universidade de Lisboa, 22–3

Gornitz, V., Lebedeff, S. and Hansen, J. 1982: Global sea level trend in the past century. *Science* 215, 1611–14

Gottschalk, M. K. E. 1971: *Stormvloeden en Rivieroverstromingen in Nederland: I, De periode voor 1400.* Assen, van Gorcum

Gottschalk, M. K. E. 1975: *Stormvloeden en Rivieroverstromingen in Nederland: II, De periode 1400–1600.* Assen, van Gorcum

Gottschalk, M. K. E. 1977: *Stormvloeden en Rivieroverstromingen in Nederland: III, De periode 1600–1700.* Assen, van Gorcum

Graff, J. 1978: Abnormal sea levels in the north west. *The Dock and Harbour Authority* 58, 366–8 and 371

Grenon, M. and Batisse, M. (eds) 1989: *Futures for the Mediterranean Basin: The Blue Plan.* Oxford, Oxford University Press

Gudelis, V. and Königsson, L.-K. (eds) 1980: *The Quaternary History of the Baltic.* Uppsala, Acta Univ. Ups. Symp. Univ. Ups. Annum Quingentesimum Celebrantis 1

Hageman, B. P. 1969: Development of the western part of the Netherlands during the Holocene. *Geologie Mijnb.* 48(4), 373–88

Hansen, J., Fung, I., Lacis, A., Lebedeff, S., Rind, D., Ruedy, R., Russell, G. and Stone, P. 1988: Prediction of nearterm climate evolution: what can we tell decision makers now? In, Climate Institute *Preparing for Climate Change. Proceedings of the First North American Conference on Preparing for Climate Change.* Rockville, Md., Government Institutes Inc., 35–47

Helmer, R. 1977: Pollutants from land-based sources in the Mediterranean. *Ambio* 6(6), 312–16

Hey, R. W. 1978: Horizontal Quaternary shorelines of the Mediterranean. *Quat. Res.* 10, 197–203

Hoffman, J. S. 1984: Estimates of future sea level rise. In, Barth, M. C. and Titus, J. G. (eds) *Greenhouse Effect and Sea-level Rise: A Challenge for This Generation.* New York, van Nostrand Reinhold, 79–103

Hofstede, J. 1990: Sea-level rise in the Inner German Bight (West Germany) since 600 AD and its implications upon tidal flats morphology. *Intl. Symp. on Transgressive Coasts: Studies and Economic Development*, Baku, USSR. Abstracts, 149–150

Houghton, J. T., Jenkins, G. J. and Ephraums, J. J. (eds) 1990: *Climate Change: The IPCC Scientific Assessment.* Cambridge, Cambridge University Press

Hutchinson, J. 1980: The record of peat wastage in the East Anglian Fenlands at Holme Post, 1848–1978 A.D. *J. Ecol.* 68, 229–49

Jelgersma, S. 1961: Holocene sea level changes in the Netherlands. *Meded. Geol. Sticht.* C.VI.7, 1–100

Jelgersma, S., de Jong, J., Zagwijn, W. H. and van Regteren Altena, J. F. 1970: The coastal dunes of the western Netherlands: geology, vegetational history and archaeology. *Meded. Rijks. Geol. Dienst* NS 21, 93–167

Jelgersma, S., Oele, E. and Wiggers, A. J. 1979: Depositional history and coastal development in the Netherlands and the adjacent North Sea since the Eemian. In, Oele, E., Schüttenhelm, R. T. W. and Wiggers, A. J. (eds) *The Quaternary History of the North Sea.* Uppsala, Acta Univ. Ups. Symp. Ups. Annum Quingentesimum Celebrantis 2, 115–42

Jongman, R. H. G. 1987: *The Rhine-Maas megalopolis and its water resources: consequences of climate-induced hydrological changes.* European Workshop on Interrelated Bioclimate and Land-use Changes, Noordwijkerhout, Netherlands

Kakkuri, J. 1987: Character of the Fennoscandian land uplift in the 20th century. *Geological Survey of Finland, Special Paper* 2, 15–20 (Proceedings of a Symposium on Fennoscandian Land Uplift at Tvärminne, 10–11 April 1986, ed. M. Perttunen)

Kidson, C. 1982: Sea level changes in the Holocene. *Quaternary Science Reviews* 1(2), 121–51

Kidson, C. and Tooley, M. J. (eds) 1977: *The Quaternary History of the Irish Sea* (*Geological Journal*, Special Issue No. 7). Liverpool, Seel House Press

Kramer, J. 1978: Coast protection works on the German North Sea and Baltic Coast. *Die Küste: Archiv für Forschung und Technik an der Nord-und Ostsee* 32, 124–39

Kwadijk, J. and Boois, H. de (eds) 1989: *Final Report. European Workshop on Interrelated Bioclimate and Land-use Changes.* 17–21 October 1987. Bilthoven, RIVM, National Institute of Public Health and Environmental Protection

Lamb, H. H. 1980: Climatic fluctuations in historical times and their connection with transgressions of the sea, storm floods and other coastal changes. In, Verhulst, A. and Gottshalk, M. K. E. (eds) *Transgressies en Occupatiegeschiedenis in de Kustgebieden van Nederland en België.* Belgisches Centrum voor Landelijke Geschiedenis No. 66, 251–84

Lamb, H. H. (1991) *Historic Storms of the North Sea, British Isles and Northwest Europe.* Cambridge, Cambridge University Press.

Lisitzin, E. 1974: *Sea-level Changes* (Elsevier Oceanography Series, 8). Amsterdam, Elsevier Scientific Publishing Co.

Long, D., Dawson, A. G. and Smith, D. E. 1990: Tsunami risk in northwestern Europe: a Holocene example. *Terra Research* 1, 532–7

Long, D., Smith, D. E. and Dawson, A. G. 1989: A Holocene tsunami deposit in eastern Scotland. *Jl. Quaternary Science* 4(1), 61–6

Louwe Kooijmans, L. P. 1974: *The Rhine/Meuse Delta: Four Studies on its Prehistoric Occupation and Holocene Geology*, Leiden, E. J. Brill

Misdorp, R., Steyaert, F., Hallie, F. and Ronde, J. de 1990: Climate, sea-level rise and morphological developments in the Dutch Wadden Sea. In, Beukema, J. J., Wolff, W. J. and Brouns, J. J. W. M. (eds) *Expected Effects of Climatic Change on Marine Coastal Ecosystems*. Dordrecht, Kluwer Academic Publishers

Mörner, N.-A. 1979: The Fennoscandian uplift and Late Cenozoic geodynamics: geological evidence. *Geojournal* 3(3), 287–318

National Research Council (US), Committee on Engineering Implications of Changes in Relative Mean Sea Level, 1987: *Responding to Changes in Sea Level: Engineering Implications*, Washington, D.C., National Academy Press

Neilson, G., Musson, R. H. W. and Burton, P. W. 1984: The 'London' earthquake of 1580, April 6. *Engineering Geology* 20, 113–41

Newman, W. S. and Fairbridge, R. W. 1986: The management of sea-level rise. *Nature* 320, 319–21

Normann, J. O. 1985: Sweden. In, Bird, E. C. F. and Schwartz, M. L. (eds) *The World's Coastline*. New York, van Nostrand Reinhold, 285–93

Oele, E., Schüttenhelm, R. T. E. and Wiggers, A. J. 1979: *The Quaternary History of the North Sea*. Uppsala, Acta Univ. Ups. Symp. Univ. Ups. Annum Quingentesimum Celebrantis 2

Paepe, R., Fairbridge, R. W. and Jelgersma, S. (eds) 1990: *Greenhouse Effect, Sea-level and Drought* (Proceedings of the NATO Advanced Research Workshop on Geohydrological Management of Sea Level and Mitigation of Drought, Fuerteventura, Canary Islands, Spain, 1–7 March 1989). Dordrecht, Kluwer Academic Publishers

Palomäki, M. 1987: Human reponse to the effects of land uplift. *Geological Survey of Finland, Special Paper* 2, 47–50 (Proceedings of a Symposium on Fennoscandian Land Uplift at Tvärminne, 10–11 April 1986, ed. M. Perttunen)

Passant, F. 1990: Leave it and Landscape it. *Nuclear Forum* (September), 20

Petersen, K.-S. 1981: The Holocene marine transgression and its molluscan fauna in the Skagerrak–Limfjord region, Denmark. In, Nio, S.-D., Schüttenhelm, R. T. E. and Weering Tj. C. E. van (eds) *Holocene Marine Sedimentation in the North Sea Basin* (Spec. Publs. Int. Ass. Sediment 5), 497–503

Pirazzoli, P. A. 1986: The Early Byzantine tectonic paroxysm. *Zeitsch. für Geomorph. Suppl. Bd.* 62, 31–49

Pirazzoli, P. A. 1987: Sea-level changes in the Mediterranean. In, Tooley, M. J. and Shennan, I. (eds) *Sea-level Changes*, Oxford: Basil Blackwell, 152–81

Pirazzoli, P. A. 1989: Present and near-future global sea level changes. *Palaeogeography, Palaeoclimatology and Palaeoecology* 75, 241–58

Pirazzoli, P. A., Grant, D. R. and Woodworth, P. 1989: Trends of relative sea-level change: past, present and future. *Quaternary International* 2, 63–71

Prescott, V. and Bird, E. 1990: The influences of rising sea levels on baselines from which national maritime claims are measured and an assessment of the possibility of applying Article 7(2) of the 1982 Convention on the Law of the Sea to offset any retreat of the baseline. In, Grundy-Warr, C. (ed.) *International Boundaries and Boundary Conflict Resolution*. Durham, International Boundaries Research Unit, 279–301

Reid, H. F. 1914: The Lisbon earthquake of November 1, 1755. *Bull. Seismological Society of America* 4, 53–80

Ritsema, A. R. and Grupinar, A. (eds) 1983: *Seismicity and Seismic Risk in the North Sea*. Dordrecht, Reidel

Robin, G. de Q. 1986: Changing the sea level: projecting the rise in sea level caused by warming of the atmosphere. In, Bolin, B. et al. (eds), *op.cit.* 323–59

Roebert, A. J. 1990: Sea level rise and artificial groundwater recharge, a study on the feasibility of geohydrologic management. In, Paepe, R., Fairbridge, R. W. and Jelgersma, S. (eds) *Greenhouse Effect, Sea Level, and Drought*. Dordrecht, Kluwer Academic Publishers, 553–64

Rohde, H. 1978: The history of the German coastal area. *Die Küste: Archiv für forschung und technik an der Nord- und Ostsee* 32, 6–29

Romariz, C. (ed.) 1989: Revista de Geologia aplicada e do ambiente. *Geolis* 3(1–2), 1–270

Rossiter, J. R. 1962: Long term variations in sea level. In, Hill, N. M. (ed.) *The Sea* 1. London, Interscience Publishers, 590–610

Schneider, S. H. 1989: *Global Warming: Are We Entering the Greenhouse Century?* San Francisco, Sierra Club Books

Schröder, P.C. 1988: Impact of sea level rise on society: a management approach. UNEP: *Report of the Joint Meeting of the Task Team on Implications of Climatic Changes in the Mediterranean*. Split, Yugoslavia, UNEP(OCA)/WG 2/25, 16–17

Sestini, G. 1989: The impact of sea level rise on low lying Mediterranean coasts. *Bollettino di Oceanologia: Teorica ed applicata* 7(4), 295–9

Shennan, I. 1987: Holocene sea-level changes in the North Sea region. In, Tooley, M. J. and Shennan, I. (eds) *Sea-level Changes*. Oxford, Basil Blackwell, 109–51

Shennan, I. 1988: United Kingdom: England: Lincolnshire. In, Walker, H. J. (ed.) *Artificial Structures and Shorelines*. Dordrecht, Kluwer Academic Publishers, 145–54

Shennan, I. and Sproxton, I. 1989: Impacts of future sea level rise on the Tees estuary – an approach using a Geographic Information System (GIS). In, *Impact of Sea Level Rise on Cities and Regions* (summaries of presentations). Venice, Marsilio Editori, 31

Shennan, I. and Sproxton, I. 1990: Possible impacts of sea-level rise – a case study from the Tees estuary, Cleveland County. In, Doornkamp, J. C. (ed.) *The Greenhouse Effect and Rising Sea Levels on the UK*. Nottingham, MI Press, 109–33

Shennan, I. and Tooley, M. J. 1987: Conspectus of fundamental and strategic research on sea-level changes. In, Tooley, M. J. and Shennan, I. (eds) *Sea-level Changes*. Oxford, Basil Blackwell, 371–90

Sissons, J. B. 1972: Dislocation and non-uniform uplift of raised shorelines in the western part of the Forth valley. *Trans. Inst. Br. Geogr.* 55, 145–59

Sissons, J. B. 1983: Shorelines and isostasy in Scotland. In, Smith, D. E. and Dawson, A. G. (eds) *Shorelines and Isostasy*. London, Academic Press, 209–25

Streif, H. 1989: Barrier islands, tidal flats, and coastal marshes resulting from a relative rise of sea level in East Frisia on the German North Sea coast. Linden, W. J. M. van der et al. (eds) In, *Coastal Lowlands: Geology and Geotechnology, Proceedings of the Symposium on Coastal Lowlands*, organized by The Royal Geological and Mining Society of the Netherlands (KNGMG). Dordrecht, Kluwer Academic Publishers, 213–23

Streif, H. 1990: Quaternary sea-level changes in the North Sea, an analysis of amplitudes and velocities. In, Brosche, P. and Sündermann, J. *Earth's Rotation from Eons to Days*. Berlin, Springer Verlag, 201–14

Tooley, M. J. 1971: Changes in sea level and the implications for coastal development. *Association of River Authorities Yearbook 1971*, 220–5

Tooley, M. J. 1974: Sea level changes during the last 9000 years in north-west England. *Geog. J.* 140, 18–42

Tooley, M. J. 1978a: *Sea level Changes: North West England during the Flandrian Stage*. Oxford, Clarendon Press

Tooley, M. J. 1978b: Interpretation of Holocene sea-level changes. *Geol. För. Stockh. förh.* 100(2), 203–12

Tooley, M. J. 1979: Sea-level changes during the Flandrian Stage and the implications for coastal development. In, Suguio, K., Fairchild, T. R., Martin, L. and Flexor, J.-M. (eds) *Proceedings of the 1978 International Symposium on Coastal Evolution in the Quaternary*. São Paulo, Universidade de São Paulo, 502–33

Tooley, M. J. 1989: Global sea levels: floodwaters mark sudden rise. *Nature* 342, 20–1

Tooley, M. J. 1990: The chronology of coastal dune development in the United Kingdom. In, Bakker, Th. W., Jungerius, P. D. and Klijn, J. A. (eds) *Dunes of the European Coasts: Geomorphology – Hydrology – Soils. Catena* Supplement 18, 81–8

Tooley, M. J. 1992: Long term changes in eustatic sea-level. In, Wigley, T. M. L. and Warrick, R. A. (eds) *Climate and Sea-level Change*. Cambridge, Cambridge University Press.

Tziavos, C. and Kraft, J. C. 1985: Greece. In, Bird, E. C. F. and Schwartz, M. L. (eds) *The World's Coastline*. New York, van Nostrand Reinhold, 445–53

UNEP 1990: *The State of the Marine Environment*, Reports and Studies No. 39. UNEP Regional Seas Reports and Studies, No. 115.

Warrick, R. A. and Oerlemans, J. 1990: Sea level rise. In, Houghton, J. T. et al. (eds) *Climatic Change: The IPCC Scientific Assessment*. Cambridge, Cambridge University Press, 257–81

Wigley, T. M. L. and Raper, S. C. B. 1987: Thermal expansion of sea water associated with global warming. *Nature* 330, 127–31

Wigley, T. M. L. and Raper, S. C. B. 1992: Future changes in global mean temperature and thermal expansion-related sea level rise. In, Warrick, R. A. and Wigley, T. M. L. (eds) *Climate and Sea Level Change: Observations, Projections and Implications*. Cambridge, Cambridge University Press.

Woodworth, P. L. 1990: A search for accelerations in records of European Mean Sea Level. *Intl. Jl. Climatology* 10, 129–43

Zagwijn, W. H. 1983: Sea-level changes in the Netherlands during the Eemian. *Geologie Mijn.* 62, 437–50. (In, Berg, M.W. Van den and Felix, R. (eds), Special Issue in honour of J. D. de Jong)

Zagwijn, W. H. 1986: *Nederland in het Holoceen*. Haarlem: Rijks Geologische Dienst

2

Relative Rise of Mean Sea-levels in the Netherlands in Recent Times

J. van Malde

2.1 INTRODUCTION

As the name itself suggests, the Netherlands (which means 'low-lying country') are vulnerable both to river floods and to storm surges. In fact their history is marked by many floods, and some of those were indeed great disasters. One of the most recent, that of 1 February 1953, struck the south-western part of the country very hard and led to the great Delta project, the aim of which was to safeguard the country according to advanced present-day standards against flooding. But floods notwithstanding, detailed water management has long since been essential to an important part of the country. It seems therefore most likely that a certain interest in water levels has existed in those areas since the distant past. In fact, probably the oldest written source kept is a sentence from 1509 imposing a certain polderdatum upon regional authorities. From the sixteenth century several references to water level observations are known, such as the mention of the tidal range at Petten (1525), of storm surge levels and floodmarks (1530, 1570) and of 'normal high water level' as reference level (for example, at Amsterdam 1556). Unfortunately, practically none of this information can be related to the Dutch Reference Datum (NAP), with the remarkable exception of the highwater level at Scheveningen (about half-way between Hook of Holland and Katwijk) during the notorious All Saints storm surge of 1570, which was either 10 or 25 cm (the two available sources differ by 15 cm) higher than the very high level of 1 February 1953 (Gottschalk 1975; for Amsterdam, until 1872 a tidal harbour on the former Zuider Zee: van der Weele 1971).

However no systematic tidal measurements are recorded from before September 1683, when in Amsterdam a start was made, and during ten out of twelve months the water level there was measured every half-hour during the night-time. The objectives of this investigation were to check whether the local reference datum valid by then was indeed the local mean high water and to fix this datum at the eight well-founded sluices and locks between the canals of the town and the tidal water called the IJ. The datum mentioned was, under the circumstances, magnificently secured by means of marble stones in the lock bays and sluice heads concerned. These 'Hudde stones', placed by order of Burgomaster Hudde, were still in place in 1851, after which they gradually disappeared (the last one in 1955). The datum was soon known as AP (= *Amsterdams Peil* = Amsterdam Datum), and since 1891 as NAP (= *Normaal Amsterdams Peil*).

From 1 January 1700 the tidal waterlevel and the wind were measured at Amsterdam as a routine; hourly during the daytime and every half-hour at night. The series of water-levels was continued until the enclosure of the Zuider Zee in 1932. After 1860 the location and equipment of the gauge were changed several times, and from the records the data for sixteen years (1750–65) are missing. Figure 2.1 shows both the graph for the three years moving average of half-tide and the corrected trend line (van Veen 1954). The latter shows a significant rise only after about 1830; this will also apply to the local mean sea-level as its annual mean values are well approximated by the corresponding half-tide values plus 0.5 cm.

Records of several other series of Dutch eighteenth-century tidal observations do exist (e.g. for Katwijk 1737–41), but they are rather short and their reduction to NAP encounters great difficulties. Though systematic tidal measurements were made here and there in the Netherlands from early in the nineteenth century, a network of relevant gauges was only established in the sixties and seventies of that century, but of course with modifications afterwards.

The locations of a number of important or relevant gauging stations are indicated in figure 2.2; the lower part of this figure also shows the high and low water levels for mean tide and the highest storm surge levels known for most of these stations. It takes ten to eleven hours for the tidal wave to move along the Dutch coast from south-west to north-east.

2.2 GENERAL TENDENCIES

For eight tidal gauges the annual values of mean sea-level (in cm and related to NAP – as are all other water level data in the Netherlands) are plotted over the period 1830–1986 in figure 2.3. Note that the revised trendline of

38   J. van Malde

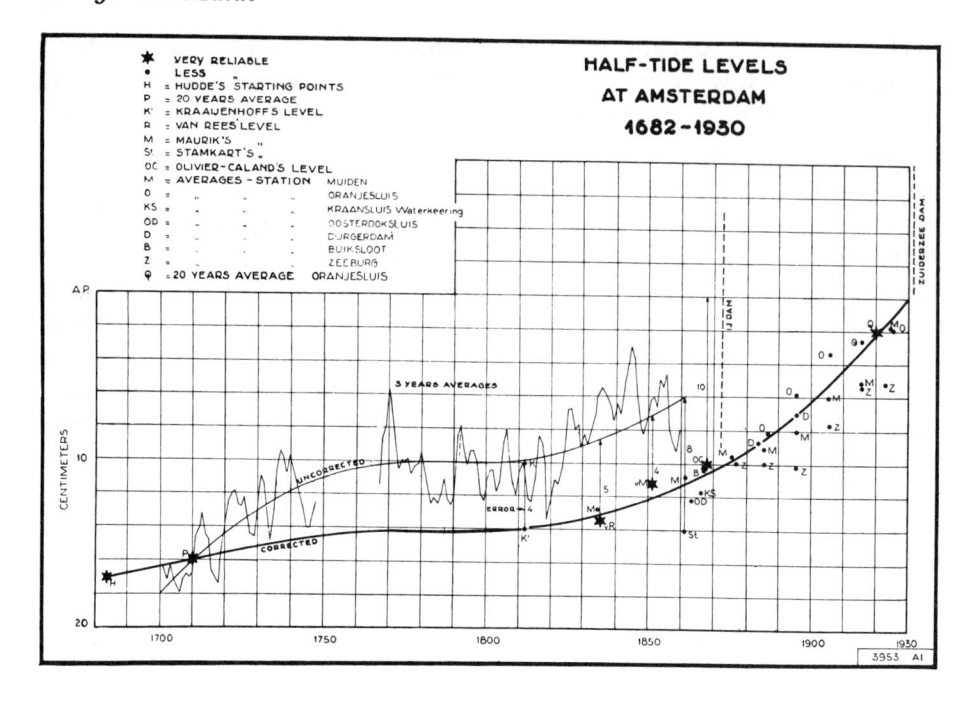

Figure 2.1   Half-tide levels at Amsterdam AD 1682–1930

figure 2.1 begins its rise just after 1830 and that the readings at the various gauges were and are utterly independent. The graphs show relative rises and falls respectively (all rises and falls discussed here are relative, as vertical movements of the reference datum NAP may not have been excluded beforehand).

Apparently there is a general rise of (annual) mean sea-level, amounting to 15–20 cm/century, over, say, the last 100 years (the very long series – Den Helder since 1832; Katwijk high tide since 1804, not reproduced here – suggests that a more or less definite rise did not start before about 1885). However this rise is far from constant. Most striking in this respect are substantial 'undulations', covering up to fifteen or even more years; some of these coincide at nearly all stations or at a number of them, whilst others do not. The considerable drop between 1877 and 1887 is especially worth mentioning here: in smoothed graphs it still amounts to 13 cm at Flushing and 9 cm at IJmuiden whilst nearly everywhere else it remains evidently present. But apart from these 'undulations' (which might partly resolve themselves to nodal fluctuations) the trend has not been a rise all the time: for quite a few gauges a period of an overall constant mean sea-level started at about 1910. This 'interruption' is most notable for Harlingen,

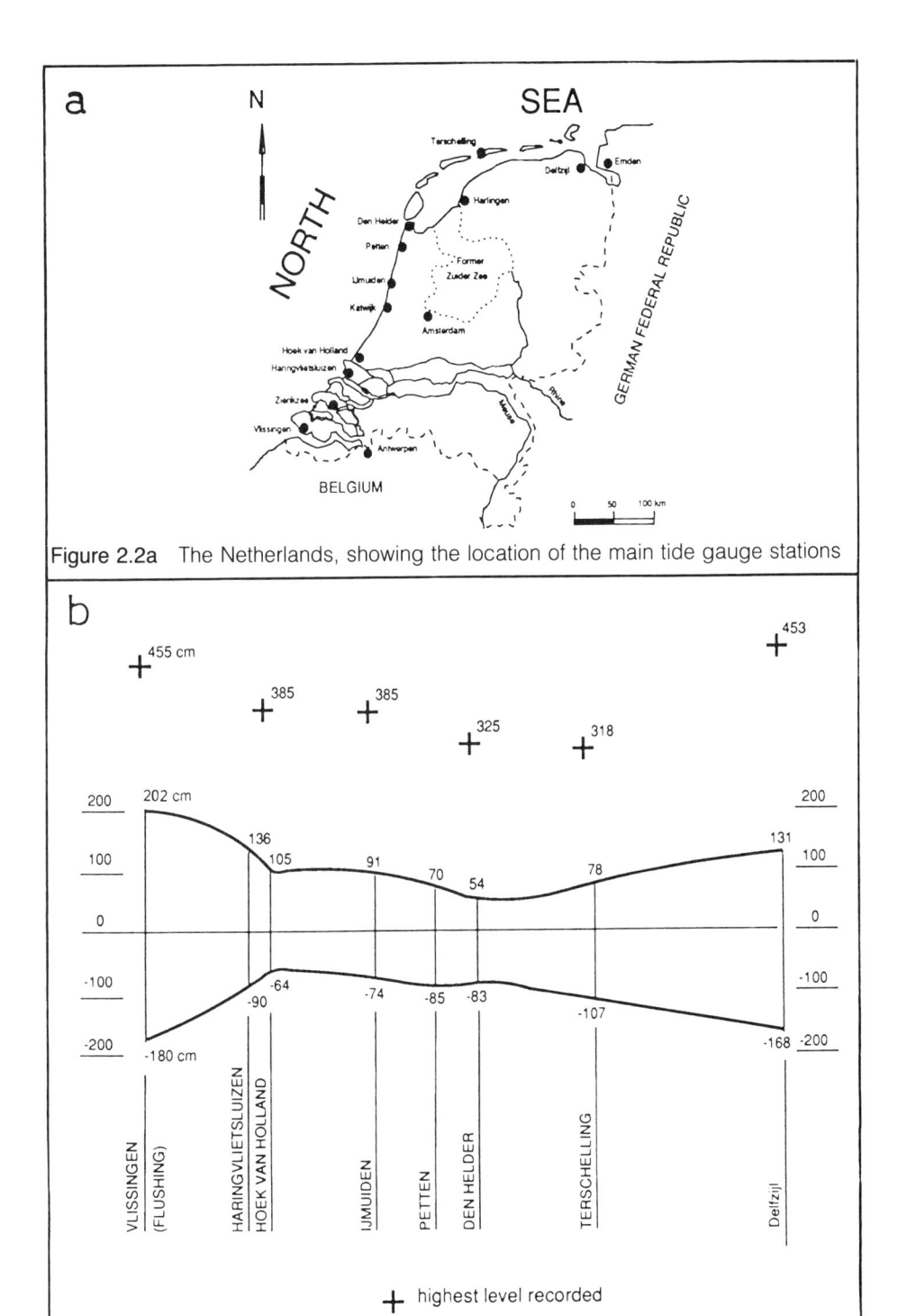

Figure 2.2a The Netherlands, showing the location of the main tide gauge stations

Figure 2.2b Mean tidal ranges along the Dutch coast on 1 January 1981 at eight tide gauge stations, together with an indication of the highest storm surge levels in relation to NAP (*Normaal Amsterdams Peil*)

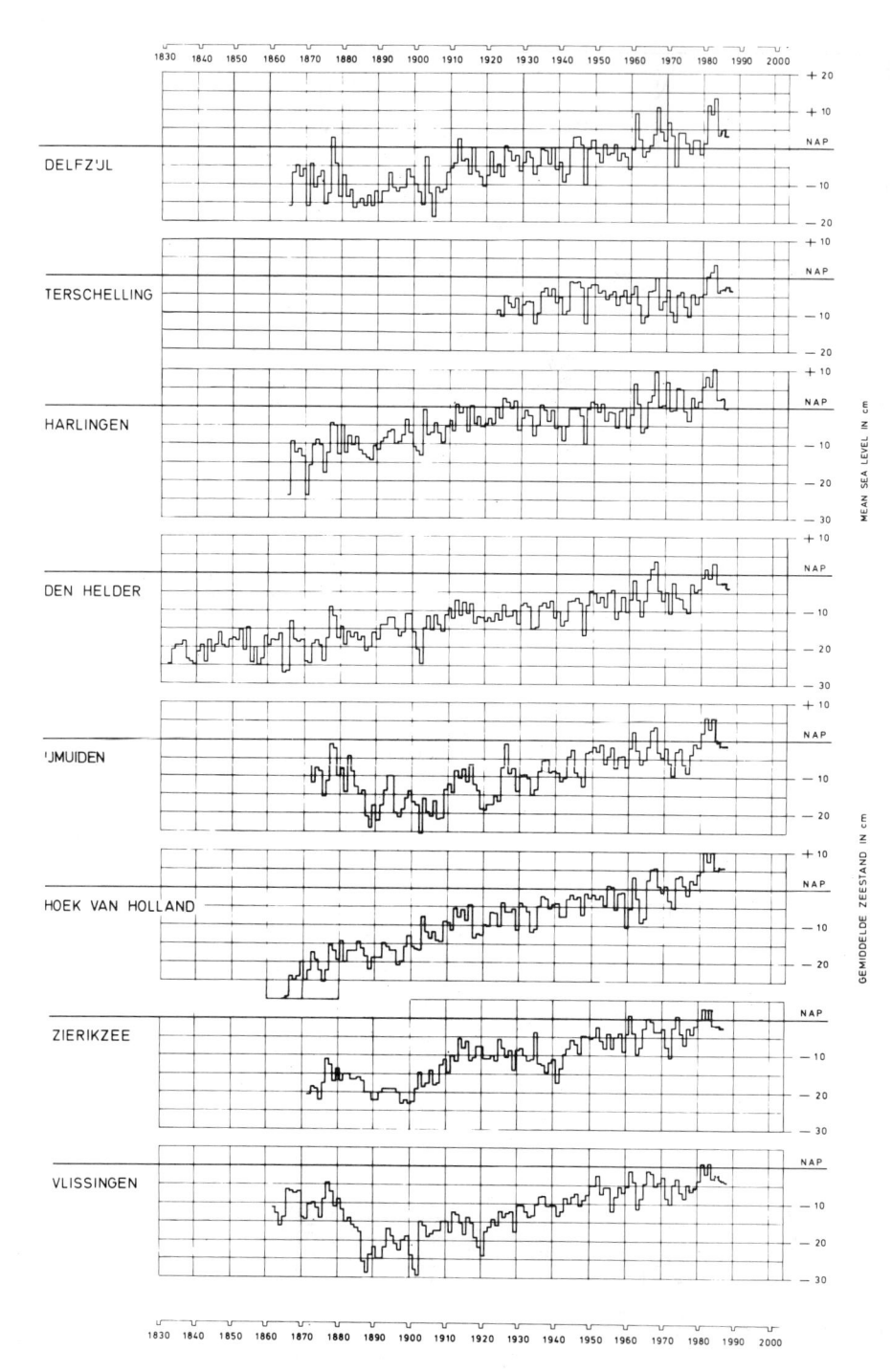

Figure 2.3 Annual mean sea-levels, related to NAP, at the main tidal stations

where it lasted some fifty years, after which the rise, prevailing over 1865–1910 (30–35 cm/century), has been resumed, maybe even at a higher speed. Finally the graphs show a third characteristic: oscillations from year to year as the result of noise.

This latter phenomenon is caused partly by natural variations and partly by human activities. Low frequency astronomical oscillations, even if not detectable, belong to the natural variations. However this first group comprises chiefly variations from year to year in: the intensities and distribution of wind set up and set down; the prevailing air pressure; the seawater density, determined by water temperature and salinity.

Of these two latter variables the annual mean values fluctuate considerably according to series for Den Helder from 1860 onward (determined from surface water samples, taken by bucket): the temperature between 8.7°C and 11.9°C, the salinity (mainly as a result of the predominantly northwards moving discharges of Rhine and Meuse) between 28.4 and 32.7$^o/_{oo}$ (van der Hoeven 1982).

From the human factor first of all the civil engineering works carried out in the Dutch tidal waters should be mentioned. Especially in the periods 1863–83 and after 1920 some of these were very large works indeed and the tidal movement at practically all Dutch tidal gauges has sooner or later been influenced by these human activities. But also levelling and data processing errors generate noise. In some cases the influences can easily be detected, as will be demonstrated later on. In other cases however such detection is far from easy, and perhaps impossible; this holds often when the interference occurred more or less gradually. Elimination of other noise factors was difficult and too laborious to be attempted in the pre-computer age.

In view of the planning of the Delta works about 1960, the conclusion was drawn with regard to the rise of mean sea level that it amounts to 15–20 cm/century along the Dutch coast. No special attention was paid to the changes of annual mean high tide (HW) and annual mean low tide (LW). Apparently it was assumed that these factors were liable to the same changes as Z (= annual mean sea level), implying that a rise of Z does not affect the annual mean tidal range (TV); this assumption seemed obvious as the rise of Z is quite small compared with the depth of the North Sea.

After activities in respect to this item had almost been stopped for about ten years, the study was resumed in 1982 and extended to the changes of HW, LW and TV (de Ronde 1982). In the following paragraphs some results of the investigations made thereafter are presented with particular reference to the gauges at Delfzijl, Harlingen, Den Helder and Flushing (Vlissingen).

2.3 CHANGING MEAN TIDAL LEVELS AT SOME DUTCH GAUGES

From each of the four gauging stations four time-series are available. As a first approach, a regression line was computed (using ordinary least squares) for each of these series, the regression line being composed of a straight trend line plus a sine with a period of 18.6 years. This sine allows for the nodal tide, being the only long-term astronomical oscillation, which according to a tentative study exerts a distinct influence on the water levels. It can be expressed in the following formula:-

$$Y = At + B \sin\frac{t + k}{18.6} \cdot 2\pi + C$$

where

Y is the quantity concerned (Z, HW etc.);
t is date;
k is a constant;
A, B, C are coefficients to be optimized.

From astronomical considerations it follows that the smallest value of k is −2 (years) and that $B > 0$ for HW and TV whilst $B < 0$ for LW.

Strictly speaking, the straight trend line is no more than a first order approximation (see section 2.2) but this is not necessarily a prohibitive objection. Furthermore it may be mentioned that in the series of TV much of the noise has been eliminated as most noise factors, mentioned earlier in section 2.2, have the same effect on both HW and LW. For the computation of the regression lines only data from 1900 onwards have been used.

The first station to be considered will be Delfzijl. Until about 1965 Z, HW and LW all show rising trends (figure 2.4), which according to the trend line computed over the period 1900–60 amount to 15 to 17 cm/century. As a consequence TV is practically constant over the same period; of all four variables this quantity shows the best accordance with the sine. But it is also quite clear that in the neighbourhood of this gauge drastic changes occurred in and after 1965, disturbing the evolution of HW, LW and TV substantially, though not influencing Z. The causes of this phenomenon are not difficult to trace: extensive harbour works locally and major dredging operations in the main tidal channel giving navigational access to Emden. In fact for the evolution up until about 1965 figure 2.4 confirms the assumption of 1960, mentioned in section 2.2.

DELFZIJL 1860-2000
COMPUTED OVER 1900-1960

$\overline{Z}$ = mean sea level
$\overline{TV}$ = mean tidal range

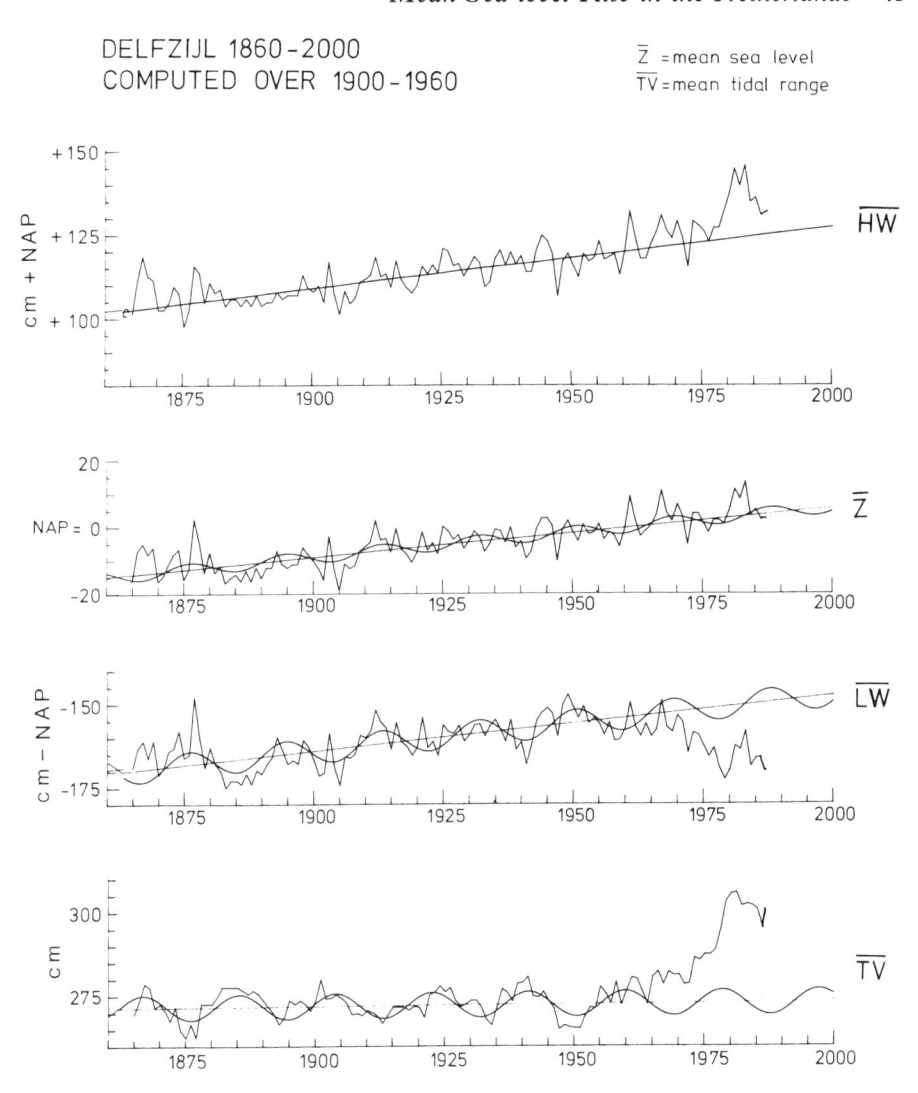

Figure 2.4 Annual mean sea-level, annual mean tidal levels ($\overline{HW}$, mean high water; $\overline{LW}$, mean low water) and annual mean tidal range at Delfzijl

The graphs for Harlingen (figure 2.5) demonstrate round about 1931/2 an even greater interference than the one for Delfzijl about 1965. Beyond any doubt the enclosure of the Zuider Zee by means of a dam of well over 30 km length, completed in 1932, caused this important change, consisting of a 16 cm rise of HW and a 34 cm drop of LW. As a result TV, which before 1932 hardly changed, increased 50 cm; moreover it started a rise at a

44 *J. van Malde*

HARLINGEN 1860-2000
COMPUTED OVER 1933-1984

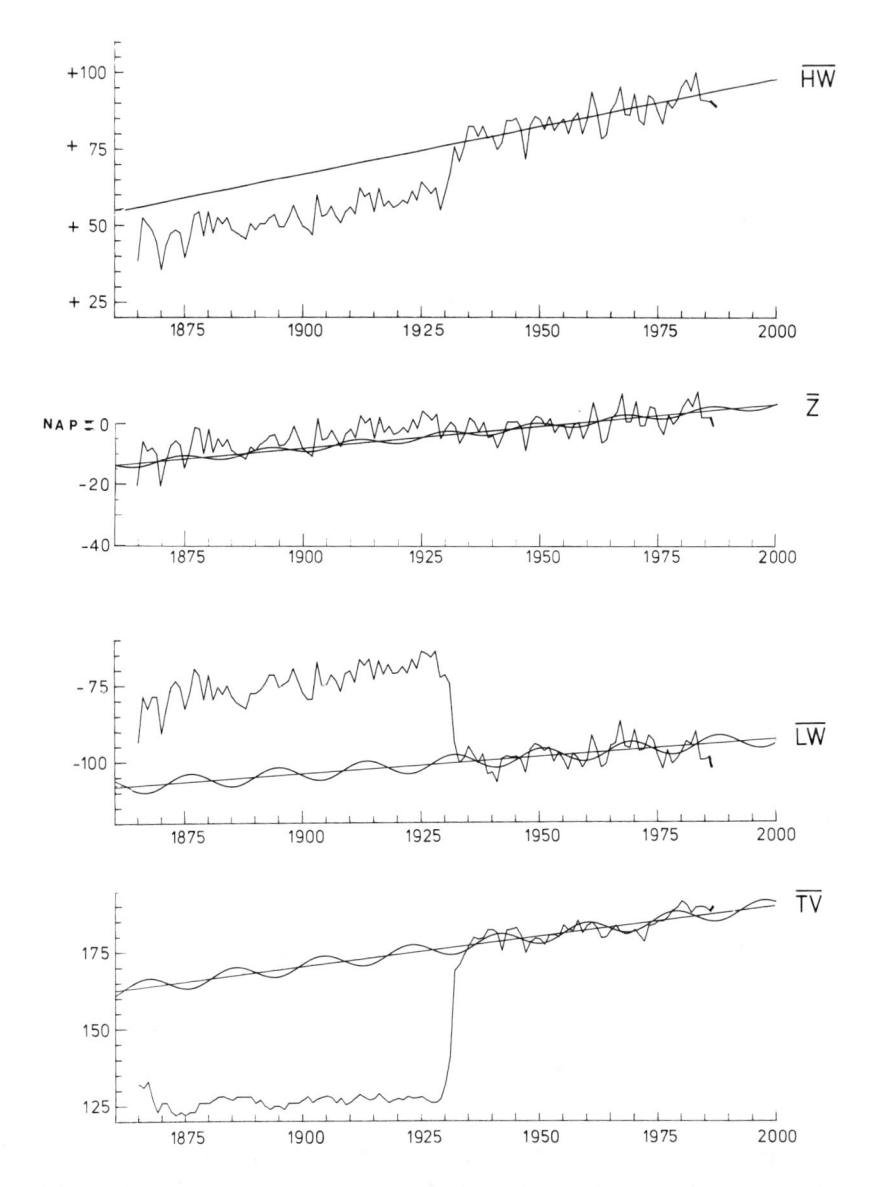

Figure 2.5 Annual mean sea-level, annual mean tidal levels and annual mean tidal range at Harlingen (units on vertical axes as in figure 2.4)

speed of 20 cm/century, mainly because of a substantial decrease of (r)LW (= speed of rise of LW). It is doubtful whether Z was influenced (by and large it has been constant between 1910 and 1950–60), but if so it would have dropped a few cm just before 1930. For the rest it should be noted that Harlingen, though as a port of some importance via a navigation channel well approachable by ships, is situated at the edge of a very shallow sea – the Wadden Sea.

At Den Helder (figure 2.6) the influence of the enclosure of the Zuider Zee is much less, HW rising 7 cm and LW dropping 8 cm here, so that TV increased 15 cm; on the other hand there is no reason to assume any change in Z round about 1932. In the period thereafter $(r)Z \approx (r)HW \approx 18$ cm/century, but $(r)LW = 8$ cm/century. As a result TV has risen 9 to 10 cm/century since; however before the enclosure of the Zuider Zee this quantity, known from 1832, had already undergone some changes in trend:

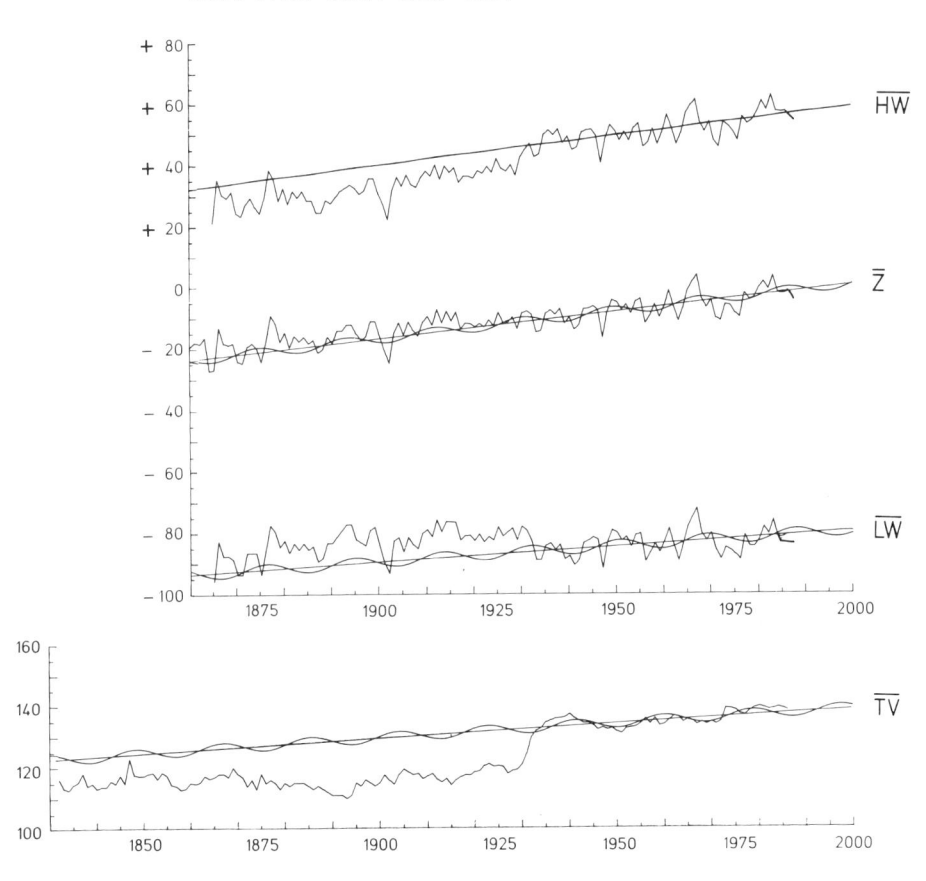

Figure 2.6 Annual mean sea-level, annual mean tidal levels and annual mean tidal range at Den Helder (units on vertical axes as in figure 2.4)

a decrease of over 5 cm between 1865–70 and 1890–5 and thereafter a somewhat greater increase.

For the three gauges at Delfzijl, Harlingen and Den Helder the causes of the considerable sudden changes in the courses of annual values of mean sea-level and mean high and low tide are quite clear. In each case the straight trend line, computed for the data from before and after the interference respectively , represents the evolution of the variable concerned satisfactorily. For all three gauges the nodal sine fits the data best for TV and is not relevant for HW (its relevance for LW and Z being doubtful). For both Harlingen and Den Helder TV has been substantially increasing over the last fifty years; for gauges along the German North Sea coast this phenomenon has also been established, in particular since 1950 (Führböter and Jensen 1985; Lohrberg 1989); this makes it the more striking that it does not manifest itself at Delfzijl. The increase of TV expressed before, is attended by (r)LW lagging behind (r)Z, whilst at Harlingen – contrary to Den Helder – (r)HW even surpasses (r)Z. Changes of TV point to changes in the tidal movement; this item needs further investigation, but from the data for Den Helder it seems likely that, contrary to the obvious view, in Dutch tidal waters changing of TV did occur before, at least locally.

Figure 2.7 shows graphs of mean tidal levels for Flushing, including graphs for mean spring and mean neap tide (the distinction between these tides is quite appropriate here in view of the relatively great differences between the relative quantities). From figure 2.7 and the graphs for mean low water (not produced here) the following conclusions may be drawn:

1   The straight trend lines fit quite satisfactorily and the sines for TV very well indeed.
2   The 'drop' of Z between 1877/1880 and 1887 (section 2.2) recurs in HW (and LW) – albeit with some delay – but definitely not in TV. However, the course of this variable TV compared to later does show a deviation before 1875.
3   (r)HW exceeds (r)LW, implying that TV has increased, certainly since 1875. On the average the increase has been 14 to 15 cm/century, falling somewhat behind however in the period 1930–50.
4   The rise of mean high water decreases in the sequence mean spring tide, mean tide and mean neap tide, as does the increase of the mean tidal range: according to the trend lines (r)HW equals successively 38.5, 33.5 and 26.0 cm/century and (r)TV 20.5, 14.5 and 6.5 cm/century.

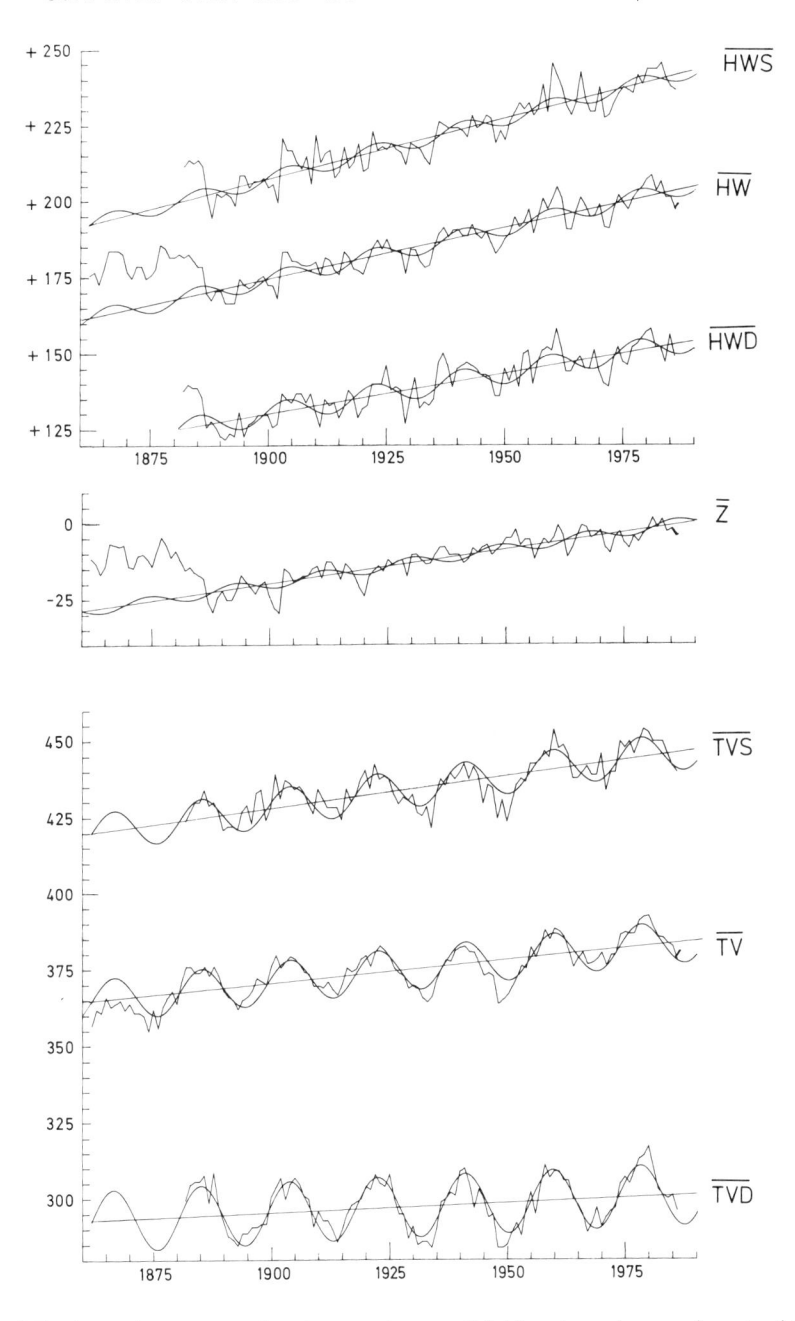

Figure 2.7 Annual mean sea-level, annual mean tidal levels and annual mean tidal range at Flushing (Vlissingen). (Units on vertical axes as in figure 2.4; D is an abbreviation for *doodtij* or dead tide or, in other words, the neap tide)

5 Even mean neap tide surpasses the speed of rise of mean high water (r)Z (= 22 cm/century).

The most striking point is the continuous increase of TV since 1875 or before, attended by (r)HW > (r)Z, and its consequence, i.e. changing tidal curves. It is unlikely that dredging in the Wester Scheldt (to improve the access to Antwerp) has been important in this respect, as such operations were up till about 1950 by and large not very extensive and were confined to the most eastern part of the estuary. Dredging in its wide mouth did not start before 1960, albeit almost at once on a large scale and indeed after some years influencing LW to some extent. According to recent model simulation, changes in the configuration of channels and banks in this mouth are responsible for this increase of TV over the period 1900–80 (Langendoen 1987), but this result should be regarded with reservation as the simulation does not represent the actual changes of HW and LW.

In view of these findings, TV for Amsterdam, the longest series available, was also investigated in spite of its location at the southern end of the shallow Zuider Zee, which evidently is far from ideal in this respect. This TV turns out to have been practically constant, both at the location before 1872 and at the location thereafter, i.e. after the enclosure of the IJ (figure 2.8). For Katwijk there are not enough data to make any statement about a trend of TV.

The widespread increase of TV might denote shifting of the amphidromic points in the North Sea, but this would just transfer the difficulty, as the relative rise of Z in the last fifty years has been very small compared to the depth of the North Sea.

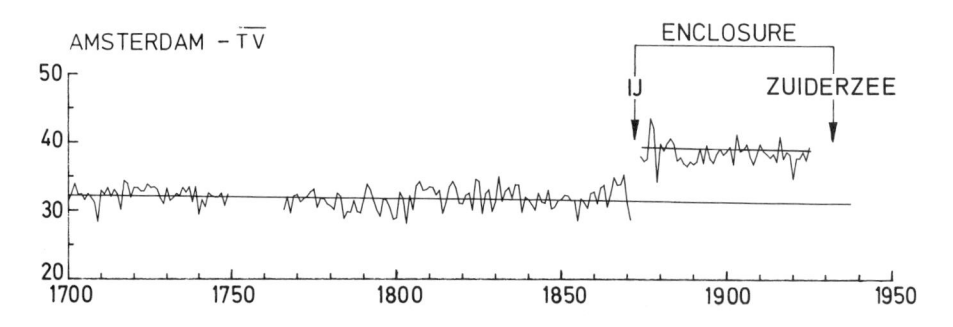

Figure 2.8 Mean annual tidal range at Amsterdam (vertical axis units = cm)

2.4 INFLUENCE OF THE NODAL TIDE

Figures 2.4 to 2.7 show that the nodal sine fits excellently to some of the quantities Y (e.g. TV Flushing), but it does not seem applicable to other quantities (HW Delfzijl, Harlingen, Den Helder). The relevance of this sine may be indicated by two factors:

1. its amplitude in its relation to the mean 'tidal amplitude' (= half the tidal range), i.e. the percentage p;
2. the ratio q between the standard deviation s_y, calculated in the computation of the regression line and the nodal amplitude, i.e. $q = (s_y:B)$ (see section 2.2). For a convincing relevancy of the nodal tide q should be less than ,say, 1 to 1.5.

Table 2.1 shows the values of q for the quantities considered and p for the most pronounced cases, being TV. From the table it follows that according to the standard just formulated the nodal sine found would be neither relevant to Z nor to HW and (at Den Helder) LW.

Table 2.1

	q for				for TV	
	Z	HW	LW	TV	p(%)	nodal ampl. (cm)
Den Helder	6.3	22.5	2.0	0.9	2.4	1.5
Harlingen	3.1	38.6	1.2	0.9	2.7	2.5
Delfzijl	2.0	45.1	1.0	0.8	2.8	4
Flushing: neap tide		1.2	0.7	0.3	6.7	10
mean tide	2.5	1.8	0.8	0.4	3.7	7
spring tide		1.8	1.1	0.6	2.3	6

It is remarkable that at Flushing p is decreasing with increasing TV whilst at the northern gauges the reverse (see also figure 2.2) seems valid for mean tide. Whether this latter tendency is of much importance might be questioned in view of the rather high values of q valid for TV at these stations.

The apparent relevance of the nodal sine to the tidal range implies that the influence of the nodal tide on the series of HW and LW cannot be negligible. Elimination of these nodal influences should improve the picture of the long term changes in these quantities. Yet in practice such elimination is no improvement when q has a high value. This may be caused by: local sensitivity for wind set-up and set-down (at the northern gauges a considerable wind set-up occurs rather frequently); inadequacy of the assumed linear trend.

It may be mentioned that, according to a maximum-entropy spectral analysis, the nodal cycle has a significant influence on Z at several west European gauges. For Delfzijl (1881–1964) and Harlingen (1865–1964) the amplitudes found are 14 and 10 mm respectively but for Den Helder (1865–1964) and Flushing (1881–1964) it could not be demonstrated (Currie 1976).

2.5 MAINTENANCE OF NAP-DATUM

The Dutch reference level NAP, established or re-established in 1684 in Amsterdam (section 2.1), was employed over large areas of the Netherlands by the eighteenth century, but the results, as far as they have been preserved, are hardly if at all usable. The first extensive 'accurate' levelling was carried out between 1797 and 1813, but many of its results were corrected after the first national precise levelling (1875–85) over almost the entire country (in view of these corrections the original designation AP was replaced by NAP). Linked to this precise levelling was the first German national precise levelling with the same reference level, however with another name (i.e. NN = Normal Null = normal zero) and established in Potsdam. Up till now four national precise levellings have been carried out in the Netherlands, of which the first three had the still available Hudde stones (section 2.1) as starting points. With each national precise levelling went the careful installation of benchmarks, of which the firm position is not easy to secure in large parts of the country, where the subsoil over a considerable depth consists of easily compressible Holocene layers. To meet this problem special 'primary benchmarks', founded on Pleistocene layers, were built as from the second national precise levelling (1926–40) onwards. Each of these primary benchmarks (numbering over 120 by now), supposed to be immovable, is in turn the starting point of periodic secondary levelling to 'secondary benchmarks'. The latter are spread all over the country in very large numbers.

Until recently the zero of a gauge in the Netherlands (being NAP) was verified by frequent levelling to 2 or 3 secondary benchmarks nearby and adjusted whenever necessary. In 1963 a programme was started to have a well-founded pile in or in the immediate vicinity of each gauging station, the benchmark of the gauge being a bolt on top of the pile. At present all tidal gauging stations are provided with such a pile, which is fully equivalent to a primary benchmark.

The first indications as to the (relative) stability of NAP are mostly from the national precise levellings (Waalewijn 1979; Waalewijn et al. 1987):

1. Relative heights of the Hudde stones:
   - in 1876 (local levelling), 5 stones: 8.0 mm difference in height between them (after more than 190 years) with a standard deviation of 3.7 mm;
   - in 1928, 2 stones, mutual difference in height: 1.1 mm.
2. Connections of Dutch to German precise levellings (v = NAP minus NN; $\bar{v}$ = mean value of v):
   - 1875–85, 4 connections; $\bar{v} = 0$ and $v_{max} - v_{min} = 60$ mm;
   - 1926–40, 5 connections; $\bar{v} = 21$ mm and $v_{max} - v_{min} = 9$ mm;
   - 1950–59, 9 connections; $\bar{v} = 13$ mm.

As the standard deviation $s_v \approx 17$ mm these results are not significant, but anyhow they give no indication that NAP has subsided in relation to NN, which is firmly founded.

The findings (1) and (2) may seem reassuring. An up-to-date approach demands also an overall comparison of the results of all appropriate precise levellings by applying advanced computation schemes. Such schemes, meant to calculate the relative vertical movements of primary benchmarks, require great computer capacity. Computations of this kind, for which Dutch data collected between 1926 and 1985 are available, were carried out by the Dutch Survey Department and were completed in 1989. It appeared that in the period, roughly speaking, 1930–80 the Amsterdam 'benchmark' did subside about 2 cm. Investigations as to the causes of this subsidence, its details and its consequences for both the definition of NAP and mean sea-level research in the Netherlands have just started.

2.6 SOME FURTHER PROVISIONAL RESULTS

The data available undoubtedly need further analysis. For that purpose some introductory efforts have been made, mainly for Den Helder, as the relevant data of this location (see also section 2.2) outnumber those of any other Dutch tidal gauge. The following series of Den Helder have been examined:

1 sea water surface temperature 1861–1981;
2 salinity of sea water surface 1936–81;
3 Z 1936–81;
4 mean monthly values of Z 1865–1984;
5 mean monthly values of TV 1944–84.

The findings of these investigations are as follows:

1 The strongly fluctuating annual mean sea surface water temperature shows between 1861 and 1900 as a trend a decrease from approximately 11°C to approximately 9.6°C, after which it rose about 0.5°C. The drop of Z between 1878 and 1887 at Flushing and elsewhere (section 2.2) may have some connection with it, but the relation is not very convincing (van Malde 1984).
1, 2 and 3 A simple correction of Z for the annual anomaly of the average atmospheric pressure does not lead to a clear reduction of noise – nor does an extended correcting procedure which also takes into account sea water temperature or sea water density.
4 The diagram of a spectral analysis, aimed to trace cycles with a period of less than two years in Z, shows the annual signal as very striking and both the semi-annual and a fourteen-month signal as less pronounced. Moreover both hardly surpass noise peaks.
5 For TV, amplitudes of cycles of certain fixed periods were computed by means of regression techniques. The sequence of clear presentation appeared to be: annual, nodal, then cycles of 8.85 years and semi-annual (both with an amplitude of just over 7 mm), 6.3 years and 4.42 years (amplitudes 3 and 2.5 mm), whilst the fourteen-month period is not relevant.

The choice of the periods considered may be elucidated as follows:

- The fourteen-month period concerns the polar tide or Chandler effect which is relatively large along the North Sea (Wunsch 1974), perhaps under the influence of meteorological effects (O'Connor 1986). Indeed the relevance of the Chandler effect for the Netherlands was first demonstrated by van de Sande Bakhuyzen in 1913.
- The 6.3-year-cycle, evidenced in 1976 for Z, has an unknown origin (Currie 1976).
- The 8.85-year-cycle is the period of the revolution of the major axis of the moon's orbital ellipse; the cycle of 4.42 years goes with this of course.

- Left out of consideration is the sun spot cycle, which may generate – whether via meteorological effects (Currie 1981) or not – an eleven-year term with an amplitude of roughly 10 mm in Europe. However this term was not traced for some of the gauges concerned, including Den Helder and Flushing, both with series up to 1964 (Currie 1976).

As from 1988 the investigation of Dutch sea-level data has been intensified as part of the contribution of the Netherlands to a European Community research project on mean sea-level changes. Within the framework of these activities special attention is paid to tracing relevant old Dutch data and to attempts to reduce noise in the series both of present and former tidal gauges. In addition, the efforts discussed above will be systematized and intensified, whilst it is also intended to apply more advanced analysis techniques such as Principal Components Analysis (with which a start has been made) and maximum entropy analysis.

One example of the work carried out so far may be mentioned here. Because of lack of data (only staff gauge readings available) or of personnel, about 80 per cent of the mean sea-level values from before 1935 were at the time determined in simpler ways than followed later. The noise, caused by the application of these simpler ways, is not negligible, in particular when mean sea-level was derived from half-tide (standard deviation 0.5 to over 1.0 cm) as was done for nearly half of these data (van den Hoek Ostende and van Malde 1989).

2.7 CONCLUSIONS

Rise of mean sea-levels in the Netherlands manifests itself clearly since 1885 and amounts on the average to 15 to 20 cm per century up until now. A complicating factor is the influence of large civil engineering works, which in some cases can be easily demonstrated, but in other cases this is not simple. Apart from the rise of mean sea-level, the mean tidal ranges have generally increased over the last fifty years (as they do at the German North Sea gauges) – at Flushing since 1875 – implying changes in the tidal movement which are not easy to explain. As this goes with a rise of mean high tide faster than the rise of mean sea-level, this development is of great importance to the mainly low-lying country. Consequently study of the long-term changes of high and low tide in Dutch tidal waters is also necessary; for this a close examination of the development of mean tidal ranges is most useful as they reflect astronomical effects best (the nodal one of course, but in any case also the 8.85-year cycle) and are much less sensitive to noise than other relevant quantities. Changes of the vertical (and the simultaneous horizontal) tidal movement as such need investigation too.

Planned further research on these items is not only required to improve our knowledge of past and present changes of mean tidal levels at Dutch gauges, but also to obtain a deeper understanding of their relationship with similar changes elsewhere and in particular in western Europe, and perhaps of their causes.

ACKNOWLEDGEMENTS

Special thanks are due to Mr J. Doekes for his valuable assistance.

REFERENCES

Currie, R. G. 1976: The spectrum of sea level from 4 to 40 years. *Geoph. J. R. Astr. Soc.* 46, 513–20

Currie, R. G. 1981: Amplitude and phase of the 11-yr term in sea level: Europe. *Geoph. J. R. Astr. Soc.* 67, 547–56

Führböter, A. and Jensen, J. 1985: Säkularänderungen der mittleren Tidewasserstände in der Deutschen Bucht (Secular changes of the mean tidal levels in the German Bight). *Die Küste 42*

Gottschalk, M. K. E. 1975: *Stormvloeden en rivieroverstromingen in Nederland*: II, 1400–1600 (Storm surges and river floods in the Netherlands, Part II). Assen, Van Gorcum (in Dutch with full annual summaries in English)

Hoek Ostende, E. R van den and Malde, J. van 1989: *De invloed van de Bepalingswijze op de berekende gemiddelde zeestand* (The influence of the ways of determining mean sea levels on their established values). Internal report, Rijkswaterstaat, Tidal Waters Division, No. GWAO–89.006 (in Dutch)

Hoeven, P. C. T. van der 1982: *Watertemperatuur– en Zoutgehaltemetingen* van het Rijksinstituut voor Visserij Onderzoek (RIVO), 1860–1981 (Observations of surface watertemperature and salinity, State Office of Fishery Research, 1860–1981). KNMI, Scientific Report W. R. 82-8 (in Dutch with detailed summary in English)

Langendoen, E. J. 1987: *Onderzoek naar de Vergroting van het Getijverschil te Vlissingen* (Investigation of the increase of the tidal range at Flushing). Delft Technological University, Faculty of Civil Engineering, report No. 5–1987 (in Dutch)

Lohrberg, W. 1989: Aenderungen der mittleren Tidewasserstände an der Nord Seeküste (Changes in the mean tidal water levels on the North Sea Coast). *Deutsche Gewässerkundliche Mitteilungen* 5/6, 166–72 (in German)

Malde, J. van 1984: *Voorlopige uitkomsten van voortgezet onderzoek naar de gemiddelde zeeniveaus in de Nederlandse kustwateren* (Preliminary results of prolonged investigation on mean sea-levels in the Dutch coastal waters). Internal report Rijkswaterstaat, Directorate of Water Management and Water Research, No. WW-WH 84.08 (in Dutch)

O'Connor, W. P. 1986: The 14 month wind stressed residual circulation (Pole Tide) in the North Sea. NASA, *Technical Memorandum* 87800.

Ronde, J. G. de 1982: Changes of relative mean sea level and of mean tidal amplitude along the Dutch coast (*Proc. NATO Advanced Research Workshop*, Utrecht, 1–4 June 1982)

Veen, J. van 1954: Tide-gauges, subsidence-gauges and flood-stones in the Netherlands. *Geologie Mijn* NS 16, 214–19.

Waalewijn, A. 1979: *De Tweede Nauwkeurigheidswaterpassing van Nederland 1926–1940* (The second precise levelling of the Netherlands 1926–1940). Rijkscommissie voor Geodesie (in Dutch)

Waalewijn, A. et al. 1987: *Drie eeuwen Normaal Amsterdams Peil* (Three centuries NAP). Rijkswaterstaat-serie No. 48, 2nd revised edn (in Dutch)

Weele, P. I. van der 1971: *De geschiedenis van het N.A.P.* (The history of NAP). Rijkscommissie voor Geodesie (in Dutch)

Wunsch, C. 1974: Dynamics of the Pole Tide and the damping of the Chandler wobble. *Geoph. J. R. Astr. Soc.* 39, 539–50.

3

Vulnerability of the Belgian Coastal Lowlands to Future Sea-level Rise

C. Baeteman, W. de Lannoy, R. Paepe and C. van Cauwenberghe

3.1 INTRODUCTION

The coast of Belgium is characterized by a low-lying plain which is situated at about mean sea level and protected from the sea by dikes and dunes. The plain, with an average width of 15 km, borders the southern North Sea coast and extends over a total length of 65 km. An important extension occurs in the western part along the river Yser The topographical level of the plain ranges from 4 to 2 m TWA, which in reality means at mean sea-level to +2 m (the Belgian reference zero represents mean low water level at spring tide, which is 2.03 m below NAP) (figure 3.1).

Only two reclaimed areas situated below mean sea-level occur; one of them, called De Moeren, is located in the far west and lies 1 to 3 m below mean sea-level; the other, called De Lage Moere, near Brugges in the eastern part of the plain, is at a level of 1 m below mean sea-level. The general topography dips southwards from the sea into the hinterland, thus causing great difficulties for drainage.

The river Yser is the only river in the coastal plain. Its mouth is located at the town of Nieuwpoort where an important sluice complex controls the drainage of a great part of the plain. In history, and not at least during the First World War, these sluices played an important role in the artificial inundation of the area.

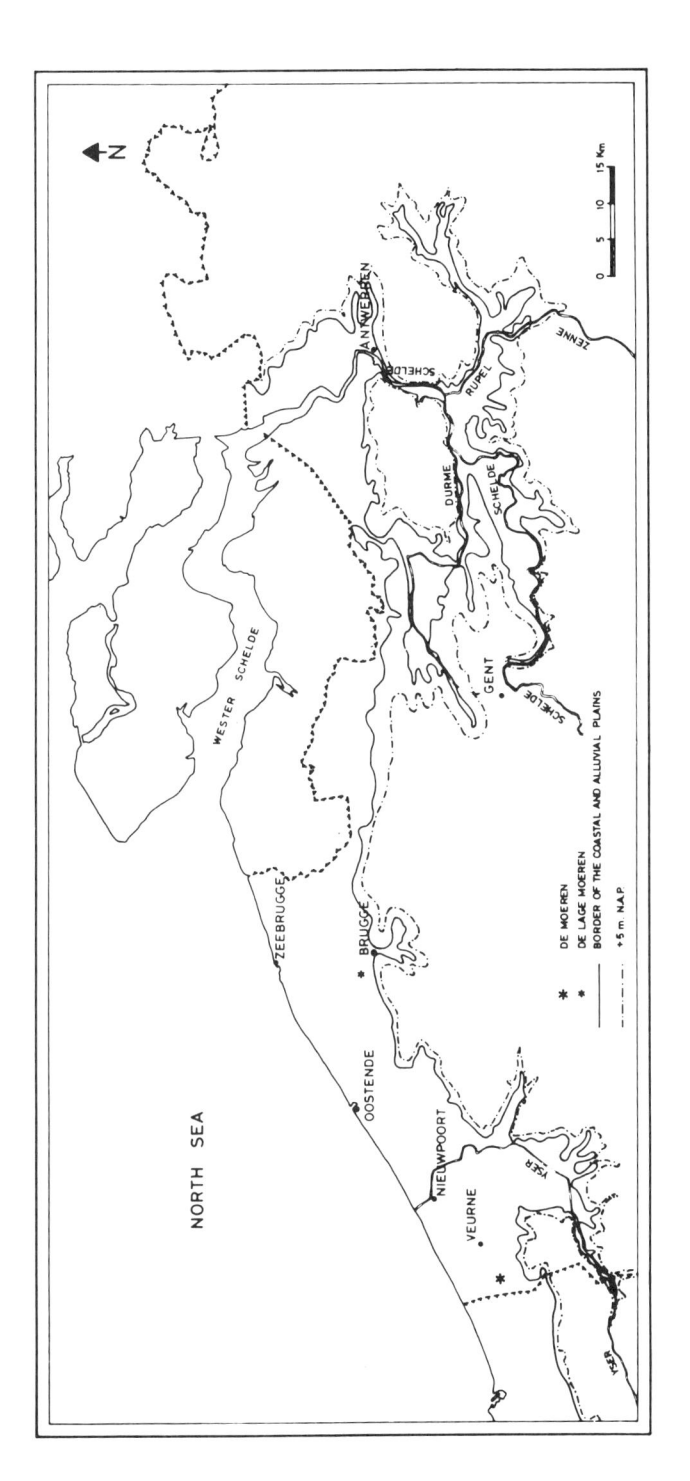

Figure 3.1 Location of the lowlands along the Belgian coast and the Scheldt estuary. The area below the +5m (NAP) contour line is considered to be a potentially hazardous region with a sea-level rise of one metre

The only two remaining natural areas not protected from the sea by man are the Zwin, a sandy salt marsh behind the dunes in the far east of the Belgian coast, and a very restricted tidal mudflat with salt marsh on the east side of the Yser mouth in Nieuwpoort.

3.2 THE SHORE AND DUNE BELT

The Belgian shoreline is a mesotidal, wind-dominated clastic shoreline with a tidal range between 4 and 5 m. The shore shows a typical runnel and ridge beach. As on the Dutch coast, wave action is generated by the prevailing westerly winds, and the coast is periodically subject to storm surges from the south-west, west and north-west. The north-westerly storms especially cause extreme high tides and high energy waves.

There is a dune belt all along the shoreline. In the western part the belt is 1.5 to 2 km wide and is very well developed, reaching elevations up to 10 to 20 m in general. The highest dune, located at Koksijde, reaches an altitude of 30 m. In the central and eastern part of the coast, the dune belt is alarmingly narrow (50 m to 800 m), although it is fairly high (ranging between 10 and 25 m, with extreme heights of about 50 m). In the far east near the Dutch border the dunes reach a height of only 5 to 10 m and are in general 1.5 km wide.

Beach erosion has been observed at different parts of the coast. De Moor (1979) described several places where severe erosion of the beach and the foredune is occurring, despite extension of defence structures and beach nourishment. The author discusses in more detail the erosion of the beach between Bredene and De Haan, where in the period between 1970 and 1979 a residual retreat of the dunefoot of about 40 m was observed. The defence structure, aiming to create an artificial beach by sand nourishment, failed completely. A survey of beach profiling during three consecutive years yielded that the main erosion consists of the following phenomena: a long-run residual lowering of the beach; a back-cutting of the back beach; a direct spring tide wave attack on the dunefoot; and a retreat of the dunefoot itself in the zones without a seawall (De Moor 1979). According to the author, the erosion is a natural phenomenon and consists of the erosional phase of coastal megaprotuberances.

The highest waterlevels ever recorded are (in m TAW, i.e. 2.03 m below mean sea-level):

along the coast:
Nieuwpoort	6.73 m (1953)
Ostend	6.66 m (1953)
Zeebrugge	6.69 m (1953)

along the Schelde:
Terneuzen	7.29 m (1953)
Doel	7.76 m (1953)
Antwerp	7.77 m (1953)
Temse	7.34 m (1976)
Dendermonde	6.94 m (1976)
Melle	6.45 m (1980)

The human impact on the shore and dune belt

In many locations, the coastline is directly bordered by high apartment buildings, thus forming an almost continuous concrete wall, sometimes called the new Atlantic Wall (plate 6). These buildings are protected from the sea by a seawall. Actually 83 per cent of the shoreline is protected by a sea defence structure.

The logic or usefulness of all these defence structures still remains an open question, as some particular situations can be observed. Seawalls are usefully constructed as a protection in areas where erosion occurs and when men and resources from the upland become endangered by inundation and erosion. But west of De Panne, for example, a 3 km-long seawall was built in front of a dune belt which is nearly 2 km wide and where neither property nor human life can be endangered, as the area is protected as a nature reserve. It is true the foredunes are subject to erosion, but it seems that the meaning of protection has been interpreted somewhat too widely, as the erosion occurs mainly in wintertime (Depuydt 1967), which is the usual situation.

It would be very interesting to leave this (restricted) erosional phenomenon to nature and observe and quantify the response of the beach and foredune to possible increased storminess and sea-level rise. Moreover due to this erosion (which is not a severe threat to this dune belt) sand would be added in the littoral system and deposited on other sand-deficient beaches.

Another special situation with respect to the planning of sea defence structures can be observed at Oostduinkerke. Large houses and apartment buildings, representing great investments, have recently been built on the foredune, about 300 m from the high water mark, without any protection at all. The private view of the beach is first-rate; the question however is for how long.

The dune belt has also been used as a zone for residential building, except for a few localities. A great number of houses and apartment blocks are built, requiring a dense road infrastructure. At Oostduinkerke a car park has been built in front of the foredune. Moreover an important part of the central dunebelt is used as a golf course. To what extent this considerable human impact on the shore and the dunes is detrimental is still to be quantified.

While there is broad agreement that seawalls are detrimental to adjacent beaches and that they are passively responsible for narrowing the beaches in front of them, controversy still remains over the question of whether seawalls play an active role in beach degradation (Pilkey and Wright 1988). Kraus (1988) in a review of the subject concluded that the majority of quantitative-type field studies indicate that seawalls do not accelerate long-term erosion of beaches if there is ample sediment or a wide surf zone exists. But if beaches are deficient in sediment or if sea-level is rising, erosion is more likely to occur on armoured beaches compared to unarmoured beaches. De Moor (1979) concluded for the Belgian coast that the classical defence structures have proved to be useful by only impeding further retreat of the dunefoot. Lowering of the beach itself still continues.

3.3 SYNOPSIS OF THE TIDAL OBSERVATIONS ALONG THE COAST: CONSIDERATIONS WITH RESPECT TO SEA-LEVEL CHANGES

In Ostend tidal observations, i.e. high water (HW) and low water (LW) records, started in 1820 (van Cauwenberghe 1977, 1985). They were established on a tide gauge near a lock in the harbour. Unfortunately, for the period 1820–34 all data are lost; it is only for 1835–52 that monthly mean values of HW and LW are found in a manuscript of Henrionet (1876). At that time, the tide gauge was linked to a reliable benchmark which was incorporated on top of a quay in the vicinity.

As only monthly mean values of HW and LW are available, mean tide level (i.e. mean HW and LW) can be calculated. The difference between this level and mean sea-level (MSL) along the coastline is fairly constant (e.g. 0.063 m + 0.006 for the period 1949–88). So, knowing the mean tide levels, it is possible to determine the MSL values for each year for the period concerned.

The tidal records of the period 1866–71 were observed by a mechanical tide gauge; however, they were lost, and those of 1878–1914 have been subject to many interruptions in the observations, so that they cannot be used for further consideration.

From 1927 until now, reliable and continued records for Ostend are available in the Coastal Hydrographic Office (the years 1940, 1942 and 1944 are very disrupted owing to the circumstances of the Second World War). For the observation sites at Nieuwpoort and Zeebrugge valuable records are available since 1967 and 1964, respectively, but they cover too short a time span for any conclusive interpretation with respect to sea-level changes.

Best-fit curve calculations on the Ostend records for the entire period, i.e. from 1835 to 1852 and from 1927 to 1988, indicate a linear increase of 0.01 m per decade for HW, LW and MSL. So far no indication of an acceleration in the increase has been shown on the moving averages of the MSL or on the graphs of the annual values.

Tidal observations are of course relative measurements. Whether the tidal observations are an indication of a rising sea-level or a record of the subsidence of the land is still to be established.

3.4 EFFECTS OF A FUTURE SEA-LEVEL RISE ON THE COASTAL PLAIN

In general the major impacts of a future sea-level rise in coastal lowlands are permanent inundation, loss of protective beaches, increased flooding and salt water intrusion (Titus 1986).

However, a rise in sea-level would seriously threaten a large portion of the coastal lowlands. The most obvious impact is that nearly the entire coastal plain of Belgium would be permanently inundated. In the case of a series of heavy storms the actual sea dikes will certainly not be high enough, and storm waves may overtop the sea walls. On the other hand, areas that are below or at sea-level have their unique characteristics of low elevations and high ground water tables which require drainage by pumping.

The impact of a rising sea-level on the coastal area is influenced to a great extent by its Holocene geological setting.

Geological setting of the Belgian coastal area

The coastal deposits represent the major infilling of the area under marine, fresh water and terrestrial conditions during the Holocene. The deposits reach their greatest thickness of about 30 m in the seaward region and wedge out towards the Pleistocene hinterland.

The western plain shows a different Holocene geological history from the eastern coastal plain, resulting from the altitude of the Pleistocene subcrop. In the eastern part the subcrop occurs at a considerably higher elevation, so that the influence of the Holocene sea-level rise started much later. The western part of the coastal plain is considerably enlarged by an important southern extension along the River Yser. As a consequence the western part contains a greater record of geological events both in time and in space (Baeteman 1981).

In the western part, the unconsolidated deposits are characterized by lateral zonation. In the seaward region, only marine and brackish clastic sediments are present overlying a basal peat layer in some places. In the central part of the western plain, the deposits consist in general of an alternation of brackish-marine sediments and peat layers. Towards the Pleistocene hinterland, the deposits are formed by only a basal peat layer overlaid by a cover of clastic brackish-marine sediments, while at the border of the outcropping Pleistocene area, the cover of brackish-marine sediments form the entire Holocene sequence (figure 3.2).

The sequence of coastal deposits is a result of the Holocene sea-level rise which initially was very rapid. During the last glacial age, sea-level was at least 100 m lower, and 9,000 years ago it was about 45 m lower than at present. On land, the very first evidence of a rising sea-level is reflected in the development of a basal peat. However, the continuous and rapid rise of

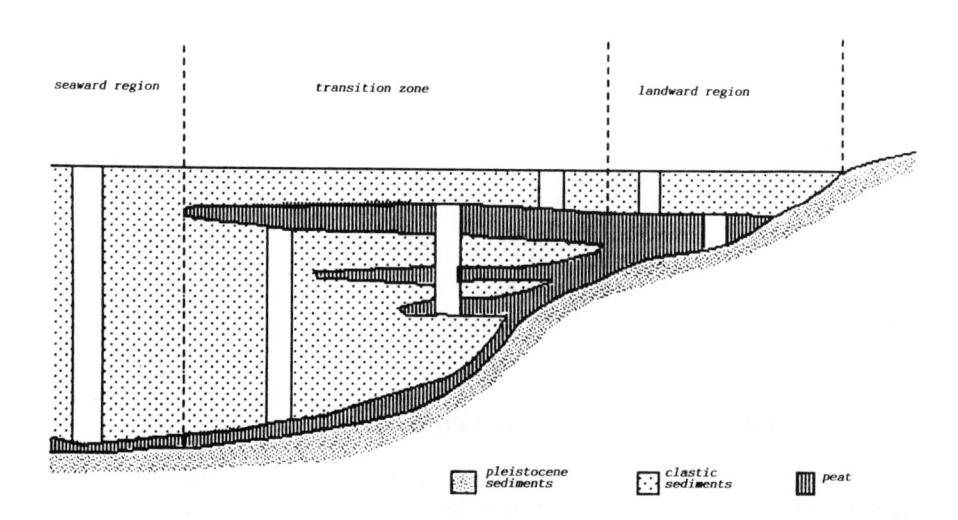

Figure 3.2 Schematic cross-section of the Holocene deposits in the western part of the coastal plain

the sea-level ultimately resulted in the deposition of sediments in the lower parts of the area.

The oldest known onset of the marine sedimentation, observed at an altitude of −16.60 m, occurred at 8440 ± 130 years BP (Baeteman 1989). From then onwards a tidal flat, and more particulary sandflats and associated tidal gullies, started to develop, characterized by a continuous deposition of clastic sediments. At the same time, the marine influence shifted landward and upward causing the end of the basal peat growth in these areas (figure 3.3).

From 7000 years BP a significant change in the general tendency took place as the rate of sea-level rise decreased (Baeteman 1987 a,b, 1989). In the residual valleys, previously infilled with sandflat deposits and fluvial deposits in the southern areas, peat growth started while in the rest of the (contemporary) plain, mudflat and salt marshes developed. From 6400 years BP, however, general peat growth is observed over nearly the entire (contemporary) plain, showing an important landward extension in both western and eastern parts (figure 3.3).

This peat layer represents the onset of what is usually called the typical cyclic formation of coastal deposits where peat repeatedly came into being alternating with the deposition of tidal sediments. The cyclic formation with the intercalated peat layers, indicating temporary regressive tendencies, generally came to an end in the time interval of 2700–2200 years BP, and in more landward areas, between 1900 and 1600 years BP.

In the eastern part, this typical cyclic formation is not so well developed. In general the Holocene sequence consists of only a basal peat that developed from 5600 years BP till 2500 years BP, overlaid by a clastic cover a few metres thick (Mostaert 1985, 1987).

The seaward extension of the coastal plain could not be established yet. Most of the evidence is situated offshore. Moreover the beach and possible dune deposits have been reworked continuously since the beginning of the Holocene. Indeed the potential for preservation of these sedimentary environments is very low in the case of a transgressive coast.

In the western part of the coastal areas, more particulary west of Nieuwpoort, older dune systems were preserved, however. An old dune group, called the Adinkerke-Ghyvelde dunes, occurs in an isolated position in the coastal plain near the French border (figure 3.4). The dunes are believed to be older than 4300 years BP (de Ceunynck 1985, 1987; Depuydt 1967). (This last remnant of older dunes is now threatened by the construction of a new highway.) From 3,000 years ago the coast (at least the western part) was characterized by progradation and a new dune system, called the Old Dunes of De Panne, developed seaward from the Adinkerke-

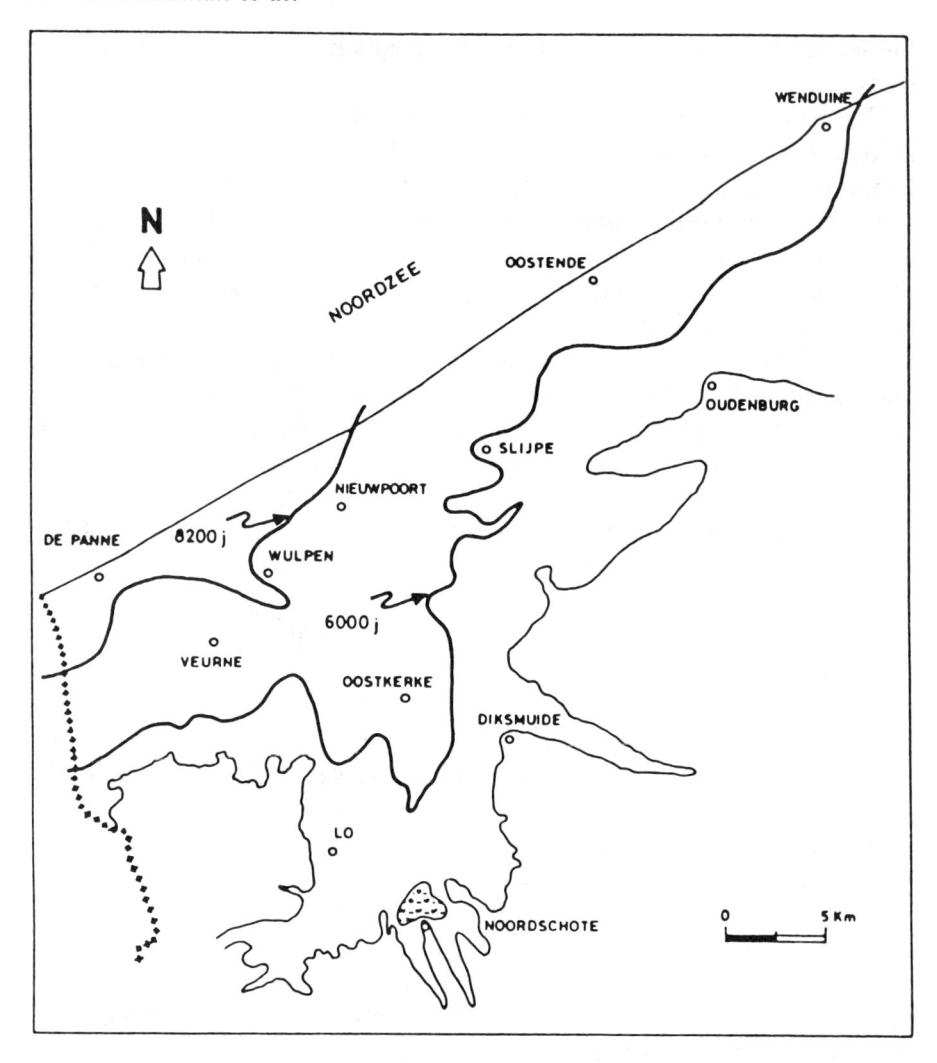

Figure 3.3 Landward extension of the coastal plain at 8200 years BP and 6000 years BP
Source: (After Baeteman 1989).

Ghyvelde dunes (de Ceunynck 1987). All evidence, and not least from archaeology, points out that the Old Dunes of De Panne were on the coast until the Roman period (de Ceunynck and Thoen 1981).

After the Roman period significant changes at the coast occurred. Severe transgression(s) took place eroding, amongst other things most of the Older Dunes of De Panne. Only from the eleventh century AD the formation of a new dune system (the Younger Dunes) was observed, which started

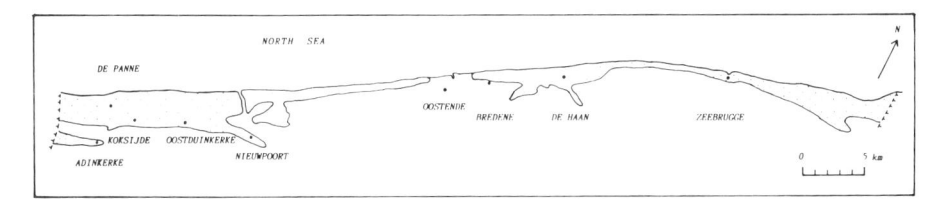

Figure 3.4 Location of the dune belt along the coast and the Old Dunes of Adinkerke-Ghyvelde. The dune belt is alarmingly narrow in the central part, and well developed only in the western part of Belgium.

probably as a result of increased storminess. From the end of the fourteenth century large parabolic dunes developed. Their formation was probably related to an increased erosion along the coast, yielding more sand to the littoral system (de Ceunynck 1985).

The vulnerability of the clay/peat area of the coastal plain

Those areas where the deposits are characterized by the alternation of peat and clay are very vulnerable. They are situated approximately at mean sea-level and the ground water table nearly reaches the surface. Hence these areas have very bad natural drainage. At present the evacuation of superfluous water is a great problem and in rainy seasons most of these areas are regularly flooded, in particular those along the River Yser. This critical situation is enhanced by the fact that the clay and peat sediments are very sensitive to compaction and as a consequence the area is subject to considerable land subsidence. In this case pumping is out of the question. Artificial drainage based on pumping certainly will lower the ground water table, but will simultaneously lower the land surface; thus the initial problem is not solved but the risk for flooding is increased. Besides the problem of land subsidence, pumping also brings the risk of intruding brackish or salt water into the upper part of the phreatic ground water.

It is obvious that (better) management for drainage is to be considered critically in the near future, if complete flooding of the area is to be avoided. It is true that general flooding of a coastal lowland and consequently the deposition of new sediments are not the result of one single storm surge only. Much more important – and critical – is the situation of the area itself with respect to the water control. In well-drained areas, superfluous water from temporary floods will be relatively quickly under control. On the other hand, in areas with deficient drainage it is clear that the evacuation of the superfluous water will become impossible. This can lead to continuous flooding from the invading sea attended with erosional incisions, resulting in considerable damage.

Proper management of the drainage system will require careful maintenance of the numerous ditches and canals and their dikes. At present many of these dikes are certainly not high enough and in too bad condition to ensure optimal evacuation of a sudden excess of water and sufficient protection.

The vulnerability of the dunes and beaches

A coastal dune belt pre-eminently forms the natural coastal defence against storm surges. But more than half of the dune belt is much too narrow to resist a period of storms. The dunes will be eroded, partly or even entirely during a heavy forty-eight-hour storm.

A rise in sea level will cause the shoreline to retreat. Beaches follow a characteristic profile. If sea level rises, the entire profile must rise by an equivalent amount. The sand necessary to raise the profile will generally be supplied by the upper part of the beach, thus resulting in a landward shift of the entire beach profile (Bruun 1962. See figure 5.14).

However in many places the first line of dunes facing the beach has been replaced by apartment buildings, hotels and concrete seawalls. Consequently there is no possibility any more for a natural sediment exchange between foreshore, beach and dunes, and during periods of high energy waves and storms there will be no sediment supply from the first dune ridge to the beach. As most of the beaches fronting the seawall are narrow, the recovery process may be absent (Kraus 1988). This lack of sediment supply can accelerate beach erosion, and the undermining of the seawall protecting the buildings is not inconceivable.

Inherent in the process of infilling of a coastal plain is the occurrence of numerous former gullies at different depths. When the gullies are post-Roman in age, they are filled with loosely packed sand saturated with water. These sandbodies are easily subject to liquefaction (quick sand flow) when the pressure on them suddenly is reduced. This situation might occur where such a former gully intersects the actual shoreline and if the pressure, formed by the fresh water pocket in the dunes, decreases considerably in the case of erosion of the dunes. This very vulnerable situation exists in particular where the dune belt is narrow, i.e in the central part of the coast.

Last but not least, the dune belt is the only natural fresh water reserve besides rivers for the coastal plain and represents the resource for drinking water for the coastal region. The broader and higher the dunebelt is developed, the bigger the fresh water reserve. But the withdrawal of fresh water from the dunes is not without danger either. The fresh water reserve forms a pressure preventing the sea water from intruding into the coastal aquifers. An excessive pumping of water, without sufficient recharge in good time, will raise the central base of the fresh water lense in the dune. As a consequence salt water infiltration will be enhanced, leading to salinization of the coastal aquifers.

It is clear that an integrated management for the dunes is required. Dunes are a very vulnerable environment and numerous influential elements can quickly cause the depletion of it. Human activity, even outside the dune area, is one of the essential causes. Some of these activities, frequently observed, are: obstructing and intercepting the sand supply (also from outside the dunes) with constructions; removal of sand which then is irrevocably lost from the littoral system in order to clear the roads and the entrances to the apartment buildings; destroying the vegetation so that the wind can easily blow up the loose sand leading to eventual erosion; lowering the ground water table in the dunes for the supply of drinking water or due to drainage in the areas adjacent to the dunes.

All these activities are to be avoided at all costs for the preservation of the dunes. It is very clear that the management of the dune area is becoming a matter of political interest.

3.5 LAND USE IN THE COASTAL AREA

Agriculture is the dominant type of land use in the coastal plain. Industry is concentrated in the port zone of Zeebrugge (which has been expanding during the last ten years) and on industrial estates in Ostend, Brugges and Veurne. The 31,000 industrial jobs account for about 25 per cent of total employment in the coastal plain. Fishing is concentrated in the harbours of Ostend, Zeebrugge and Nieuwpoort, which dominate the national fish trade.

The population of the coastal plain stands at approximately 360,000 inhabitants with important urban concentrations in Brugges (118,000 inhabitants) and Ostend (69,000). Population densities of the municipalities of the littoral fringe range from 187 inhabitants per km^2 in Middelkerke to 1830 inhabitants per km^2 in Ostend. This situation is entirely different during summer holidays when on a peak day about 250,000 day trippers join the 300,000 resident tourists and the 180,000 local residents.

The Belgian coast is indeed the longest most concentrated area of major seaside resorts along the Channel and the North Sea. In terms of tourist activity the coastal zone is the most important region of the country. In 1982, 20.4 per cent of all vacations (at least four nights) of the Belgian population took place on the Belgian coast (68 per cent of the vacations of Belgians were taken abroad). In 1985 tourists (Belgians and foreigners) spent 15.3 million nights on the Belgian coast, representing 49 per cent of all (registered) tourist nights in Belgium. Fifty-one per cent of all day trips in 1982 had the coastal zone as a destination. All these activities take place along a coastline only 65 km long, so that congestion is the main problem on peak days (Boerjan 1985; Vanhove 1973).

Tourist activity on the Belgian coast has been growing for more than 100 years. The influence of Leopold II (Belgium's second king) was important; he compared the Belgian coast to a gold mine waiting to be exploited. After the Second World War the evolution in the tourist activity grew very rapidly. The increase of free time and income in the 1950s and 1960s and growing motorization brought about an explosion in demand for tourism. The demand for accommodation causes a rapid increase of land prices, which makes it easy to understand that many hotels and houses are replaced by apartment buildings. Apartment tourism soon becomes the dominant type of accommodation. A building boom took place especially after 1955. Project-developers were on the hunt for decayed hotels and houses and for new building sites (often located on land of great value from an ecological point of view). *Belle époque* houses were replaced by monotonous high-rise buildings which formed a new Atlantic Wall. Seaside resorts are losing their identity and are expanding towards each other, leaving few open coastal spaces and natural dune areas. Holiday villages and new camping grounds are built in the polders. Camping tourism expanded considerably and also consumed much of valuable space in the dunes and polders. Many camping grounds are concentrated along the littoral fringe, in particular in those localities where the dune belt is alarmingly narrow. The type of accommodation on camping grounds changed from tents to caravans and solid constructions. Camping sites became overcrowded and disorderly and sometimes lacked basic amenities. The infrastructure for recreation (swimming pools, tennis courts etc.) took additional space.

The explanation of the destruction of the natural landscape of the coast resides in the objectives of the tourist sector, which are purely economically oriented. The tourist supply is simply determined by its demand. The coast is considered as a 'tourist product' that has to be sold. Real estate agents invite the public to buy an apartment as an excellent investment. The ecological value of dunes and polders gets little attention unless as an element of the tourist product. Even in the seventies a buffer zone around the beautiful nature reserve of De Panne was sacrificed to an upper-class residential development. The Belgian coast lost about 300 ha of dunes between 1965 and 1985 mainly through the extension of residential areas.

The government continues to invest heavily in tourism. Many harbours for sailing yachts have been enlarged or newly built; Nieuwpoort for example is becoming one of the largest in north-west Europe. Recently numerous breakwaters were built along the shore in an attempt to offer tourists broader beaches.

3.6 THE ALLUVIAL PLAIN ALONG THE SCHEDULT ESTUARY

The polders and low wetlands along the Rivers Scheldt, Rupel and Durme are situated at a level of -0.5 to 3 m TAW and belong to the lowermost regions of Belgium. The main problem of these areas is the discharge of superfluous water.

The tidal range in the estuary increases from about 4 m at the mouth of the Scheldt to more than 5 m at the mouth of the Rupel. In periods of heavy rain and high tides, discharge of surface water becomes impossible and several times already this situation has given rise to extensive flooding of the alluvial plains. The highest tide ever recorded reached 7.77 m in 1953 in Antwerp.

A serious improvement of the drainage system, a precise water level control and a maintenance and elevation of the dikes will be essential. However, intensive drainage and pumping is increasing the salt water intrusion in the aquifers, which is already occurring in some parts of these polders.

The planning of navigation and port facilities will have to take rising sea-level into account. The sea-level rise will gradually decrease clearance under bridges, requiring drawbridges to be opened more frequently.

The Scheldt estuary is characterized by the dominant position of the industry of the harbour of Antwerp. The harbour of Antwerp forms the most important industrial centre of Belgium (chemical, petrochemical sector, petroleum refinery, motor industry, electricity production). A nuclear plant is located along the Wester Scheldt at a short distance from Antwerp. The harbour itself is spread over a 13,800 ha area; docks and quays reach a length of about 80 km. Yearly 85 to 90 million tons of merchandise are exchanged and 16,000 to 17,000 units of the merchant service arrive at the port. Tremendous private investments have been applied to industry; nearly 70 per cent of investments were realized by foreign firms and joint ventures. About 75,000 persons are involved in the harbour activities (industry included), which means that (with their families) no less than 200,000 Belgians are dependent on it.

3.7 FINAL CONSIDERATION

The coastal and alluvial low-lying wetlands are important areas for the country as an industrial and recreational zone. The areas are characterized by a dense population, and great investments have been made to develop industry and tourism.

However these areas are very vulnerable. Already nowadays flooding is occurring regularly and an eventual sea-level rise in the near future will severely damage these entire plains. The drainage system is insufficient and in a bad condition. The dikes along the canals must be elevated, and discharge feasibilities must be reconsidered as well as the increasing salt water intrusion. The coastal dune strip is for its greatest part far too small to act as a natural defence against the power of the sea, and the huge human impact on that environment makes its situation even more vulnerable.

REFERENCES

Baeteman, C. 1981: *De Holocene ontwikkeling van de Westelijke Kustvlakte (België)*. Proefschrift Vrije Universiteit Brussel, 297 pp.

Baeteman, C. 1987a: Ontstaan en evolutie van de kustvlakte (tot 2000 jaar voor heden). In, Thoen, H. (ed.) *De Romeinen langs de Vlaamse Kust*. Brussels, Uitgave Gemeentekrediet, 18–21

Baeteman, C. 1987b: De Westelijke kustvlakte in de Romeinse Tijd. In, Thoen, H. (ed.) *De Romeinen langs de Vlaamse Kust*. Brussels, Uitgave Gemeentekrediet, 22–3

Baeteman, C. 1989: Radiocarbon dates on peat from the Holocene coastal deposits in West Belgium. In, Baeteman, C. (ed.) *Quaternary Sea-level Investigations from Belgium* (Professional Paper, 241, 1989/6), 59–91

Boerjan, P. 1985: De Belgische kust toeristisch doorgelicht. In, *Ruimtelijke Planning*, II. E. 2.e. Antwerp, Van Loghum Slaterus

Bruun, P. 1962: Sea-level rise as a cause of shore erosion. *J. L. Waterways and Harbors Div. ASCE* 88, 117–30

Cauwenberghe, C. van 1977: Overzicht van de tijwaarnemingen langs de Belgische kust. Periode 1941–1970 en 1959–1970. *Tijdschr. Openbare Werken van België* 4, 339–49

Cauwenberghe, C. van 1985: Overzicht van de tijwaarnemingen langs de Belgische kust. Periode 1971–1980. *Tijdchr. Openbare Werken van België* 5, 437–57

Ceunynck, R. de 1985: The evolutions of the coastal dunes in the western Belgian coastal plain. *Eiszeitalter u. Gegenwart*, 35, 33–41

Ceunynck, R. de 1987: Ontstaan en ontwikkeling van de duinen. In, Thoen, H. (ed.) *De Romeinen langs de Vlaamse kust*. Brussels, Uitgave Gemeentekrediet, 26–9

Ceunynck, R. de and Thoen, H. 1981: The Iron Age settlement at De Panne-Westhoek; ecological and geological context. *Helinium*, 1981/1, 21–42

Depuydt, F. 1967: Bijdrage tot de geomorfologische en fytogeografische studie van het domaniaal natuurreservaat De Westhoek. *Dienst Domaniaal Natuurreservaat en Natuurbescherming, Werk*. 3, 1–100

Henrionet, J. 1876: *Notice sur les travaux topographiques, exécutés au Dépôt de la Guerre de Belgique*. Brussels, Archief Nat. Geogr. Inst.

Kraus, N. C. 1988: The effects of seawalls on the beach: an extended literature review. *Journal of Coastal Research* SI 4, 1–28

Moor, G. de 1979: Recent beach erosion along the Belgian North Sea coast. *Bull. Belg. Ver. Geol.* 88(2), 143–57

Mostaert, F. 1985: *Bijdrage tot de Kennis van de Kwartairgeologie van de Oostelijke Kustvlakte op basis van Sedimentologisch en Lithostratigrafisch Onderzoek*. Proefschrift Rijksuniversiteit Gent, 351 pp.

Mostaert, F. 1987: De Oosteljike kustvlakte in de Romeinse tijd. In, Thoen, H. (ed.) *De Romeinen langs de Vlaamse kust*. Brussels, Uitgave Gemeentekrediet, 23–5

Pilkey, O. H. and Wright III, H. L. 1988: Seawalls versus beaches. *Journal of Coastal Research* SI 4, 41–64

Rottier, H. and Arnolus, H. 1984: *De Vlaamse Kustvlakte: van Calais tot Saeftinge*. Tielt, Lannoo

Titus, J. 1986: Greenhouse effect, sea-level rise and coastal zone management. *Coastal Zone Management Journal* 14(3), 147–71

Vanhove, N. 1973: *Het Belgische Kusttoerisme – Vandaag en Morgen*. Bruges, Westvlaams Ekonomisch Studiebureau

Vermeersch, C. 1986: De teloorgang van de Belgische kust. In, *Ruimtelijke Planning*, II. E. 2.f. Antwerp, van Loghum Slaterus

4

Impacts of Sea-level Rise on The Wash, United Kingdom

I. Shennan

4.1 BACKGROUND

The stability of the coastline is a function of sediment accretion, sediment erosion and relative sea-level change for the specified temporal and spatial scales over which each is considered. Human perception of stability of the coastline is complicated to assess and is often based on observations over a short time period. Yet it is perhaps the most significant factor in determining a management response to coastline change and coastal zone land use change. Future changes in the environment of The Wash, and of the Fenland, will be dependent on the management response to social, political, economic and environmental conservation arguments as well as the seemingly more 'natural' changes. The main change to be considered here is the potential change in sea-level and its impact over the next century. 'Natural' is not really a suitable adjective since it is likely that future sea-level rise may well be induced by man-made emissions of carbon dioxide, nitrous oxide, chlorofluorocarbons and other radiatively active gases. It is expected that these emissions will result in a warming of the earth's atmosphere, which may be exaggerated in high latitudes, leading to enhanced ablation of icesheets and glaciers. This will have a global effect, which is a baseline upon which to judge the local response in The Wash.

4.2 CARBON DIOXIDE AND TRACE GAS EMISSIONS, CLIMATIC AND SEA-LEVEL CHANGE

The amount of carbon dioxide within the atmosphere was *c.* 200 ppmv at the maximum of the last glaciation, increased rapidly to *c.* 300 ppmv during the period of deglaciation and remained relatively stable until the industrial revolution, since when levels have risen to over 340 ppmv (Neftel et al. 1982, Rycroft 1982) and by 1990 had reached 353 ppmv with a current rate of change of 1.8 ppmv/year (Houghton et al. 1990). The 'greenhouse model', involving more complex processes than this popular name suggests, describes the response of net warming of the atmosphere coupled to rising carbon dioxide levels and other trace gases (e.g. Seidel and Keyes 1983). The greater warming effect in high altitudes may enhance partial melting of the ice sheets and ice shelves but could be offset by increased snowfall. A series of positive feedback mechanisms involving a slight initial sea-level rise, seasonal variations in meltwater discharge, ocean temperature, density, and salinity, precipitation and iceberg calving may induce a relatively rapid rise in sea level (e.g. Mercer 1978, Ruddiman and McIntyre 1981, Barth and Titus 1984). To address these problems, a research initiative called SeaRISE has been established, and will focus on marine ice sheets, such as the West Antarctic ice sheet, grounded well below sea-level, the catastrophic collapse of which would raise sea-level by six metres (Bindschadler 1990). The need for empirical data is paramount not only to establish trends, but also to refine models. Global climate models simulate a linked rise in global air temperatures and increasingly the models are being refined to reveal significant regional variations in temperature response. But there are scientists now questioning the empirical evidence for a global rise in atmospheric temperature (e.g. Harris 1985). Climatic data are statistically noisy, and if the aim is to identify the response in climate to an increase in atmospheric carbon dioxide levels then local effects must be removed from the data. These consist, for each station, of any changes in the location of the instruments, changes in the instruments themselves, changes in the observation time, changes in the computational procedures used on the data, and most importantly, changes which have taken place in the surroundings of the station. These include building construction, grass replaced by concrete or tarmac, growth of trees, and are called the urban effect. If the urban effect is removed, it has been argued, changes in surface air temperature in the last 100 years are not statistically significant (e.g. Wood 1988, Karl et al. 1989, Wijn-Nielsen 1989).

In a similar fashion the interpretations of global tide gauge data are disputed. Many tide gauge trends are dominated by local or regional effects, both oceanographic and geological (Pirazzoli 1989). While some authors have claimed to have filtered out these effects (e.g. Gornitz and Lebedeff 1987), others have concluded that there is no observable increase in global sea-level associated with the 'greenhouse effect' (Stewart 1989).

It should also be noted, based on empirical evidence, that there remains the possibility of a very rapid rise in sea-level, decimetres in a few weeks or even metres in a few years. However, the explanations of the causative processes, the accumulation and catastrophic release of subglacial meltwater, have yet to be tested adequately (Tooley 1989).

4.3 SEA-LEVEL RISE SCENARIOS

The various factors involved in modelling future sea-level rise are very varied, and as yet the models have not been universally accepted. One of the most publicized approaches has been by the US Environmental Protection Agency (EPA), and this is taken as the example here (figure 4.1 and table 1.1 in chapter 1). There has been a convergence of scientific opinion since the publication of the EPA scenarios with a tendency to limit predictions to shorter periods and scale down values of expected sea-level rise. These are summarized in Warrick and Oerlemans (1990) who give as best estimates for the 'business-as-usual' scenario sea-level rise of 18 cm for AD 2030 and 44 cm for AD 2070 (see figure 1.1 in chapter 1). The EPA approach was first to model fossil fuel consumption and other factors influencing the release of carbon dioxide and trace gases into the atmosphere up to AD 2100 (e.g. Barth and Titus 1984). Atmospheric composition, volcanic aerosol emissions and variations in solar radiation are also considered. The model predictions of climate sensitivity to such changes over the past eighty years appear quite a good fit to climate records, and upon this basis global warming to AD 2100 has been modelled, using a range of estimates for each of the factors that will influence atmospheric composition.

The sea-level response to global warming was modelled as the sum of two factors. Firstly, by estimating rates of heat absorption by the upper layers of the oceans, the increase in sea level is calculated. From this model between one third and one half of the observed global sea-level rise this century could be accounted for. The contribution of ice and snow discharge was not modelled, but it was assumed that the remaining one half or two-thirds was attributable to these factors. Thus by modelling the thermal

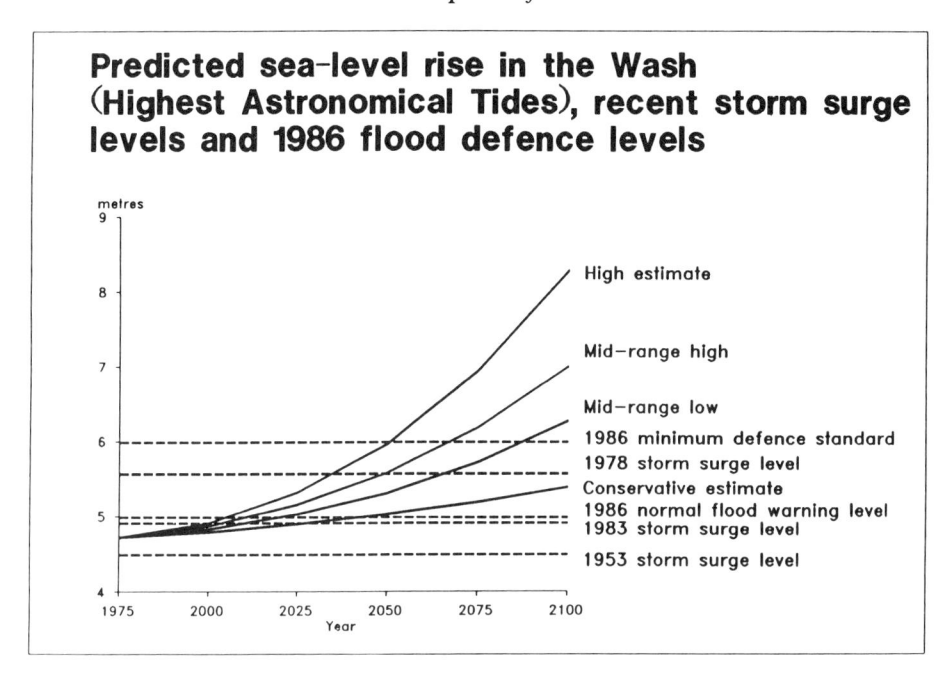

Figure 4.1 Sea-level rise scenarios (adapted from Barth and Titus 1984) and applied to The Wash. The altitudinal origin for the scenarios is the High Astronomical Tide (4.8 m OD). More recent predictions of future sea-level rise tend to fall within the band bounded by the conservative and mid-range estimates.

expansion factor, total sea-level rise to AD 2100 could be estimated with either a 1:1 or 2:1 ratio for thermal expansion: ice and snow discharge. This is a weakness in this and subsequent models, since such a linear relationship could not have held for the whole of the time since deglaciation. The positive feedback mechanisms discussed by Ruddiman and McIntyre (1981) are not accommodated by this assumption.

A series of global sea-level scenarios has been produced using various combinations of high, mid-range and low assumptions for each factor that determined atmospheric composition, global warming, thermal expansion and ice and snow discharge (Barth & Titus 1984). However if the majority of the snow and ice discharge is from Antarctica then changes in gravitational forces would increase the amount of sea-level rise around the UK to 20 per cent above the global average arising from that cause (Clark and Primus 1987).

The global scenarios, excluding the gravitational factors, have been modified for The Wash to take into account local tidal conditions and crustal movements and are summarized in figure 4.1. The range of the scenarios shown poses some problems for future planning, particularly after AD 2025, when the divergence between the scenarios increases rapidly.

4.4 ENVIRONMENT RESPONSE TO PAST SEA-LEVEL RISE

Since at least 6,500 years ago the Wash/Fenland embayment (figure 4.2) has been influenced by sea-level-controlled sedimentation, ranging from sub-tidal and inter-tidal clastic sequences to brackish and fresh water semi-terrestrial peat deposits. Throughout this time the coastline has not been static, but has undergone major shifts in both horizontal and vertical dimensions (Shennan 1982, 1986a,b). Analyses of the sedimentary data allow an explanation of relative sea-level change (figure 4.3) and coastline movements to be made (Shennan 1986b). Past sea level can only be estimated within an error band of c.1 m due to factors such as sediment compaction and assumptions about past tidal regimes and sedimentation. Nevertheless the reconstruction of Mean High Water of Spring Tides between 6,500 and 2,500 years ago reveals a variable rate of change. During the periods of low, static or negative rates of sea-level change peat accumulation and extension seawards across previously intertidal sediments was characteristic.

If the tidal range in the past is assumed constant then relative sea-level is the sum of the regional eustatic factor (essentially the level of the oceans controlled by the volume of water in the ice sheets) and crustal movements affecting The Wash area. By comparison with other relative sea-level curves and a reliable estimate of the regional eustatic factor it is possible to calculate the crustal factor. An average rate of 0.91 m/1,000 yr crustal subsidence, with possibly a greater rate for the last 5,000 years, has been calculated (Shennan 1986b, 1987a).

The regional eustatic factor (Mörner 1976) shows more fluctuations between positive and negative rates of sea-level change than those presently observed for The Wash (figure 4.3). The combination of this eustatic factor and the rate of crustal subsidence can be used to investigate the relationship between these numerous eustatic fluctuations and the stratigraphic and sea-level records for The Wash area (figure 4.4). It seems that the small eustatic fluctuations in isolation had little effect on the major observed altitudinal and horizontal movements of the coastline. A more stepwise relationship existed, where only major changes in the rate of sea-level rise would bring to an end a period of peat development and landward movement of the coastline. Similarly a short period of reduced sea-level rise

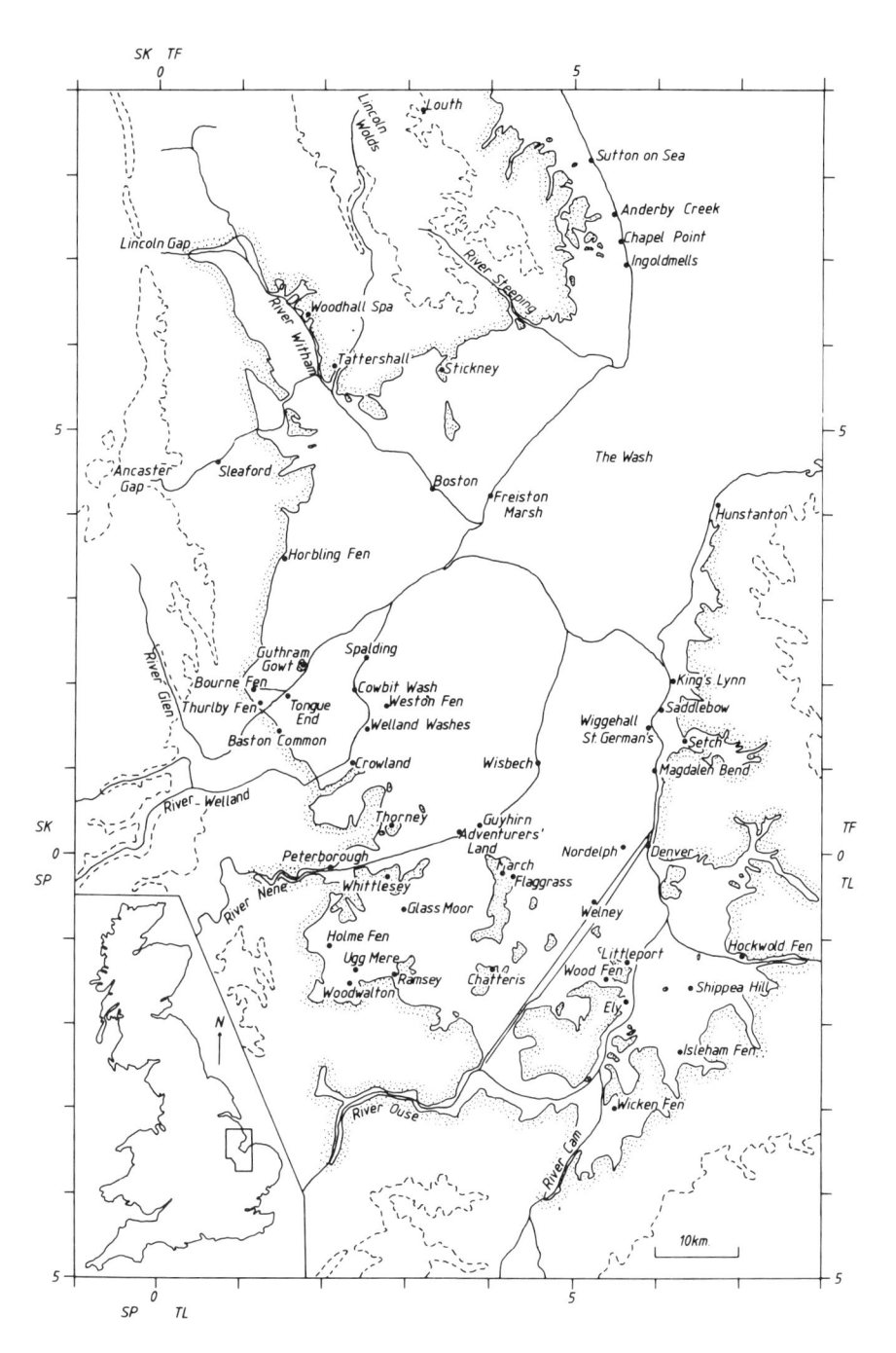

Figure 4.2 Map of The Wash and the Fenland. The stippled line marks the approximate edge of unconsolidated sediments of Holocene Age, and between this line and the present coastline most of the ground lies below +5.0 m OD. The dashed line is the 60m contour.
Source: (Shennan 1986a, b)

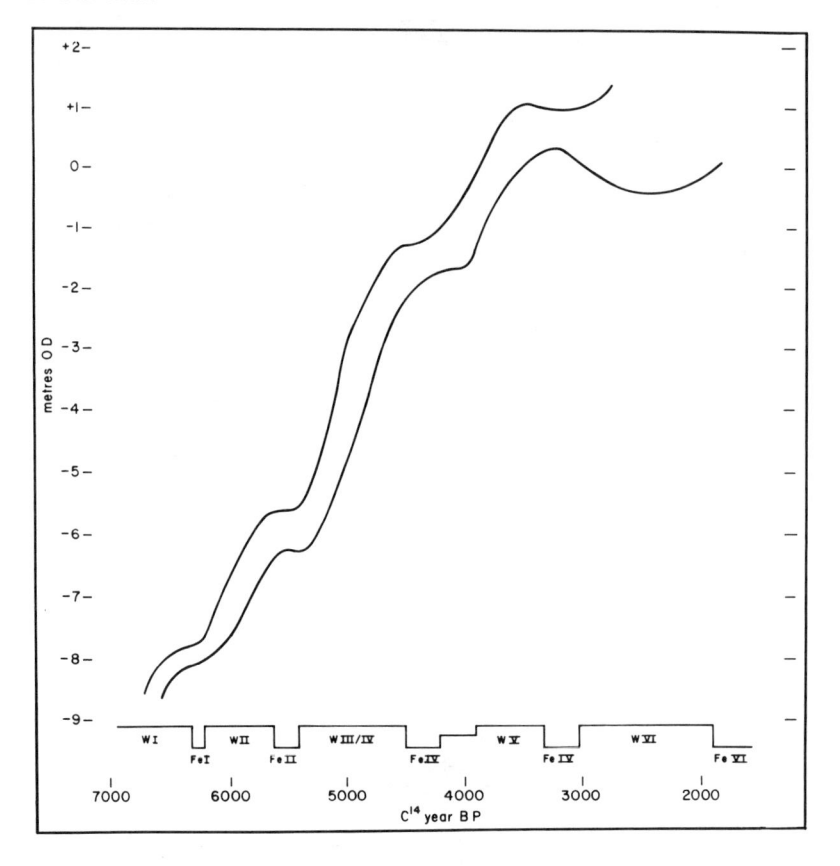

Figure 4.3 Error band of Mean High Water Spring Tides for the past 6,400 years, estimated from stratigraphic data from the Fenland. Present MHWST intersects the coastline at +3.80 ± 0.07 m OD. The Wash/Fenland chronology of tendencies of sea-level movement is shown at the foot of the diagram: Wash (W)I to VI are positive tendencies of sea-level movement and Fenland (F)I to VI are negative tendencies of sea-level movement.
Source: (Shennan 1987b)

or actual fall would not always halt a landward movement of the coastline. Thus a sustained period of sea-level change in one direction or another is required to change the balance between coastline advance and retreat. The stratigraphic evidence shows that once the rate of sea-level rise is sustained at 5 mm/year, net coastline retreat occurs, with maximum rates of rise being slightly higher (table 4.1).

The stratigraphic data cannot be easily used to quantify rates of sea-level change after c. 2000 BP since the surface sediments have been too disturbed. Some useful data can be obtained by accurate measurement of

Impacts of Sea-level Rise on The Wash 79

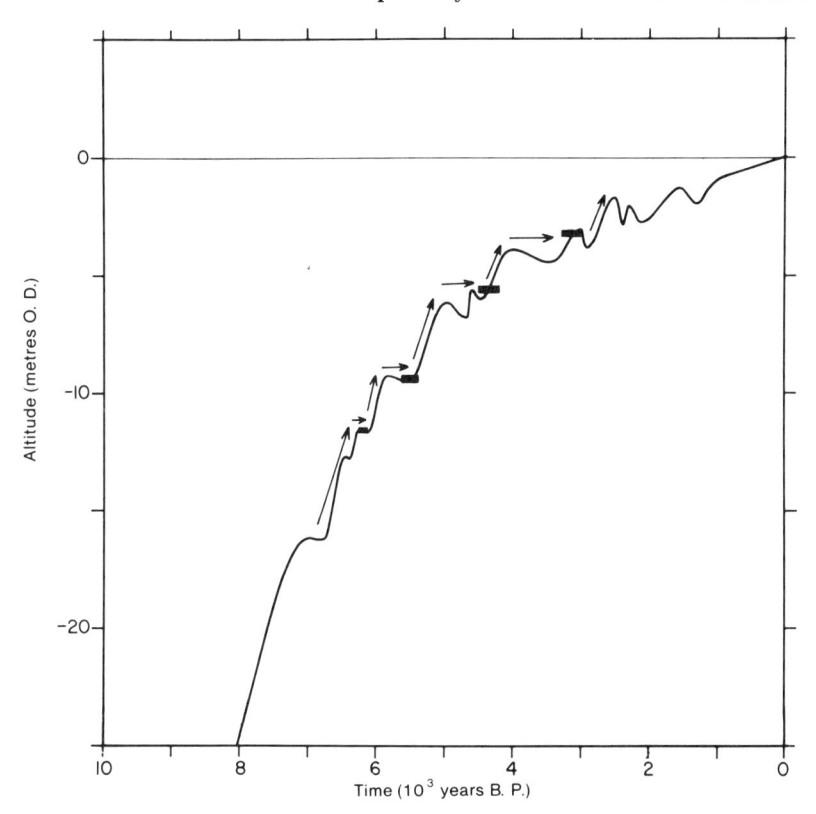

Figure 4.4 Predicted relative sea-level change for The Wash, showing the relationship between eustatic fluctuations and periods of coastline advance, represented by black boxes.
Source: (Shennan 1987a)

Table 4.1 Maximum rates of sea-level change in mm/year (50-yr mean), associated with the end of periods of regressive overlap. These are the maximum rates following the end of coastline advance, which occurred when the rate reached 5 mm/year, and continued to rise in most cases (from figure 4.4). Eustasy models from Shennan (1987a), figures 4.3 and 4.7.

Age (BP)	Eustasy model 1	Eustasy model 2
6200	14.5	14.5
5400	10.0	9.6
4200	8.0	7.5
3000	5.2	2.9
2500	8.3	6.3

saltmarsh reclamations of different ages (figure 4.5). Preliminary results (table 4.2) reveal a recent average rate of sea-level rise of 1.4 mm/year.

Direct measurements of changes in sea-level are currently obtained by tide gauge observations, and will be enhanced by satellite altimeter data in the future. The nearest stations for which digital data are available from the Permanent Service for Mean Sea Level are Immingham and Lowestoft where records were started in 1959 and 1955 respectively. It is very difficult to obtain a clear sea-level signal from such short records, for example the record from Immingham reveals an estimated linear trend of +0.3 mm/year, but with a large standard deviation, thus the trend is not significantly different from zero (Pugh and Faull 1982). This observation should be considered in the context of the longer records from Aberdeen, North Shields, Sheerness, Southend and Newlyn which all reveal a flattening-off

Table 4.2 Surface altitudes of reclamations – Butterwick transect

Butterwick to 13th-century sea bank	2.73–3.07 m OD
13th-century sea bank to 1809 bank	3.09–2.87 m OD
1809 sea bank to 1972 bank	3.32–3.06 m OD
Present marsh beyond borrow-pit	3.27 m OD
Average rate of rise c. 1250–1809	0.0 to 0.64 mm/year
Average rate of rise 1809–1972	1.41 mm/year

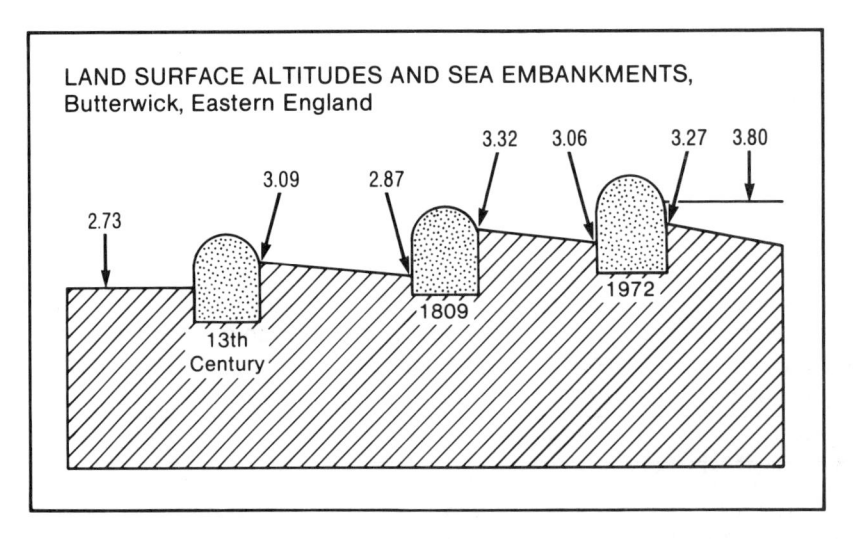

Figure 4.5 Altitudes (in metres above OD, Newlyn) of reclamations at Butterwick, near Freiston Marsh, Lincolnshire. The present-day Mean High Water Mark of Spring Tides intersects the 1972 sea embankment at +3.80 m OD.

of sea-level rise over the past twenty years. For example the 1929–70 average linear trend at Southend is 3.4 mm/year, but for 1929–79 this is reduced to 1.4 mm/year.

Thus the direct observation of sea-level change from tide gauges and the reconstruction of sea-level change from stratigraphic and other data all reveal the significant interaction of long-term and short-term fluctuations. Longer-term changes, over 100 years or so, cannot be reliably observed from monitoring on a scale of one or two decades because the signal-to-noise ratio is too low. Furthermore the stratigraphic record indicates that some short-term variations in sea level will not have a major effect on coastline development; only periods of sustained change are significant. The magnitude of change apparently significant is shown in figure 4.4 and table 4.1.

4.5 MAN'S RESPONSE TO PAST SEA-LEVEL RISE

Reclamation and sea defence have been successfully achieved in the past with relatively simple engineering methods. The response to rising sea levels has been to build higher structures, either by modifying existing banks or by increasing the standard of banks built with new reclamations (figures 4.5 and 4.6). The timescale of rise has been long enough to allow such responses. Some 438 square kilometres have been reclaimed from The Wash between Saxon times and AD 1980. Of the present Lincolnshire coastline (154 km), 14 km are sand dunes, 26 km are dunes protected by some artificial revetment, 74 km are embankments of unconsolidated sediments. Part of the coast between Mablethorpe and Skegness (22 km) is protected by groynes, which have been increased from 82 in 1953 to 263 in 1979/81 (Shennan 1988). For a period arguments were centred not on sea defence but on the continued seaward shifting of the coastline for reclamation at the possible expense of the conservation resource of the coastal zone (Lincolnshire County Council 1982), but sea defence is again the prime concern.

In addition to reclamation and sea defence, land drainage in the Fenland has been a major response by man. The overall effect has been to increase agricultural productivity, but at the cost of enhanced rates of topsoil loss (Burton and Seale 1981), a decrease in land surface altitudes due to consolidation, wastage and erosion (Hutchinson 1980), and the need for major financial investment in drainage and protection against fresh water flooding.

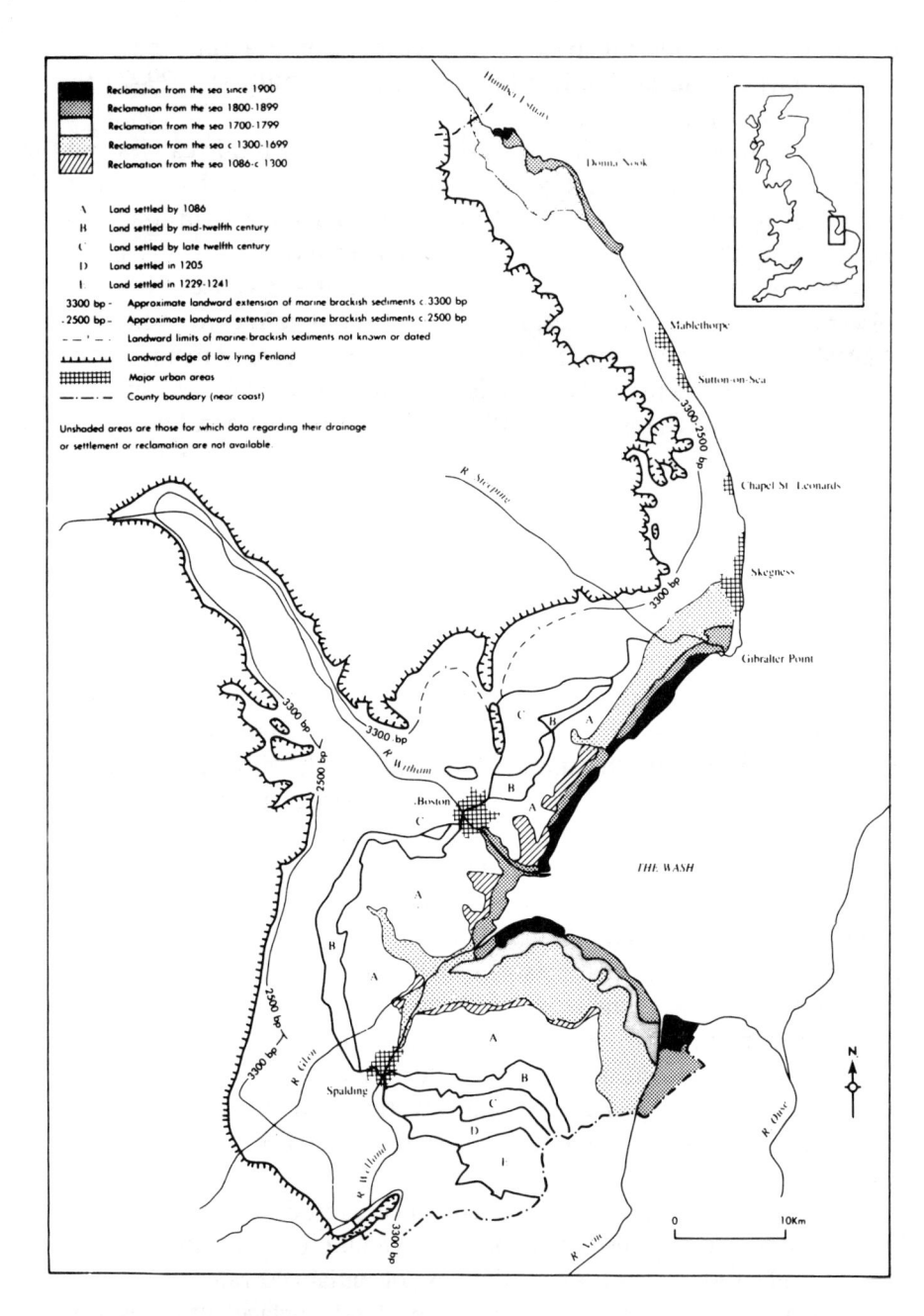

Figure 4.6 Map of The Wash and part of the Fenland in Lincolnshire showing the stages of reclamation.
Source: (Shennan 1988)

4.6 ENVIRONMENT RESPONSE TO FUTURE SEA-LEVEL RISE

Prior to significant modification of the coastline by man the major periods of coastline advance were each halted by a sustained rise of sea level. Short term fluctuations (<50 years) had little effect on the whole embayment although locally could be important (see section 4.4 and Shennan 1986a,b). The estimated rates of sea-level rise associated with the termination of coastline advance were mostly 5 mm/year (taken as a fifty-year average) rising to a slightly higher maximum rate as the coastline retreated (table 4.1). Rates of rise for each scenario (table 4.3), other than the conservative, indicate that a landward migration of the intertidal vegetated and sediment zones would be the most likely response, i.e. a transgressive sediment sequence would be established resulting in a narrower saltmarsh zone.

The sea-level record for the period during which active reclamation has occurred is not well established since it falls in the gap between direct observation from tide gauges and the reliable part of the stratigraphic record. During the past two centuries sea-level has risen at a rate considerably less than 5 mm/year (table 4.2 and tide gauge data), and the sedimentary response has been accretion in front of new sea banks. It appears that the same response could not be guaranteed in the future, but to test the relationship between reclamation, continued accretion and rates of sea-level change more research is urgently required.

Two further points should be noted but are not discussed in further detail. Firstly, the predicted increase in carbon dioxide may increase photosynthesis of the saltmarsh vegetation and thus counteract the decrease in the area of the saltmarsh zone. The effect of this may be significant and the computer simulation models developed in earlier studies (Natural Environment Research Council 1976) could be enhanced to predict the net balance.

Table 4.3 Rates of sea-level rise in mm/year (50 year average) for each scenario (figure 4.1)

1975–2025	2000–2050	2050–2100	
2.6	3.8	4.6	conservative
5.2	8.2	18.4	mid-range low
7.8	12.8	27.6	mid-range high
11.0	20.0	45.6	high

Secondly, North Sea coasts are profoundly affected by storm surges, the frequencies of which vary considerably (Lamb 1980) and may be affected by sea-level change (Rossiter 1962). Linke and others (e.g. Linke 1984) indicate a strong correlation between sea-level rise, storminess and coastal retreat but the cause–effect link is not clear. This factor should nevertheless be borne in mind.

4.7 MAN'S RESPONSE TO FUTURE SEA-LEVEL RISE

Management options regarding future sea-level rise fall essentially into four main categories :

1. flood protection, flood proofing and flood zoning;
2. abandonment of the coastline and relocation inland;
3. programmes to reduce sea-level rise;
4. doing nothing different from present policy.

Option 3 is a response to a global problem and cannot be effected solely by national or local action. Some reduction of both sea-level rise and carbon dioxide-induced environmental changes can be made (e.g. Seidel and Keyes 1983; Newman and Fairbridge 1986) but would require substantial international planning. Even so, the overall effect is likely to delay sea-level rise by one or two decades. Local plans should be made on the assumption that the international cooperation is not forthcoming.

Option 2 at first appears difficult to conceive, yet is a serious proposal for much of the sandy shoreline of eastern USA (e.g. Kerr 1981) and has been costed (Titus 1987). In countries without the necessary financial reserves this may be the only option for their low-lying coastal zones. For The Wash and the Lincolnshire coast the decision will be based on some form of cost–benefit analysis and the necessary data bases should now be collected to aid such decisions if they are ever to be required. What is the amount of sea-level rise and the necessary level of financial input that would render such an option defendable?

Option 1 has been the favoured and successful response for at least the last thousand years in the Fenland. But what for the future? Will option 1 suffice, i.e. continue building structures higher, allowing for a fifty-year lifespan for the structure, with a fifty-year sea-level rise prediction built into the specifications? This option is attractive if the perception is that the problem is no greater than it has been in the past and that current procedures offer an adequate solution, i.e. option 1 is the same as option 4.

The alternative is that society should provide a flexibility and reserve capacity to deal with the range of changes that are precedented in the past, and this should be increased due to possible man-induced change (c.f. Schneider 1985). Because the range between the various scenarios cannot be narrowed at present this is reason to be cautious now rather than complacent. Because sea-level variations on a short time scale, say ten- -fifteen-year trends, are such poor indicators of longer-term trends of fifty to a hundred years, it may not be until AD 2025 that a clear signal of which scenario seems correct is obtained. Woodworth (1990) has concluded that clear evidence for accelerations of sea-level at the 95 per cent confidence level will be observable by AD 2010 and at the 99 per cent confidence level by AD 2020. When a wide range of uncertainty exists the worst dangers remain unknown and there is greater risk than if more is known about the subject.

Monitoring programmes

Direct observation of change can be undertaken at local, national and international scales and early establishment of the required range of monitoring programmes and integrated databases is crucial if rapid and efficient management responses are to be made should sea-level induced changes occur rapidly during the next century. Local monitoring programmes and integrated databases for The Wash will be considered further.

Firstly, all existing tidal data should be accumulated in computer-compatible form for analysing trends and the local variability of extreme events. Secondly, the balance between erosion and deposition at the coastline should be annually reviewed, not only for The Wash but also the whole coastline of eastern England since this may influence sediment supply. A cost-effective method of monitoring annual change may be remote sensing. Preliminary results reveal that the main saltmarsh vegetation communities and surface sediment zones can be identified using multispectral data (Hobbs and Shennan 1986). Landsat Thematic Mapper (TM), airborne multispectral and SPOT remotely-sensed imagery have been used successfully to identify saltmarsh vegetation communities and sediment variation around The Wash (Donoghue and Shennan 1987). A classified image has been converted to pc ARC INFO format and forms the first part of an operational monitoring GIS system, which can integrate various classified images and other geo-referenced data. Annual analysis of satellite data, assuming the availability of suitable cloud-free data, and/or aircraft-borne multispectral scanner data will provide an essential database for assessing rates of change if net accretion and seaward expansion of the vegetation and sediment zones in front of reclamation embankments is to be replaced by the movement of these zones towards the present sea banks as the rate of sea-level rise increases.

Decision-making

The first decision, and probably the easiest, will be to continue to protect the shoreline of The Wash by increasing the height of sea banks as sea-level rises. The highest costs will be incurred by the modification of structures such as river and drain outfalls, harbour facilities and transport structures.

In conservation terms a rate of sea-level rise in excess of 5 mm/year sustained over twenty-five years or more is likely to alter substantially the intertidal environment affecting the distribution of plant and animal communities. The net movement of the vegetation and sediment zones towards the sea bank would lead to the narrowing and possible removal of some ecologically valuable environments.

The alternative courses of action are controversial – firstly, flooding of recently reclaimed land, or secondly, the construction of embankments around existing valuable intertidal areas. These pose the question of whether the intertidal environments are sufficiently valuable to require conservation at relatively large cost. Organizations such as the Nature Conservancy Council need to develop now a strategy for implementing such schemes since a relatively rapid series of decisions will have to be made if the rate of sea-level rise increases as envisaged in the first part of the next century. Both options suggested will require financial investment in land purchase and engineering structures. For the second option potential areas for conservation can be identified. The most mature marshes in The Wash not used for military purposes include the 'Old Marsh' at Gibraltar Point, the area between the outfalls of the rivers Welland and Witham, and Leverton Marsh. An earth embankment with sluices for the main creeks would provide the means for managing tidal flow, the balance between erosion and deposition, and the vegetation succession. Tidal environments are currently managed in a similar fashion in the Easter Scheldt estuary (Groenendijk 1984). Depending on the alignment of the banks such a scheme may provide a solution to both flood protection and conservation requirements. For example, an embankment following the marsh front along the western side of the Welland channel, with three or four sluices across the major tidal creeks, may provide a conservation scheme in which the saltmarsh environment could be managed. An embankment extending from Tab's Head to the southern shore of The Wash, crossing the Welland channel with a major sluice which would allow ships to pass through to reach Fosdyke, would provide a wider range of environments to be managed (saltmarsh and intertidal flats) and provide a flood protection solution for the Welland estuary. It would require a different level of technology to build a structure so far out in the intertidal zone than is currently employed in sea defence programmes in the area.

Plans for the cost, legal, engineering and ecological requirements for a range of schemes to cover the whole coastline of The Wash and the range of environments considered essential for conservation should be developed as soon as possible.

In addition to flood protection, flood alleviation and land drainage must be considered. Monitoring of land levels, drainage requirements, saltwater intrusion in ditches and ground water, and fresh water run-off inland of the coast form part of the monitoring requirements. A 1 m rise of sea level will cause serious difficulties for the evacuation of fresh water by gravity discharge during the period of low tide, which will of course be one metre higher. Flood alleviation costs are a significant part of the decision-making process, along with population numbers and property values in any cost–benefit analysis. The databases required to contribute to such procedures are found in a very disaggregated form but procedures to integrate these within a computer-based information retrieval system are currently being developed (*e.g.* Shennan 1987b, Shennan and Tooley 1987 and figure 4.7). These procedures are being developed for national use and will need to be evaluated for specific local requirements. The Geographical Information System has been extended to comprehend the coast of eastern and southern England, from Northumberland to Kent (Shennan and Sproxton 1990). It has also been enhanced so that the topographic data used for the definition of flood risk in a coastal lowland are not confined to ward boundaries, which overestimate the area of flood risk (Shennan and Sproxton 1990). The GIS approach is currently being developed for cost-benefit analyses. (An example of one of the range of maps produced is shown in figure 4.8). Preliminary analysis of data for the whole Fenland indicates that the relationship between surface morphology, census data and sea-level rise scenarios is non-linear and characterized by altitudinal thresholds at 5 m and 7–8 m. These thresholds indicate suitable parameters for the development of flood planning strategies.

Investment

The investment requirements can be divided into three :

1. low cost: to establish suitable databases (figure 4.7) and monitoring schemes (satellite, tide gauge, field surveys, simulation models to

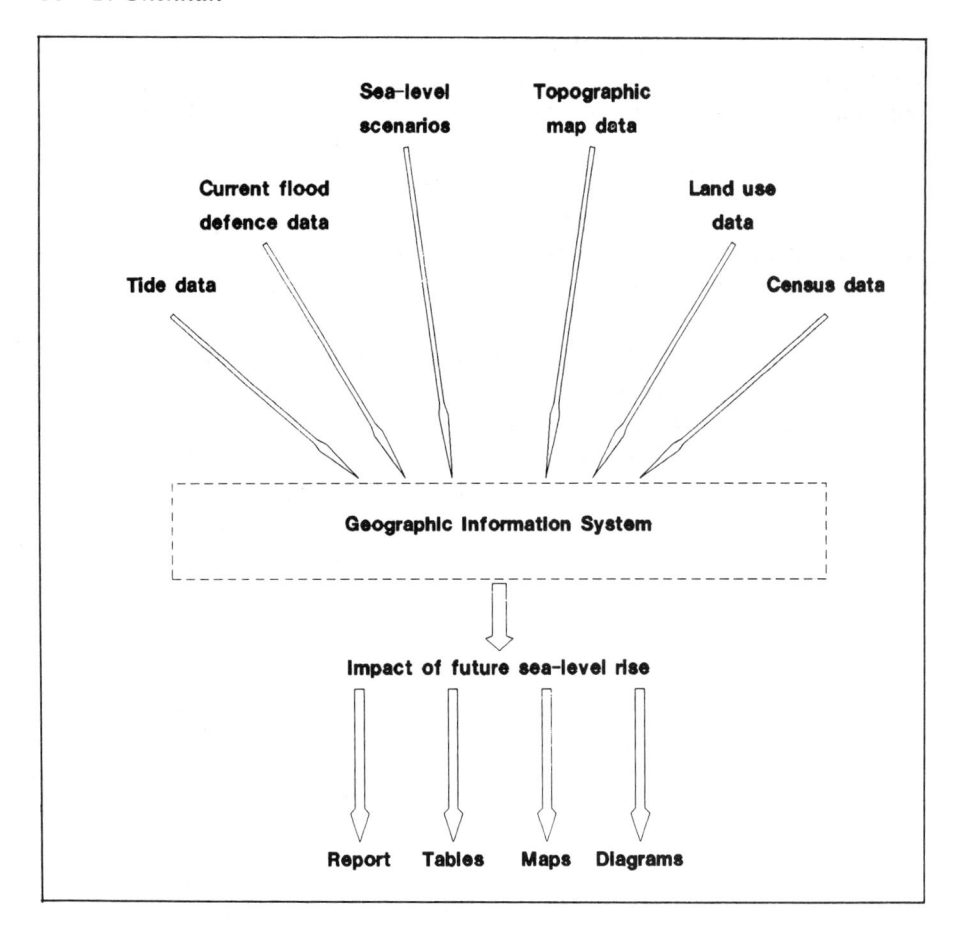

Figure 4.7 Organization of a Geographic Information System used to provide data for future management of the impact of sea-level rise.
Source: (Shennan and Tooley 1987)

compute the effect of sea-level rise on vegetation and sediment balance, tidal range and storm surges);
2 medium cost: to raise and widen the base of existing embankments as sea-level rises at a rate similar to the current rate;
3 high cost: to provide harbour facilities, bridges, outfalls, sluices, installation of new pumping stations, protection of embankments and sea defences against erosion if the protection of the saltmarsh is removed, and major new flood protection and conservation schemes.

Figures 4.8a–d Examples from Lincolnshire, United Kingdom, of map output from the GIS being developed to integrate census data, topographic data and sea-level rise scenarios

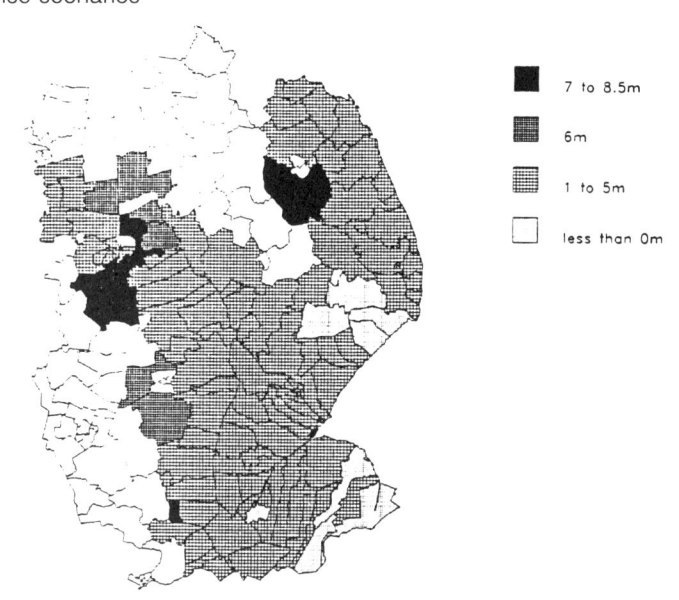

Legend:
- 7 to 8.5m
- 6m
- 1 to 5m
- less than 0m

Figure 4.8a Minimum altitude in the ward based on spotheight data

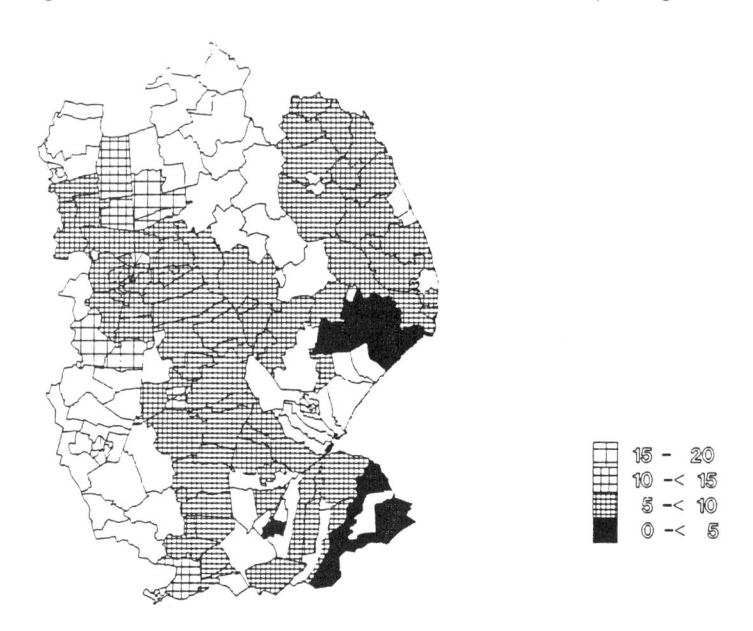

Legend:
- 15 – 20
- 10 –< 15
- 5 –< 10
- 0 –< 5

Figure 4.8b Altitude in metres in the ward based on minimum contour values recorded in the ward

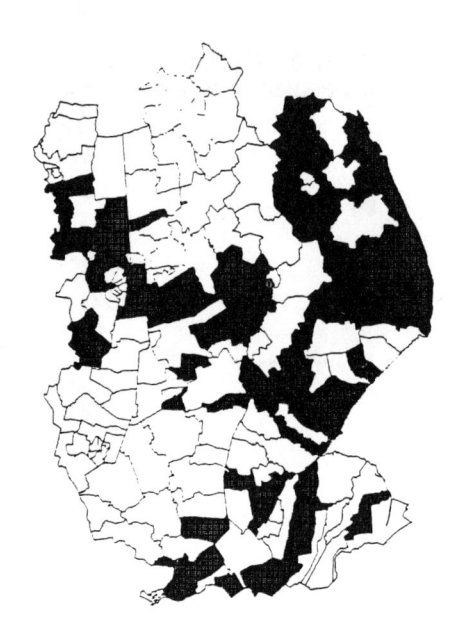

Figure 4.8c Presence of industry in a ward

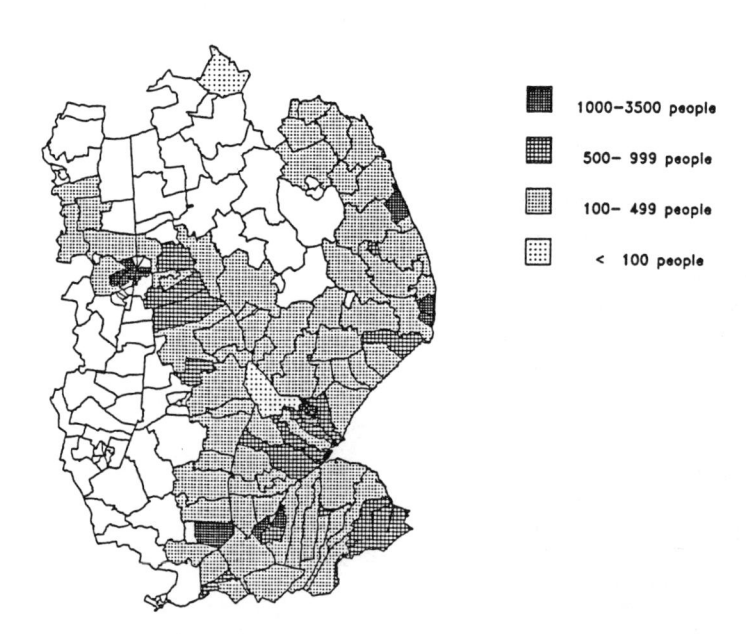

Figure 4.8d Number of people by ward in households with no car for wards with a minimum altitude less than +5 m OD

ACKNOWLEDGEMENTS

The author thanks the student groups from Durham University Geography Department who carried out the detailed levelling of the Butterwick transect and the landowner, John Saul, for access to the land, his cooperation and interest in this work.

This chapter is a revised and expanded version of a paper published by the Nature Conservancy Council in: P. Doody and B. Barnett (eds) *The Wash and its Environment, Research and Survey in Nature Conservation*, No. 7. Permission to reproduce this paper here has been given by the Nature Conservancy Council and is gratefully acknowledged.

REFERENCES

Barth, M. and Titus, J. (eds) 1984: *Greenhouse Effect and Sea-level Rise : A Challenge for This Generation*. New York, van Nostrand Reihnold

Bindschadler, R. A. 1990: *Sea RISE: A Multidisciplinary Research Initiative to Predict Rapid Changes in Global Sea Level Caused by Collapse of Marine Ice Sheets*. NASA, Goddard Space Flight Centre, Greenbelt, Md.

Burton, R. G. O. and Seale, R. S. 1981: Soils in Cambridgeshire 1. Sheet TL18/28 (Stilton). *Soil Survey Record*, 65

Clark, J. A. and Primus, J. A. 1987: Sea-level changes resulting from future retreat of ice sheets : an effect of CO_2 warming of the climate. In, M. J. Tooley and I. Shennan (eds) *Sea-level Changes*. Oxford, Basil Blackwell, 356–70

Donoghue, D. N. M. and Shennan, I. 1987: A preliminary assessment of Landsat imagery for mapping vegetation and sediment distribution in The Wash estuary. *Intl. Jl. Remote Sensing* 8, 1101–8

Gornitz, V. and Lebedeff, S. 1987: Global sea-level changes during the past century. In, Nummedal, D., Pilkey, O. H., Howard, J. D. (eds) *Sea-level Fluctuation and Coastal Evolution*, (SEPM Special Publication 41). Tulsa, Okla., Society for Economic Paleontologists and Mineralogists, 3–16

Groenendijk, A. M. 1984: Tidal management : consequences for the saltmarsh vegetation. *Water Science Technology* 16, 79–86

Harris, R., 1985: Variations in the Durham rainfall and temperature record, 1847–1981. In, Tooley, M. J. and Sheail, G. M. (eds) *The Climatic Scene*. London, George Allen & Unwin, 39–59

Hobbs, A. J. and Shennan, I. 1986: Remote sensing of saltmarsh reclamation in The Wash, England. *Journal of Coastal Research* 2, 181–98

Houghton, J. T., Jenkins, G. J. and Ephraums, J. J. 1990: *Climate Change: The IPCC Scientific Assessment*. Cambridge, Cambridge University Press

Hutchinson, J. A. 1980: The record of peat wastage in the East Anglian Fenlands at Holme Post, 1848–1978 AD. *J.Ecol.* 68, 229–49

Karl, T. R., Tarpley, D., Quayle, R. G., Diaz, H. F., Robinson, D. A. and Bradley, R. S. 1989: The recent climate record: what it can and cannot tell us. *Reviews of Geophysics* 27, 405–30

Kerr, R. A. 1981: Whither the shoreline? *Science* 214, 428

Lamb, H. H. 1980: Climatic fluctuations in historical times and their connexion with transgressions of the sea, storm floods and other coastal changes. In, Verhulst, A. and Gottschalk, M. K. E. (eds) *Transgressies en occupatiegeschiedens in de kustgebieden van Nederland en België*. Belgisches Centrum voor Landelijke Geschsiedenis No. 66, 251–84

Lincolnshire County Council 1982: *Development on the Lincolnshire Coast : Subject Plan*. Lincoln

Linke, G. 1984: Arensch (south of Cuxhaven). *INQUA Subcommission on Shorelines of Northwestern Europe. Field conference 1984*. Hanover, NLFB, 22–7

Mercer, J. H. 1978: West Antarctic ice sheet and CO_2 greenhouse effect : a threat of disaster. *Nature* 295, 220–3

Mörner, N.-A. 1976: Eustatic changes during the last 8000 years in view of radiocarbon calibration and new information from the Kattegatt region and other northwestern European coastal areas. *Palaeogeography, Palaeoclimatology, Palaeoecology* 19, 63–85

Natural Environment Research Council 1976: *The Wash Water Storage Scheme Feasibility Study: A Report on the Ecological Studies*. NERC Publications Series C, No. 15

Neftel, A., Oeschger, H., Schwander, J., Stauffer, B. and Zumbrunn, R. 1982: Ice core sample measurements give atmospheric CO_2 content during the past 40,000 yr. *Nature* 295, 220–3

Newman, W. S. and Fairbridge, R. W. 1986: The management of sea-level rise. *Nature* 320, 319–21

Pirazzoli, P. A. 1989: Present and near-future global sea-level changes. *Palaeogeography, Palaeoclimatology and Palaeoecology* 75, 241–58

Pugh, D. T. and Faull, H. E. 1982: Tides, surges and mean sea level trends. In, *Shoreline protection*, London, Thomas Telford, 45–55

Rossiter, J. R. 1962: Longterm variations in sea level. In, Hill, M. N. (ed.) *The Sea* 1. London, Interscience Publishers, 590–610

Ruddiman, W. F. and McIntyre, A. 1981: The mode and mechanism of the last deglaciation : oceanic evidence. *Quaternary Research* 16, 125–325

Rycroft, M. J. 1982: Analysing atmospheric carbon dioxide levels. *Nature*, 295, 190–1

Schneider, S. H. 1985: Science by consensus: the case of the National Defense University study 'Climatic change to the year 2000' – an editorial. *Climatic Change*, 7, 153–8

Seidel, S. and Keyes, D. 1983: *Can we Delay a Greenhouse Warming?* Washington, D.C., US Environmental Protection Agency

Shennan, I. 1982: Interpretation of Flandrian sea-level data from the Fenland, England. *Proceedings of the Geologists' Association*, 93, 53–63

Shennan, I. 1986a: Flandrian sea-level changes in the Fenland I: the geographical setting and evidence of relative sea-level changes. *Journal of Quaternary Science*, 1, 119–54

Shennan, I. 1986b: Flandrian sea-level changes in the Fenland II: tendencies of sea-level movement, altitudinal changes, and local and regional factors. *Journal of Quaternary Science*, 1, 155–79

Shennan, I. 1987a: Holocene sea-level changes in the North Sea. In, M. J. Tooley and I. Shennan (eds) *Sea-level Changes*. Oxford, Basil Blackwell, 109–151

Shennan, I. 1987b: Impacts on The Wash of sea-level rise. In, Doody, P. and Barnett, B. (eds) *The Wash and its Environment*, (*Research and Survey in Nature Conservation* No. 7), 77–90. Peterborough, Nature Conservancy Council

Shennan, I. 1988: UK; England, Lincolnshire. In, Walker, H. J. (ed.) *Artificial Structures and Shorelines*. Dordrecht, Kluwer Academic Publishers, 145–54

Shennan, I. and Sproxton, I. W. 1990: Possible impacts of sea-level rise – a case study from the Tees estuary, County Cleveland. In, Doornkamp, J. C. (ed.), *The Greenhouse Effect and Rising Sea Levels in the U.K.* Nottingham, M1 Press, 109–33

Shennan, I. and Tooley, M. J. 1987: Conspectus of fundamental and strategic research on sea-level changes. In, M. J. Tooley and I. Shennan (eds) *Sea-level Changes*. Oxford, Basil Blackwell, 371–90

Stewart, R. W. 1989: Sea-level rise or coastal subsidence question. *Atmosphere-Ocean* 27, 461–77

Titus, J. G. 1987: The greenhouse effect, rising sea level and society's response. In, Devoy, R. J. N. (ed.) *Sea Surface Studies: A Global View*. London, Croom Helm, 499–528

Tooley, M. J. 1989: Global sea levels: floodwaters mark sudden rise. *Nature* 342, 20–1

Warrick, R. and Oerlemans, J. 1990: Sea level rise. In, Houghton, J. T. et al. (eds), *Climate Change: The IPCC Scientific Assessment*. Cambridge, Cambridge University Press 257–81

Wijn-Nielsen, A. C. 1989: The greenhouse effect: a review of data- and model-studies. In, Frankenfield Zanin, J. (ed.) Summaries of presentations: First International Meeting, *Impact of Sea Level Rise on Cities and Regions*, Venice, 7–13

Wood, F. B. 1988: On the need for the validation of the Jones et al. temperature trends with respect to urban warming. *Climatic Change* 12, 297–312

Woodworth, P. J. 1990: A search for accelerations in records of European Mean Sea Level. *Intl. Jl. Climatology* 10, 129–43

5

Vulnerability of the Coastal Lowlands of the Netherlands to a Future Sea-level Rise

S. Jelgersma

5.1 GENERAL DESCRIPTION OF THE LOWLANDS

Introduction

About half of the Netherlands is characterized by a plain mainly situated below sea-level and protected by dikes and dunes against the sea. This picture of the coastal lowland is mainly due to the important rise of sea-level after the last glaciation, a slight tectonic subsidence and the sediments derived from the floor of the North Sea and supplied by rivers. Especially in the western and south-western part the tidal and deltaic plain of the Rivers Rhine, Meuse and Scheldt have been of great importance to the pattern of sedimentation in the Holocene. The geological map (figure 5.1) shows the so-called lower part of the Netherlands covered by these young deposits. Due to the slope of the underlying Pleistocene surface seawards, the Holocene sediments vary in thickness. The lithology shows also great variation: in the coastal area mostly sand (coastal barriers and dunes) and landwards the Holocene sediments are built up of clays, peat and fine sands (figure 5.2).

To protect the low-lying areas from flooding, dike building as a method of flood defence has been applied by the inhabitants of the Low Countries throughout the centuries with varying degrees of success. The central part

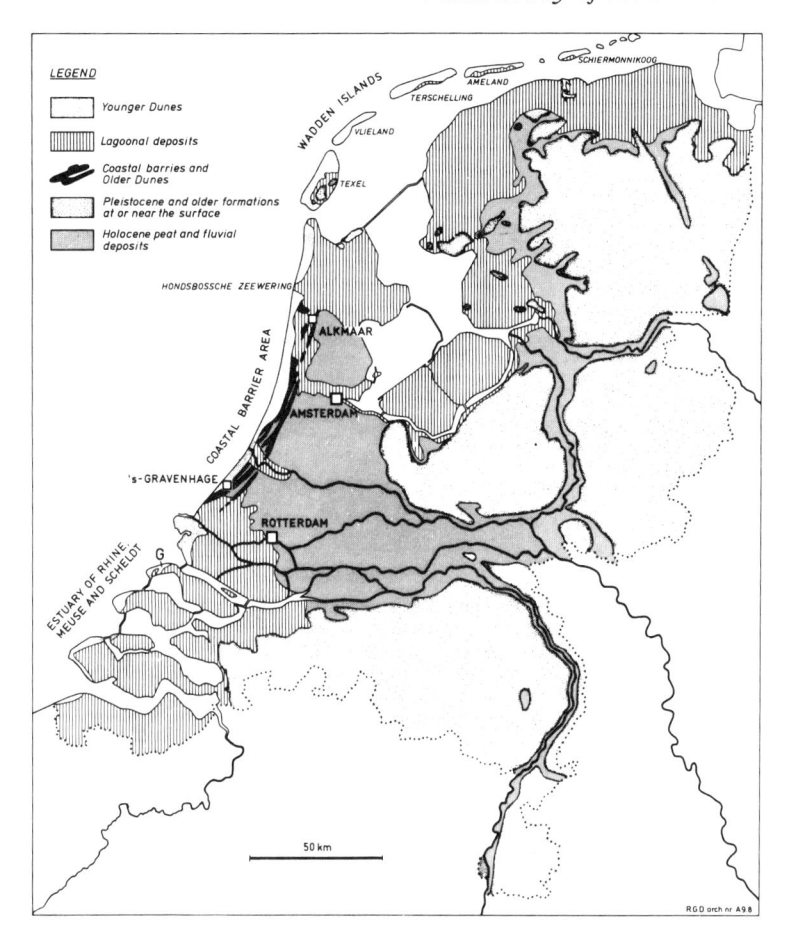

Figure 5.1 Simplified geological map of the Netherlands

of the Low Countries has a high economic value: most of the important industrial activities are present here. Beside important agricultural areas the major part of the population (8 million of the 14 million) has settled in the big cities like Rotterdam, The Hague and Amsterdam.

The shoreline

The coastline of the Netherlands is part of the low sandy shore bordering the south-west part of the North Sea between the French–Belgian border and the north of Denmark.

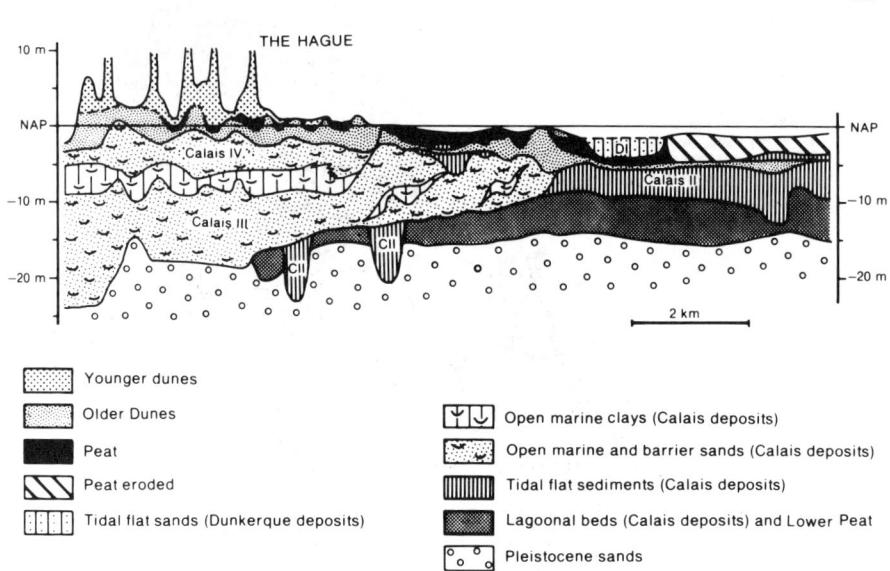

Figure 5.2 Section across the coastal barrier complex near The Hague.
Source: (After Jelgersma et al. 1979)

The total length of the Dutch coast line is more than 400 km, in which the following three units can be distinguished: the Rhine, Meuse and Scheldt estuary in the southwestern part; the coastal barriers in the central part; and the Wadden Islands bordering a large tidal flat area in the northern part of the country (figure 5.1).

The tidal range on most of the Dutch coast is 1.50–2.00 m, but in the south-western part it may be as much as 3.00–4.00 m.

Sand drift along the coast is of great importance along the Wadden Islands (from West to East), of lesser importance between the Belgian border and the Old Rhine (from south-west to north-east) and is rather ineffective between the Old Rhine and the island of Texel (van Straaten 1961).

Wave action on the Dutch coast is generated by the prevailing westerly winds. It is periodically subject to storm surges from the south-west, west and north-west. The last direction is the most dangerous one, as the configuration of the southern North Sea causes extreme high tides during north-west storms.

The whole littoral area can be classified as a micro- to mesotidal wind-dominated clastic shoreline.

The Rhine, Meuse and Scheldt estuary is located south of the Hook of Holland. For centuries this area has fought a battle against the storm floods of the North Sea. Only the west side of the islands has dunes; the major part of the coastline on the inlets and outlets is protected by dikes against the sea. After the disastrous inundation in 1953, the so-called Delta plan was initiated for better protection of the land and its inhabitants. This plan, now almost completed, is designed to shorten the line of defence against floods by closing off the larger inlets and outlets with dams and to reconstruct and to heighten all sea dikes throughout the Netherlands. After completion only the Nieuwe Waterweg to the harbour of Rotterdam and the southernmost inlet giving access to the harbour of Antwerp will be left open.

The coastal barrier area lies in the western part of the Netherlands. In this part of the coast two units can be distinguished: a series of coastal barrier ridges overlaid by low dunes (Older Dunes), part of which is overlaid by Younger Dunes. The latter are much higher (up to 40 m) than the Older Dunes. The width of both landscapes varies between 1 and 9 km. The area overlaid by Younger Dunes is smaller: between 300 m and 5 km (Jelgersma et al. 1970, van Straaten 1965).

The Wadden Islands and the coastline north of the Hondsbossche dike (figure 5.1: plate 4) comprise six islands bordering a large tidal flat area. On the side facing the North Sea, the islands have a dune coast; on the south side the islands are embanked marshes and tidal flats. There are some indications that the coastal dunes (Younger Dunes) are overlying Older Dunes and coastal barriers.

The whole area of the Frisian barrier islands is highly dynamic: there is a lateral displacement of the inlets due to strong littoral drift.

The tidal flat area, the 'Wadden Sea', is rather young. From the twelfth century the sea invaded a lagoon and swamp area, transforming it into the present tidal flat. This area is one of the last natural landscapes left in the low-lying countries. It consists of high tidal flats, partly marshes, and low tidal flats. As a natural reserve the area is of importance to bird life and as a nursery for the fishery offshore. The dynamic process of the tidal currents in the tidal inlets and creeks gives rise to erosion and sedimentation. Because of the asymmetrical tidal current sedimentation is more important than erosion in this tidal flat area (van Straaten 1961). It must be mentioned that sedimentation in this area has kept up with the present sea-level rise of 15 cm/century, measured by tide gauge readings, during the last 100 years.

The backswamp

The alluvial valley and the deltaic plain of the rivers Rhine, Meuse, IJssel and Scheldt is built up out of alternating layers of peat, clay and sand. These layers have been formed in different environments: sand in the channels; clay and peat in the flood basins; and sandy clay on the banks. This area is bordered seawards by the tidal plain where brackish and marine deposits like tidal flats of silts and sands, lagoons with clays and swamps with peats are present. The rest of the coastal plain consists of alternating layers of sand, silt, clay and peats formed in lagoons, marshes, tidal flats, creeks and swamps (Zagwijn 1986).

Bordering the Pleistocene outcrop a zone of peat is present. These unconsolidated Holocene deposits have a maximum thickness of 24 m. The compaction in these layers can be very high. This is mentioned below as human-induced land subsidence due to the drainage of land (compaction and oxidation of peat clay layers) has occurred.

Sea-level changes and subsidence

The observed sea-level changes are the combined effect of eustatic sea-level changes and tectonic movements. The eustatic factor is controlled by the changes in the ice volume on the earth. The building up of land ice caps during the last glacial age caused a drop of sea-level of more than 100 m. Accordingly the whole southern North Sea became dry land. After the last glaciation the enormous land ice caps melted down, resulting in a rise of sea level. The tectonic movements in the area under discussion are more complicated. There is a continuous tectonic downwarping related to the North Sea basin. Figure 5.3 gives a contour of the base of the Tertiary which indicates that the area of maximum tectonic subsidence is situated in the North Sea. In the Netherlands the rate of downwarping is supposed to be 1.5 cm/century (Jelgersma 1966, Zagwijn 1983).

The other tectonic component is caused by isostatic movements arising from the accumulation and the melting of the Scandinavian and British ice caps. The load of the Scandinavian and British ice caps caused a depression of the crust followed by a slow updoming after the withdrawal of the ice. This special case of isostasy is proved by the heights of the numerous raised beaches in the Scandinavian and Scottish areas. Most geophysicists agree that the depression of the crust by the ice loads should cause an uplift in the marginal areas, the so-called peripheral bulge.

The isostatic recovery after the retreat of the ice should on the other hand result in a subsidence of adjacent areas. These ideas have never been

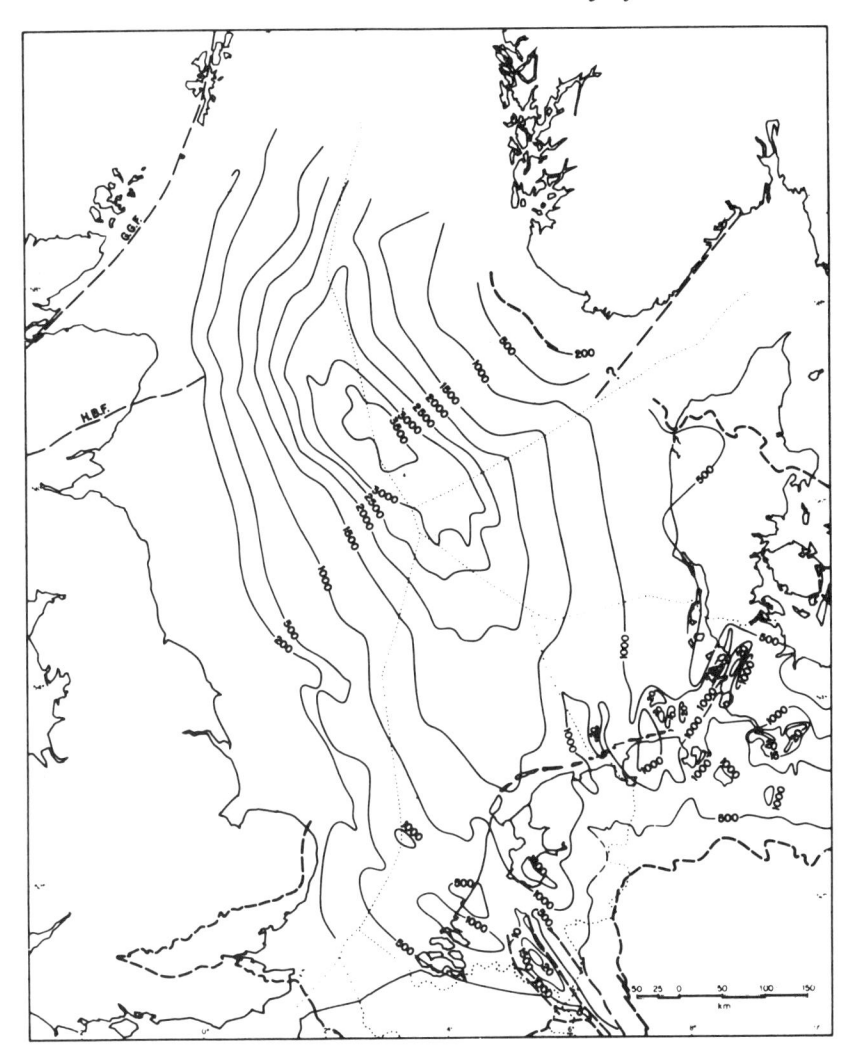

Figure 5.3 Depth contours in metres of the base of the Tertiary.
Source: (After Ziegler and Louwerens 1977)

confirmed empirically by field studies, and the effects in the marginal areas, such as the Netherlands, are still a much disputed subject.

In figure 5.4 the present uplift and downwarping of north-west Europe is given. The figure suggests a relation between the present subsidence in the Netherlands and the adjacent part of North Germany, and the isostatic uplift in Scandinavia and Scotland. As discussed above this relationship is questionable, but we accept that the observed downwarping is related to the tectonic subsidence of the North Sea basin.

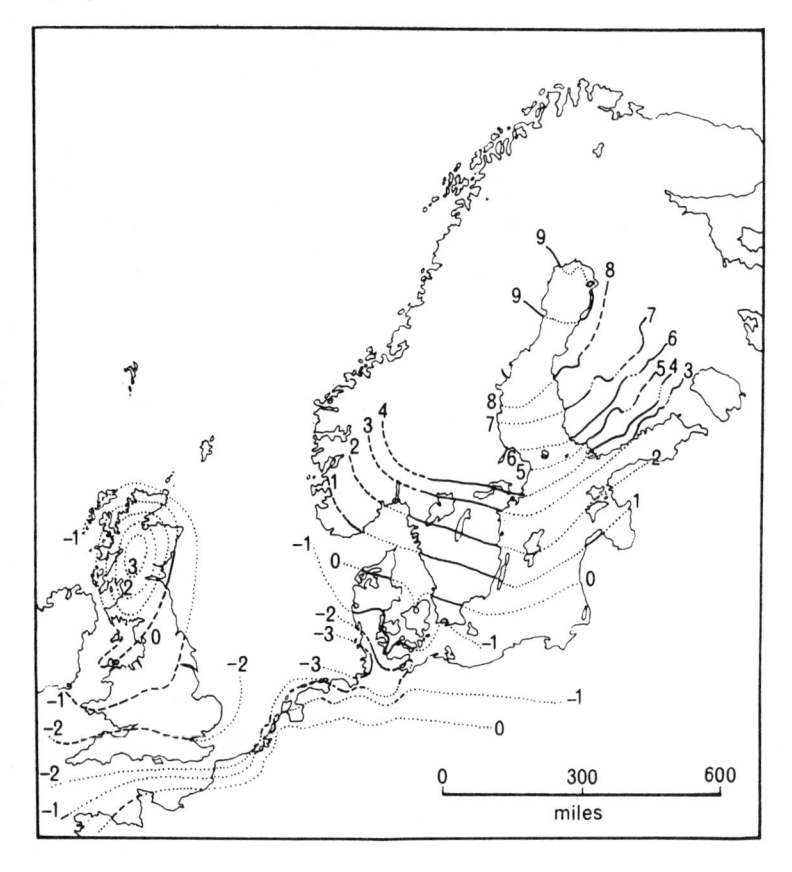

Figure 5.4 Present-day uplift and downwarping in north-west Europe. The isobases represent the rate of change (+ or −) in mm/year, determined from tidegauge records. Dashed isobases are less certain: the dotted isobase lines are interpolations.
Source: (West 1968)

Another important factor contributing to land subsidence is due to human activity in the low-lying areas. These activities like drainage for reclaiming land, ground water extraction and mining of gas and oil cause compaction in the sediments. In the Netherlands Bennema et al. (1954) and Schothorst (1982) have done extensive research on compaction of reclaimed land. The history of reclamation of the Dutch fenlands has been described by van der Linden (1982).

Clays with 35 per cent finer than 2 μm have a porosity of more than 80 per cent and they will be compressed to about half of their original thickness after reclamation over a hundred-year period. Peats will be compressed much more: to about one ninth of their original thickness.

Holocene sands, because of their lesser porosity, are not subject to much compaction.

Striking compaction differences can be encountered in land reclaimed centuries ago where peat layers have been eroded by creeks. The latter have been filled up by sand and silt and now emerge up to 2 m above the adjacent part of the landscape owing to the compaction of the adjacent peat and clay.

Significant compaction was found in some land reclaimed at the end of the nineteenth century. In an old creek filled up with peat mud a compaction of 2.5 m was measured over seventy-five years. Outside this creek the surface lowering in the polders amounts to 1–1.5 m over the same period of time.

After the Second World War new polders were reclaimed in the Netherlands in the former Zuider Zee area. To forecast subsidence of the surface after reclamation, consolidation constants were derived from the pore space and the thickness of the sediment (de Glopper 1969, 1986) before reclamation. Before the polder was pumped dry permanent measuring sites were installed to observe the regional pattern of subsidence after reclamation. In twenty-five years the observed subsidence varied between 10 and 100 cm. It is related to the thickness of the Holocene layer and its composition. Because of poor drainage conditions, compaction during the first ten years is much lower than in the following fifteen years.

The curve for the relative rise of sea-level during the last 8,000 years is given in figure 5.5 (no compaction is included in the curve). The curve shows a continuously rising sea-level with a gradual levelling off after 6000 BP. Data about the position of sea-level during the last 2,000 years are not available. If the level in Roman times is compared with the present one, the relative rise of sea-level during this period will not have been more than 1 m (about 5–6 cm/century). This very slow relative rise is in contradiction with the observed tide gauge measurements of the last century. These observations indicate a rise of 15–20 cm/century (van Malde, this volume). It cannot be excluded that this current rate of rise of sea-level is caused by global warming due to the greenhouse effect.

5.2 FLOOD CONTROL, WATER MANAGEMENT AND LAND USE

Land reclamation and water management

Land use in the low-lying areas had already started in the neolithic, Bronze and Iron Age on the natural levees of tidal creeks and rivers in the Rhine and Meuse area. The occupation of some of these settlements is thought to be restricted to the season of low risk from flooding. Permanent settlement

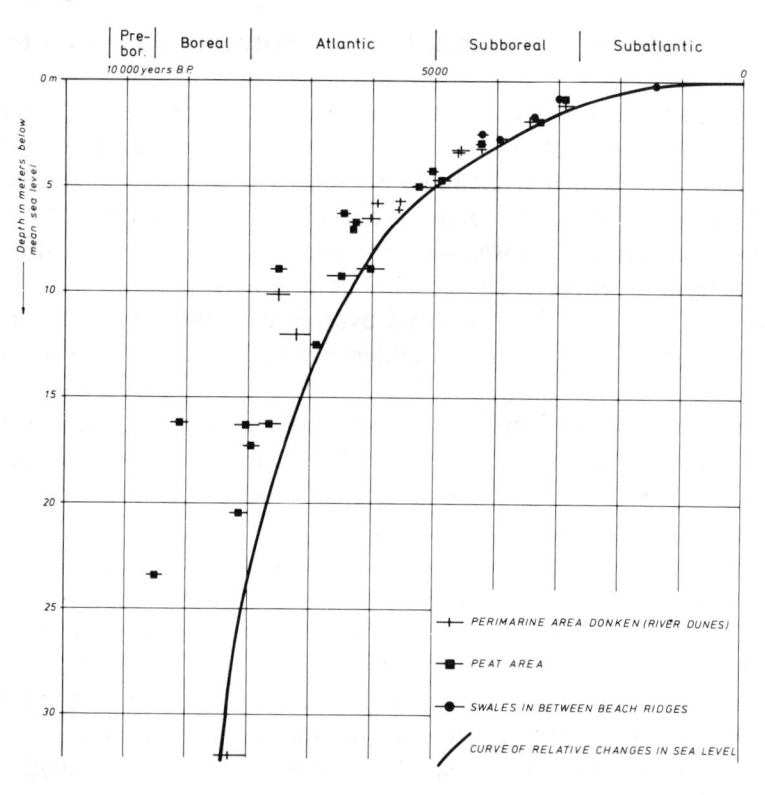

Figure 5.5 Curve of relative sea-level rise during the Holocene, based on data from the Dutch coastal plain.
Source: (Jelgersma 1979)

would be in higher areas: in the coastal dunes and on the Pleistocene outcrops. The activities in the settlements were first restricted to hunting and cattle farming. Later on agriculture was practised. At the beginning of the first century AD settlements are known in the marine area of the northern part of the Netherlands. Shortly after that time, because of the increased danger of flooding, dwelling mounds had to be constructed to protect the settlements from flooding (van Giffen, 1940).

It was not until the twelfth century that the first embankments were made to protect and allow the reclamation of the low-lying coastal areas. From this time onwards the Dutch people have found that the method of flood defence by dike building is the most successful method that can be applied.

An optimum water level in the reclaimed land was maintained by discharging excess water by sluices with gates which opened at low tide. Owing to compaction of the fine-grained sediments, the accompanying surface lowering has caused the tidal drainage to be replaced by windmills.

This subsidence was already noticeable and discussed in 1570 in the Court of Holland. In 1650 the tidal drainage of large areas of Holland had to be replaced by windmills. At the end of the sixteenth century the windmill was also used to reclaim lakes. This method was initiated by rich merchants from the port of Amsterdam who wanted to invest money in new farmlands.

The larger and deeper lakes, like the Haarlemmermeer and parts of the IJsselmeer, could not be reclaimed until the nineteenth and twentieth centuries when steam, oil and electric pumps were developed. The earlier history of reclamation of the western fenlands has been described by van der Linden (1982).

From the twelfth century onwards more than 7,000 square km of land were reclaimed. This amount includes 2,500 square km of reclaimed lakes. The reclaiming of land from the sea was not always so successful, as large storms frequently inundated the inhabited low areas; but during the periods of relatively low storm frequency the battle restarted to reclaim the land.

One of the last severe storm surges in 1953 inundated large parts of the south-west Netherlands as a result of dike breaching. Nearly 2,000 inhabitants lost their lives and thousands of square kilometres were flooded. This last disaster has given rise to the Delta Plan described earlier.

The area drained by the Rivers Rhine and Meuse has also been periodically subject to flooding due to high river discharge. Many times in the history of the Netherlands the embankments were not high and strong enough to constrain the extremely high river levels. Figure 5.6 gives a map indicating the front-line barriers, a total of about 400 km. In addition to these main sea dikes about 200 km of coastal dunes – the Younger Dunes – present a natural defence against the sea. In the low land itself there are also thousands of kilometres of inland river and polder dikes.

Nearly the whole lower part of the Netherlands is made entirely of polders. The water level of the polders, lying below mean sea-level, is maintained by pumping the excess water into storage canals. Figure 5.7 represents several maps indicating the area of the low lands situated 5 m below N.A.P. In figure 5.8 the infrastructure of the Dutch water system is given. The sluices discharge the water into the North Sea at low tide. A map of the Delta project with the distribution of fresh and salt water is given in figure 5.9.

Land use in the backswamp

During the last century and especially during the last forty years the land use of the low areas has changed. This is due to industrialization and the

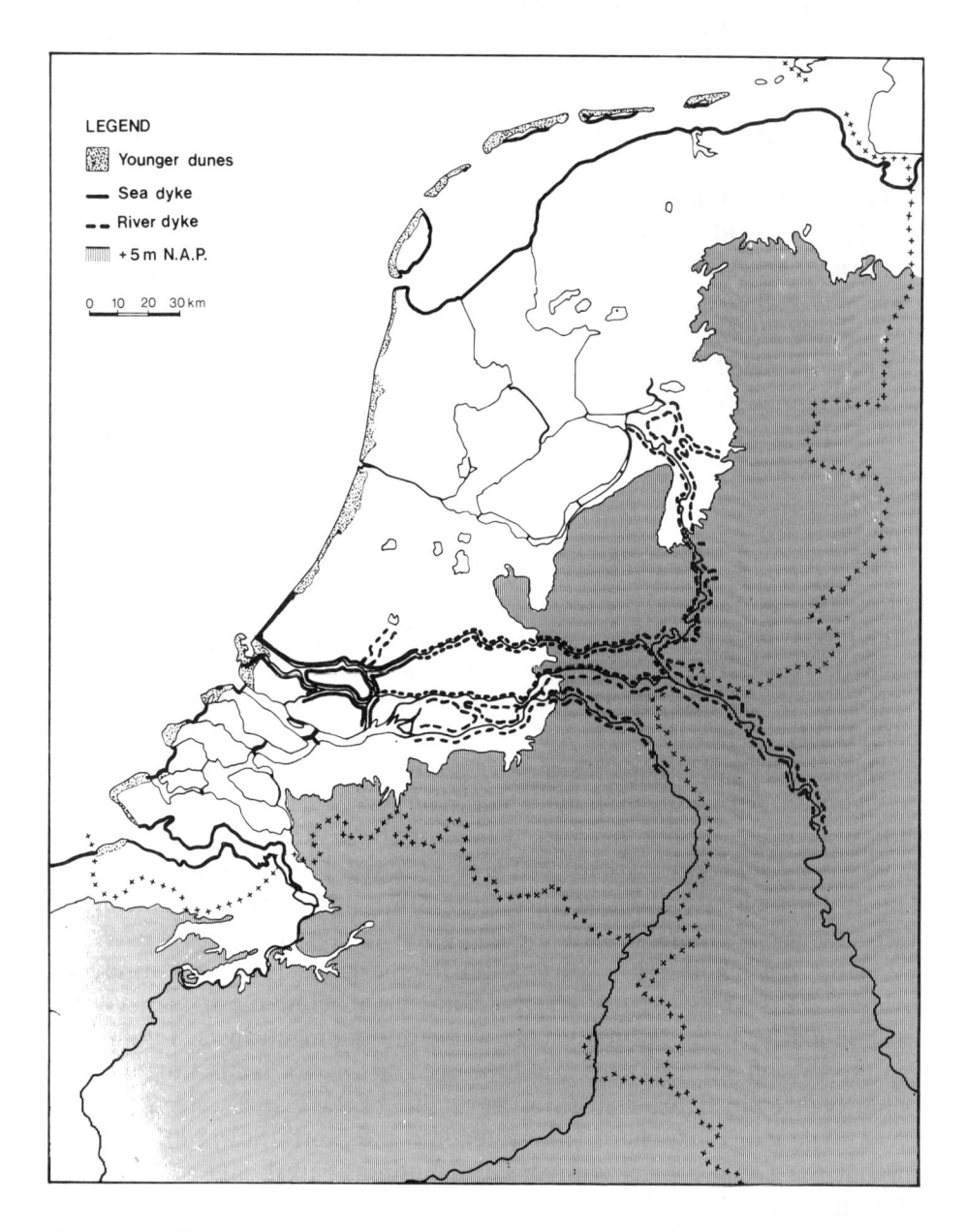

Figure 5.6 The system of sea and river defences against flooding, including the natural defence of the Younger Dunes

increase of population, which has considerably reduced the land under agriculture. At the present time, 8 million of the 14 million people of the Netherlands are living below mean sea-level, and the most important industrial activities can be found in this region. Rotterdam with Europoort has become the biggest harbour in the world. Together with the airport and the industrial activities near Amsterdam the whole area has a dominant position as a gateway to Europe.

In this context the oil harbour and oil refineries in the port of Rotterdam are of vital importance to the heavy industry of the Ruhr area in Germany. The most important oil and gas fields are situated in the lower part of the Netherlands. These mineral resources are not only important for domestic use but also important export products. In figure 5.10 the oil fields, gas fields and the +5 m contour are given. Their relationship to oil and gas fields in the southern North Sea is shown on figure 1.5c.

A general problem related to land use is the pollution of ground water. This is caused by the fact that the low lands are a densely populated and heavily industrialized part of the country with important agriculture.

Another problem is the pollution of the River Rhine. River water supplies about two-thirds of the total fresh water needed in the Netherlands, and accordingly the pollution of the river water is a serious problem.

Another problem related to the industrial activity in the port of Rotterdam is the increased salt water intrusion. Deeper entrance channels and harbours have caused the salt water intrusion to move about 35 km further upstream during the last sixty years. This causes problems for the adjacent agriculture which in the dry season needs fresh water from the river.

About land use for agriculture the following outline can be given. On the well-drained marine soils in the south-west, in the north and in the recently reclaimed polders the land is used for arable farming. In the clay and peaty soils dairy farming is practised.

Land use in the coastal zone

In the western part of the country the area of the low coastal dunes has been levelled and is used for bulb culture.

The higher coastal dunes have become during the last 100 years important areas for drinking water supply and recreation. Before that time the only activity in these areas was hunting; the region was named the 'wilderness'. Owing to overexploitation of the ground water in the dunes an important artificial recharge with river water has become necessary.

Figure 5.7 The areas of the Netherlands below (−), at or above (+) the Dutch Ordnance Datum (NAP = *Normaal Amsterdams Peil*)

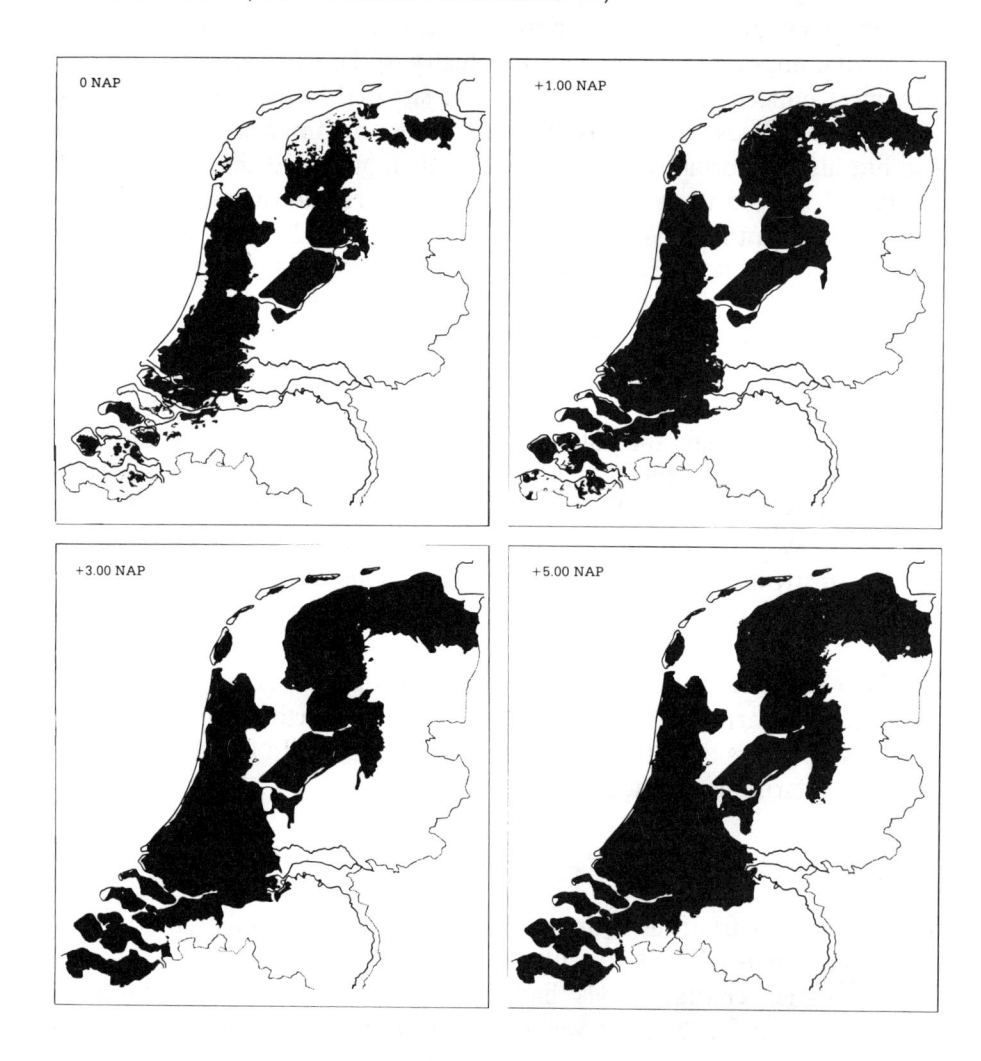

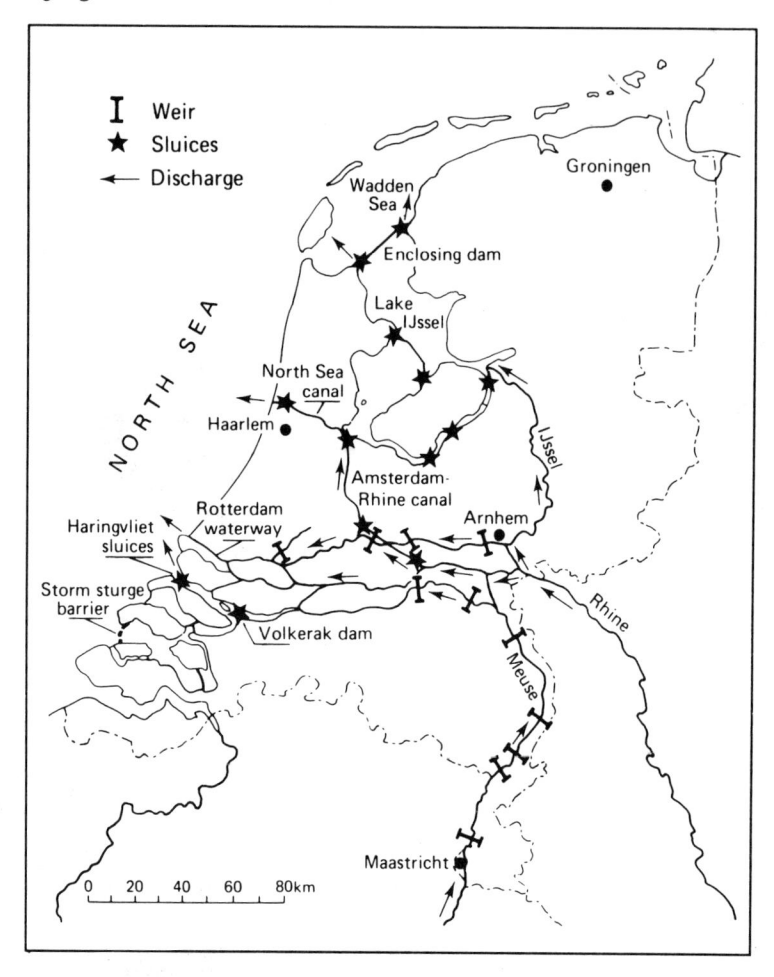

Figure 5.8 Infrastructure of the Dutch water system showing the position of weirs, sluices and discharge points into the North Sea and Wadden Sea.
Source: (Colenbrander 1986)

It must be mentioned that in the lower part of the Netherlands the ground water is brackish; accordingly the only source of fresh ground water is situated in the coastal dune area. The increasing demands of inhabitants, tourists and industry during the last forty years have caused the overexploitation of ground water.

The coastal area is an important recreation area for national and international tourism. The same can be said for the open water areas of the south-west, Lake IJssel and the Wadden Sea. These open water areas are,

Vulnerability of the Netherlands 109

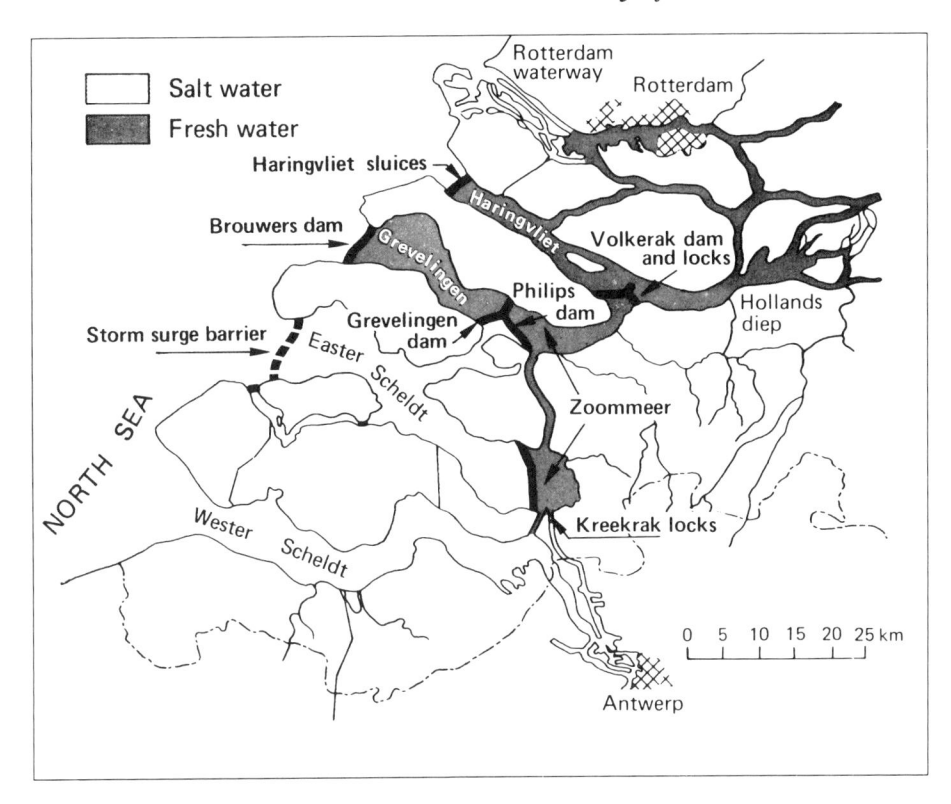

Figure 5.9 Map of the Delta project. The total discharge of the Scheldt and Meuse and 90 per cent of the Rhine discharge is in these distributaries. The Rotterdam Waterway and the Wester Scheldt were excluded from the project because they give access to the great European harbours of Rotterdam and Antwerp.
Source: (Colenbrander 1986)

beside recreation, of vital importance to bird migration. The Wadden Sea, itself a unique natural wetland reserve, and the shallow water offshore zone is an important nursery of several fish species.

5.3 EFFECTS OF A FUTURE SEA-LEVEL RISE ON THE COASTAL LOWLAND

General remarks

The starting point in the following discussion is a rise in sea-level of 1 m in the next 100 years. This is the value agreed at Noordwijkerhout in 1987 (see chapter 11, section 11.5). However, many estimates have been published

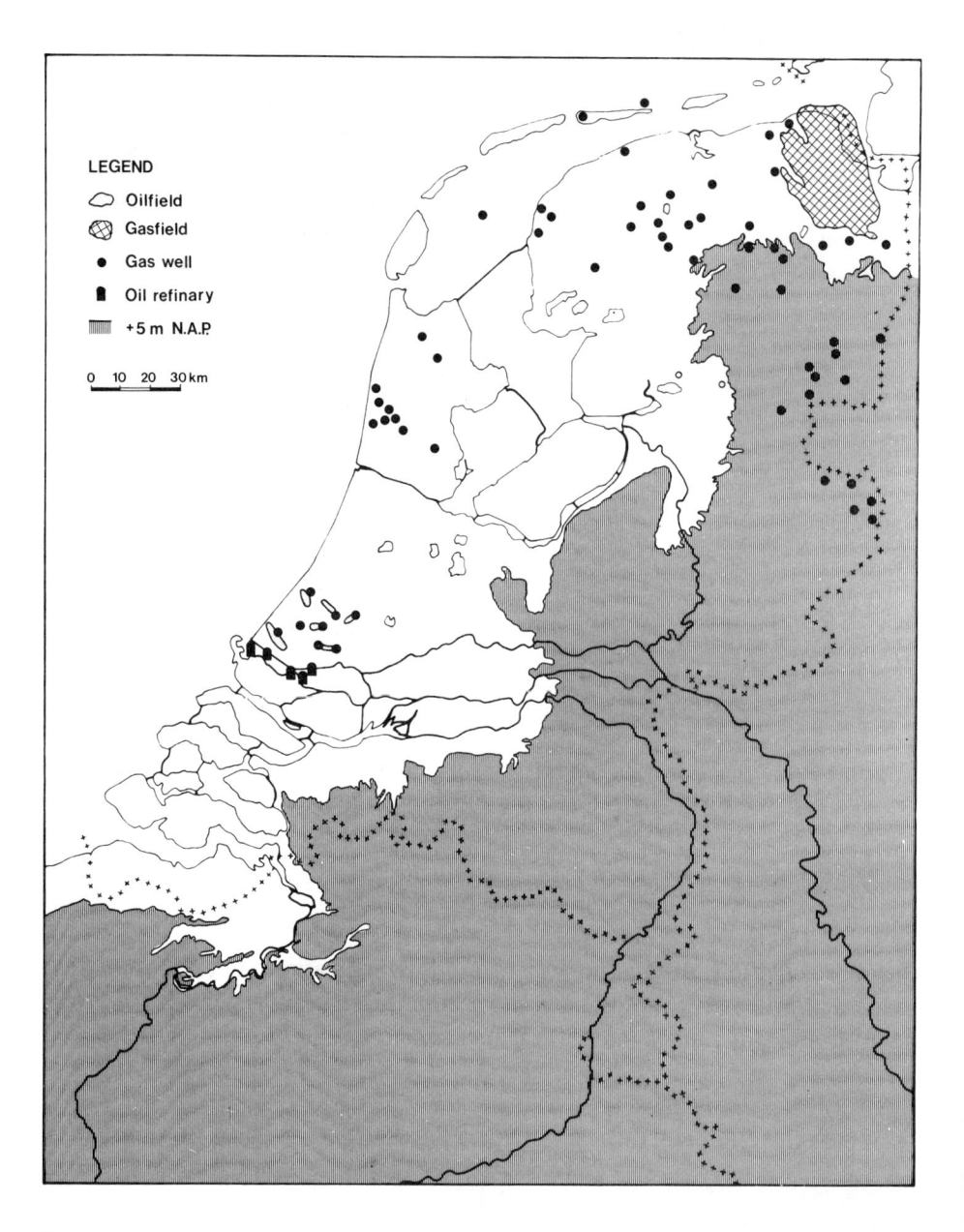

Figure 5.10 The location of oil and gas fields, gas wells and oil refineries in relation to the +5 m NAP contour (see also figure 1.5c)

(e.g. Robin 1986, Titus 1986, van der Veen 1986. See also the introductory discussion in chapter 1). Another important factor that can have great influence on the coastal area is climatic change due to the predicted rise in temperature. It seems likely that precipitation will change; less in summer and more in winter. The increased discharge in the river area can give problems with the embankments. The decreased discharge in summer can cause salt water intrusion. The latter will be a problem for agriculture since fresh water is needed during the dry season. Another unknown aspect is that owing to a change in sea-level the tidal range along the coast may change.

A change in climate can result in an increase in storminess along the coasts in the southern North Sea basin. It is well known that the most serious damage on a coastline occurs during storm surges at the time of high tide.

In the discussion of the impact of future sea-level rise on the coastline, the backswamp and the riverplain these impacts will be assigned a risk factor. The evaluated area will be classified into one of three zones: Extreme Risk, High Risk and Moderate Risk.

The coastal zone shown in figure 5.1 will be subdivided into three sectors: the south-west area, the coastal barrier area and the Wadden Sea.

In the present situation, the main distributaries of the estuaries of the Rivers Rhine, Meuse and Scheldt have been closed off by dams. Only the Wester Scheldt and the entrance to the harbour of Rotterdam are left open (figure 5.9). The 1953 flooding established a new philosophy with respect to the design water level for the country's sea dikes. The heights of new dikes of the Delta plan are based on the decision that the probability of exceedence is set at 1/10,000 years.

This idea was introduced by Van Wemelsfelder who constructed, in relation to the probability of exceedence of storm surge levels, probability exceedence lines. He used a linear scale for the water level and a logarithmic scale for frequencies of occurrence. The tidal range along the Dutch coast varies between 1 and 4 m; accordingly this has to be taken into account for the exceedence line in the given location (figure 5.11). Thus the height of a given dike, the so-called delta level, is derived from the following factors:

storm surge level	5.00 m above sea-level
wave run up	9.90 m
seiches and gusts	0.35 m
sea-level rise	0.25 m
settlement	0.25 m
total height	15.75 m

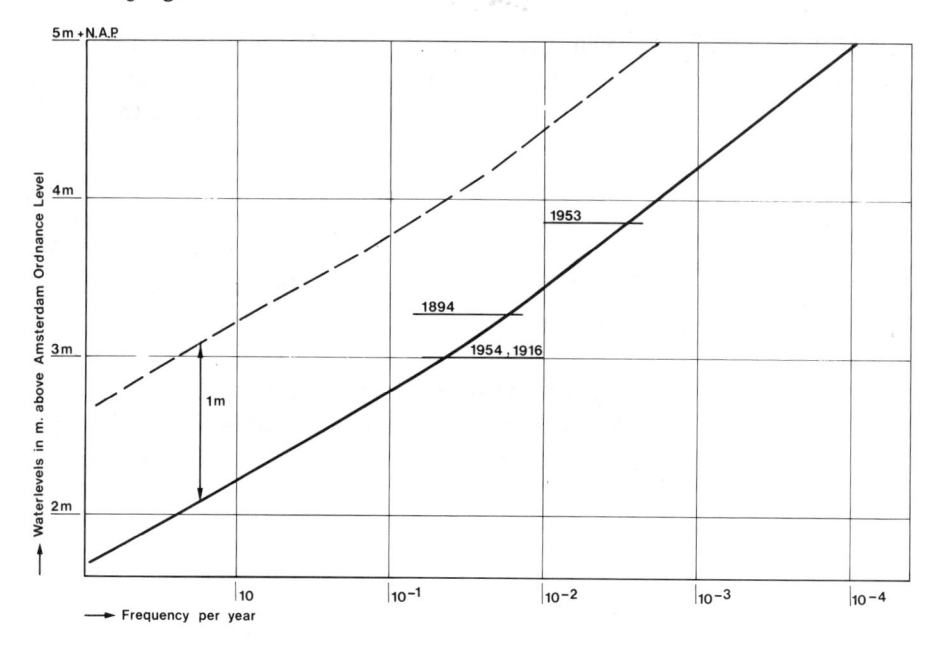

Figure 5.11 Present frequency and risk of storm surge levels at Hook of Holland (continuous line) and for a 1 m rise of sea-level (dashed line). Extreme water levels in relation to Amsterdam OD (NAP = mean sea level) are shown for AD 1894, 1916, 1953 and 1954).

If this safety norm is to be maintained the delta dikes throughout the Netherlands should be elevated, not only for an increase in sea level but also for an increase in storm surge level and wave run-up.

The south-west area

The same applies to the dikes surrounding the Wester Scheldt and the Nieuwe Waterweg to Rotterdam. To safeguard the hinterland, either sluices or higher dikes should be the answer to the future rise of sea-level. These safety measures would have a serious impact on the present water management system in this area. To protect the city and harbour of Rotterdam from future sea-level rise, a storm surge barrier in the Meuse Waterway has been designed, proposed and accepted by the Dutch Parliament (Third Policy Document on Water Management, 1989).

The coastal dune strip on the western part of the former islands in the Delta area is not very high and small in area. This natural defence against the sea has shown erosion and progradation in the past. It seems likely that

1 The Tees coastal lowlands, North-east England.
2 Dungeness nuclear power station in Kent, South-east England.

3 Bergen aan Zee, North Holland.
4 The Hondsbossche Sea dyke beyond the high dunes north of Bergen aan Zee.

5   Callantsoog, North Holland.
6   Nieuwpoort, Belgium.

7 Aigues-Mortes on the Rhône Delta in France.
8 The Po Delta, Italy near the village of La Pila.

Vulnerability of the Netherlands 113

the trend of erosion and sedimentation will change after the completion of the Delta works.

Impact on the coastal barrier area

The coastal dunes between Monster and the Hondsbossche dike are between 300 m and 5 km wide. Their heights show a variation between 8 and 40 m above mean sea-level. On figure 5.12, a map of the sandy littoral zone is presented and indicates the amount of retreat and progradation during the last 100 years. On a larger scale, on the former island of Schouwen (Figure 5.13), the pattern of erosion in relation the high and low dunes is shown. The conclusion that a rise in sea-level will accelerate erosion and stop or reverse progradation seems to be likely. The theoretical modelling of the amount of beach erosion caused by a rise in sea-level has been developed by Bruun (1962, Bruun and Schwartz 1985). The Bruun Rule states that a beach (figure 5.14) which has attained equilibrium with coastal processes (Profile 1) will respond to a rise in sea-level by losing sand from the upper part of the profile and gaining it in the nearshore area until a new equilibrium (Profile 2) is established. The coastline will thus retreat from A to B and then C, as the direct result of the transference of the sand seaward (from V_1 to V_2). It is possible to predict the extent of coastline recession where the conditions proposed by Bruun apply, but it should be noted that other factors also influence the changes that will occur on a beach as a result of a sea-level rise (for example, diminution of fluvial sand supply to the coast as a result of reduced run-off).

Another approach developed by geographers is the analysis of old shoreline maps. In the Netherlands important work has been done by Schoorl (1973). Leatherman (1984) has given this method new attention in using it to forecast shoreline erosion in the analysis of several old shoreline maps of a given area. According to this approach the amount of historical retreat is assumed to be directly correlated with the rate of sea-level rise. The date of the maps indicate the period under consideration. The amount of relative sea-level rise is known from tide gauges.

If the rise of sea-level in the coming 100 years is five times as great as at present, the retreat of the shoreline will be at least five times as much as observed in foregoing 100 years. If we apply this method on the coastline with the narrow low dunes north of the Hondsbossche dike where the retreat of the coast during the last 100 years has been more than 200 m, the following can be said. The calculated sea-level rise in the last 100 years has been ±20 cm/century which caused a coastline retreat of a 200 m. If sea-level rise is 100 cm, five times as great in the coming 100 years,

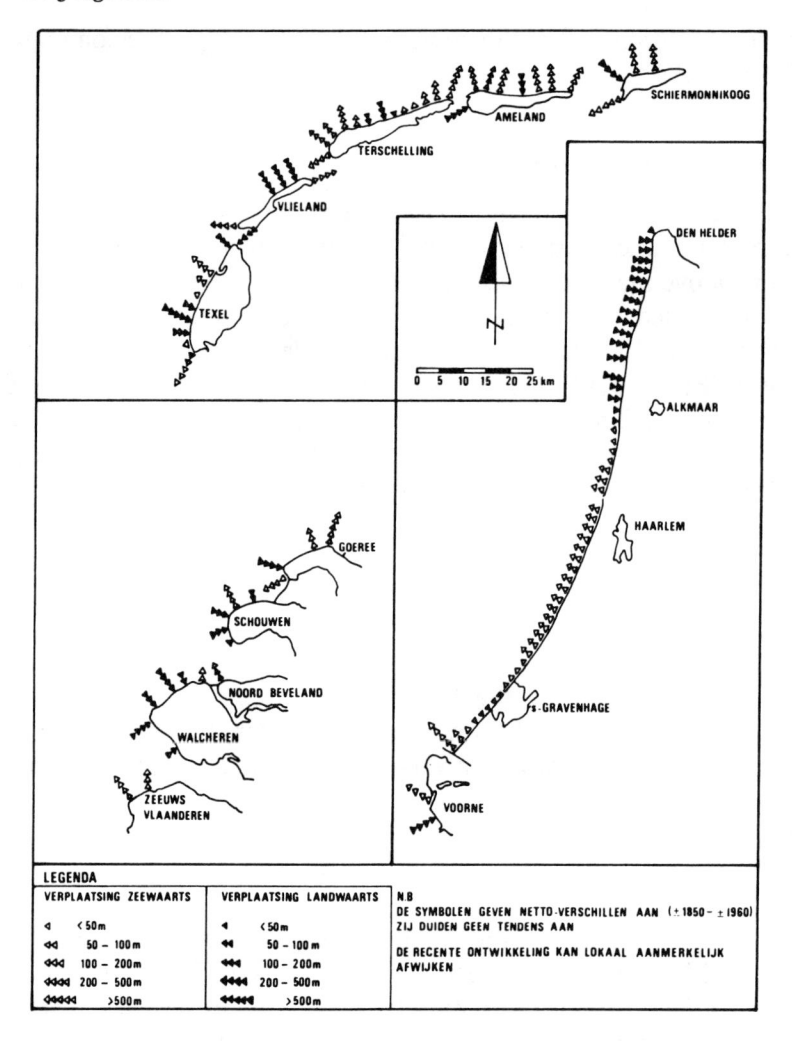

Figure 5.12 Shift in the dune foot along the Dutch coast, 1850–1860 and 1960–1970. Open triangles indicate progradation and amount; black triangles indicate coastline retreat and amount. NB the symbols indicate net values from c. 1850 to c. 1960
Source: (Klijn 1981)

shoreline retreat will be at least 5×200 m = 1000 m. This could give rise to a very dangerous situation due to the very narrow dune strip in this region.

The following can be remarked about the coastal dunes south of the Hondsbossche dike. If coastal erosion takes place the situation is not too

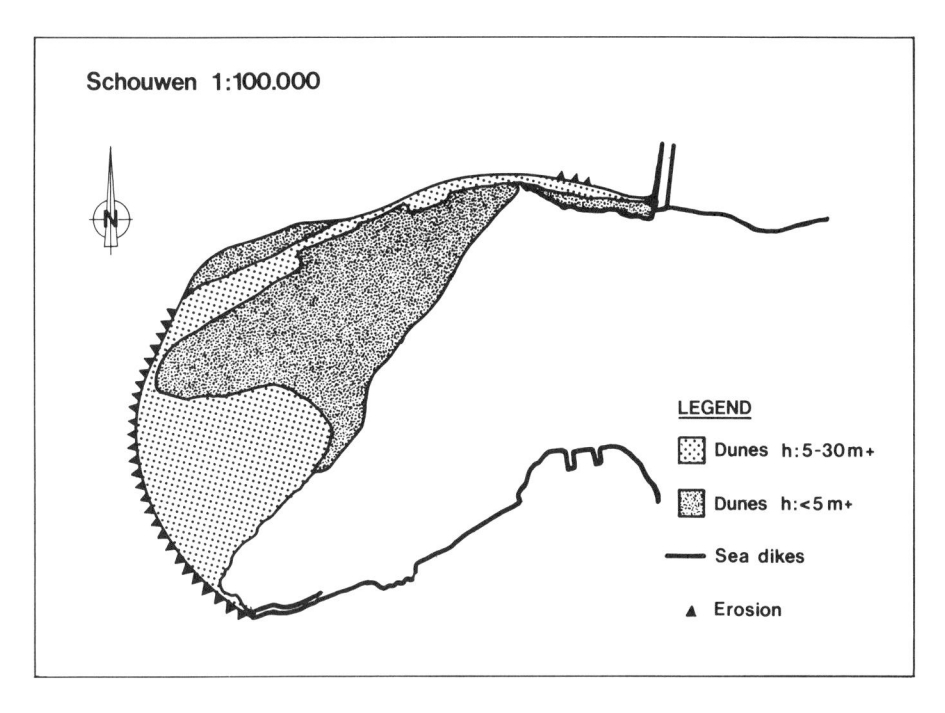

Figure 5.13 Map showing the elevation of the dune areas of the former island of Schouwen in the delta area in relation to the pattern of erosion. Schouwen is located on figure 5.12.

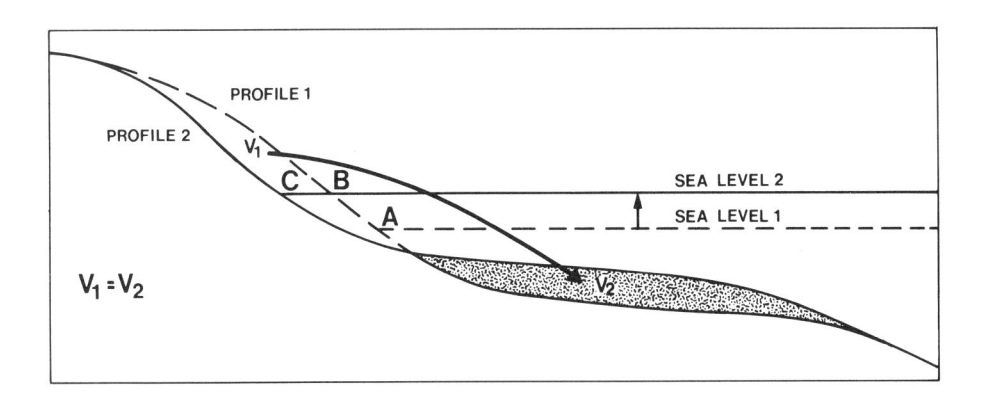

Figure 5.14 The Bruun rule of beach erosion due to sea-level rise (V_1 is the volume of sand eroded, as a consequence of sea-level rise from level 1 to level 2, and is deposited in the near-shore zone as V_2)

dangerous in relation to the hinterland. The dune belt is wide and the dunes are high. Only two points are more risky: the former outlet of the River Old Rhine where the dunes are very narrow; and the levelled area of the harbour of IJmuiden.

If erosion occurs the sand will be removed in the near-shore zone and accordingly the eroded material gives an input to the sand budget. The effect of erosion is likely to be different north of the Hondsbossche dike and on the Wadden Islands. There the dunes are low and will be moved landwards by means of wash-overs. In this case no eroded material will be transported to the near-shore zone.

In general it would be recommended to subdivide the shoreline into several sectors and analyse historical shoreline movements. Together with wave exposure, longshore transport and the local relative sea-level rise during the time under consideration, a trend line in the behaviour of the shoreline to a future sea-level rise should be derived.

The extensive dune area in the coastal barrier region is of great importance for the hinterland because of the occurrence of fresh water. Erosion of this dune area will cause a lowering of the ground water table resulting in a decrease of the underlying fresh water reservoir which also opposes seepage of salt ground water in the polder area.

Coastal erosion would also have a serious impact on the coastal resorts located on or near the shoreline. Damage would result in an important loss of property and income through the decrease of tourism.

The Wadden Islands and the coast north of the Hondsbossche dike

The coast north of the Hondsbossche dike has already been discussed, and the conclusion may be given that this area can be classified in the Extreme Risk category.

The Wadden Islands are located on the north side of a great tidal flat area which is southwards bordered by the sand dikes of Groningen and Friesland. In 1932 the Zuider Zee was closed off from the tidal flat area by a great dike (enclosing dam).

The Wadden Islands lie in a dynamic area. There is important sand drift from west to east. Figure 5.12 indicates the rate of erosion and progradation during the last 100 years. The supposed rise in sea-level of 1 m in the next 100 years will certainly increase the rate of erosion. It does not seem unlikely that the islands will be moved landwards by means of wash-over. The tidal inlets will increase in size, whereas the islands will decrease in size.

The following can be said about the effect on the tidal flats of the rise in sea-level. At present, the sedimentation in the tidal flat keeps up with the rise in sea-level. Because of the asymmetrical tidal curve, the current vectors result in a net transport and deposition landwards (lag effect between the vertical sediment distribution and the governing tidal cycle with an asymmetrical tidal current, Postma 1961, 1967, 1981).

It seems to be questionable that sedimentation can keep up with the expected rise in sea-level which is five times as great as at present. Secondly, owing to the rise in sea-level, hydrodynamic conditions can change. If the change results in a reversal of the tidal asymmetry curves a significant amount of sediment from the Wadden Sea would be transported to the North Sea. If this happens the whole environment of the present tidal flat area of the Wadden Sea will change. This will have a serious impact on the ecology: the function as a nursery for fish species, and for migrating birds. The whole area can be classified as an Extreme Risk region, including the dikes of the mainland.

The backswamp, the rivers and lake areas

The whole low-lying area behind the dunes and the sea dikes is drained by an intricate system of waterways that are carefully managed. Precise water level control is essential in the low-lying areas for many purposes. In the south-west, the delta area coastal reservoirs (figure 5.9) have been made by closing the estuaries. In the north, the Zuider Zee has been closed off by the more than 30 km long enclosing dam; which brought Lake IJssel into being. As mentioned before, the surplus of drainage water is discharged at low tide to the North Sea by drainage sluices.

It may be clear that a rise of sea level of even 0.50 m will upset this whole water management system. Drainage at low tide by means of sluices will be impossible, and drainage will have to be done either by pumplift or by setting up the water level (figure 5.8). In the IJssel the higher water level would make it necessary to elevate and adapt the surrounding dikes. Both methods will be costly.

In the delta area, storm surge barriers would be closed more often than calculated, and it is questionable whether the salt water environment of the Easter Scheldt can be maintained. Change would have a serious impact on oyster culture and other related activities. The low areas, especially the polders, will have to be more intensively drained. This action will increase salt water intrusion and cost more in energy in increased pumping.

A rise in sea level will also have consequences for river run-off. It is expected that sedimentation will shift further upstream. This will elevate the river levels and have consequences for dike heights, shipping and bridges. Salt water intrusion in the rivers will also increase, and this will cause great problems for agriculture when the intake of fresh water is needed during the dry season.

As mentioned before, a change in climate is also thought to be likely owing to the greenhouse effect. This climatic change will have an effect on river discharge rate and groundwater recharge, and hence rate of salt water intrusion.

In winter the precipitation may increase, and during the summer decrease. These changes in precipitation indicate a lower river discharge during the summer which would increase the salt water intrusion much more. This could have a serious side-effect on agriculture. Figure 5.15 represents a risk hazard map of the low area of the Netherlands.

Economic impact

A background for the consideration of the impact of sea-level rise on the economy of the Netherlands can be given by reference to figure 5.16. This shows the area that would be flooded without dikes and the value of these areas in terms of the gross national product.

Some preliminary calculations have been done about the amount of money needed to keep the low part of the Netherlands dry. Goemans (1986) has given some indication of the costs for water management and shoreline protection of a 1 m rise of sea-level. The costs would be 10 billion florins to be spent on:

1. change from sluices into pumplift stations;
2. existing pumplift stations operating for longer, requiring more energy, and in some cases adaption of the pumplift stations;
3. reconstruction of dikes, bridges and weirs in the coastal zone and in the river area;
4. measures to reduce salt water intrusion.

Adjustments to harbour and dock facilities on the coast and along the rivers will cost 1 billion florins. Adjustments to the water management in the low parts about 3 billion florins.

The raising of the dikes and nourishment/supplementation on the sandy shore will amount to 6 billion florins. According to Goemans (1986) the raising of the dikes must be done in steps of 0.50 m. Each step will require at least twenty years in time. The total costs, even with a rise in sea level of 2 m, will amount to a yearly expenditure of 250 million florins, 0.1 per cent

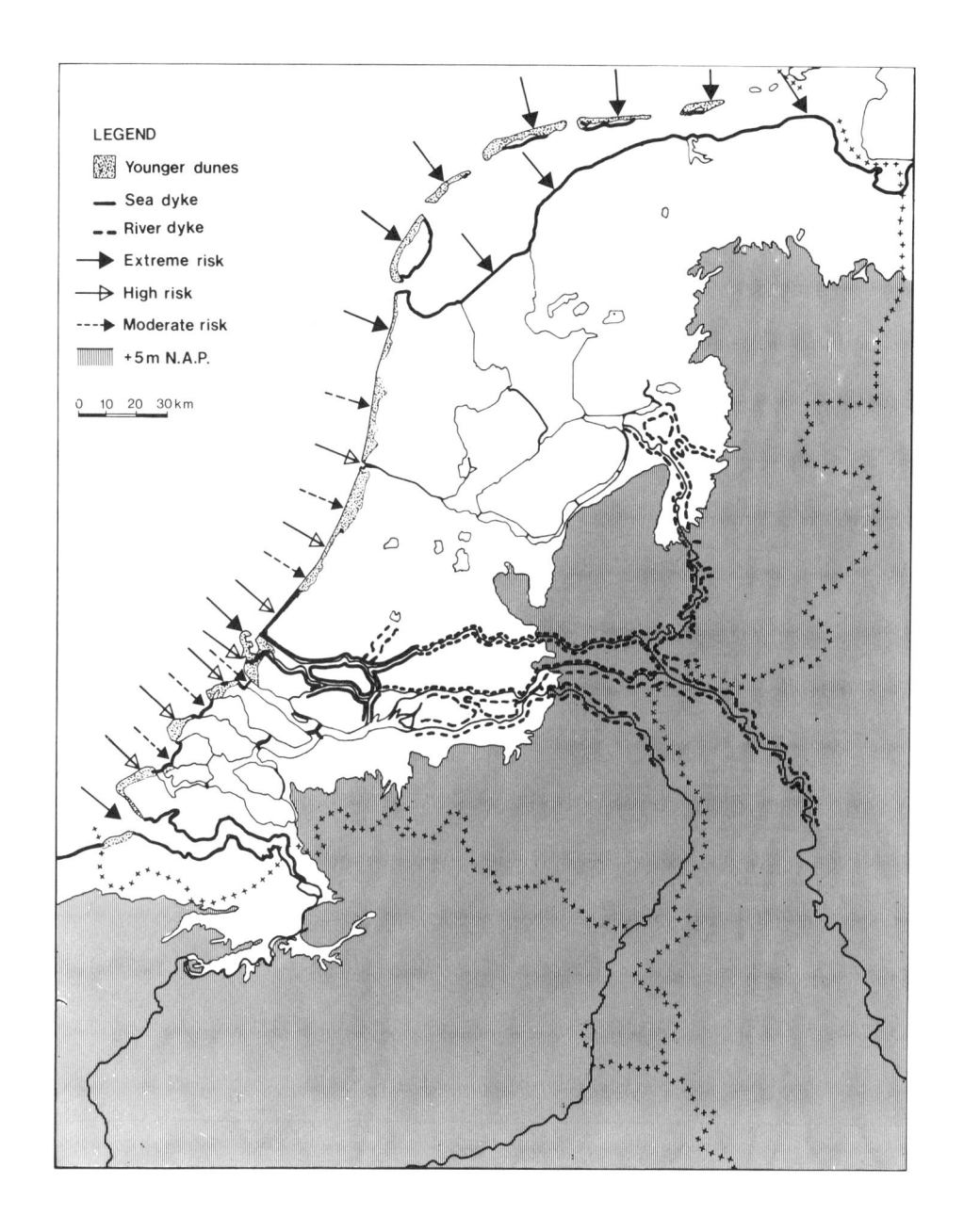

Figure 5.15 Risk hazards map of the Netherlands related to a rising sea-level

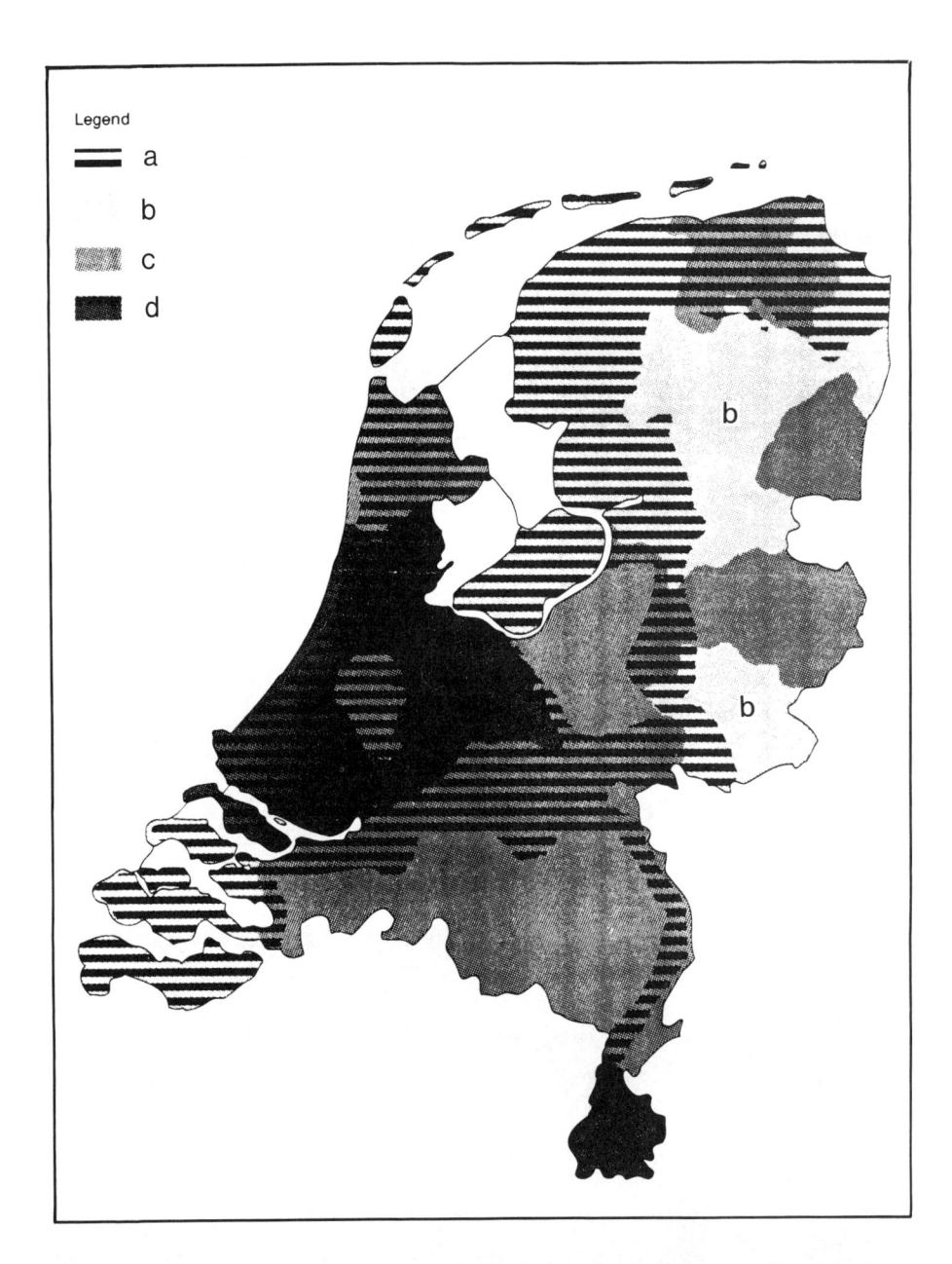

Figure 5.16 Map of the Netherlands showing the economic value of the low part of the country in terms of the distribution of present gross national product: a, area flooded without dikes; b, value less than 5 million Dfl./km^2 (GNP); c, 5–15 million Dfl./km^2 (GNP); d, more than 15 million Dfl./km^2 (GNP). (Ministry of Transport and Public Works 1989)

of the Dutch GNP. These figures do not include extra expenditure in energy and enlarging pumplift installations.

The impact on crops, the fisheries of the tidal flat area of the Wadden Sea, recreation, storm damage to coastal resorts and the fresh water resources in the dunes have not yet been studied and translated into costs.

Some more recent calculations have been done in extensive reports for the government of the Netherlands by the Ministry of Transport and Public Works (1989). Several scenarios different in time and sea-level rise are given which are only for dike reconstruction and sandy shoreline erosion protection. For the dune coast and sandy shoreline there are 4 possibilities.

1 retreat;
2 selective maintenance;
3 maintenance;
4 seawards building.

One of the sea-level rise scenarios is 85 cm in the coming 100 years. It is calculated that the total costs of coastal defences, the raising of dikes and to maintain the position of the sandy shoreline, will be about 90 million florins a year for the coming 100 years. The building costs of a storm surge barrier in the entrance channel to the harbour of Rotterdam, to protect against high water and to slow down salt water intrusion, is calculated at 1 billion florins.

REFERENCES

Bennema, J., Geuse, E. C. W. A., Smits, H. and Wiggers, A. J. 1954: Soil compaction in relation to Quaternary movements of sea-level and subsidence of the land, especially in the Netherlands, *Geol. Mijn.* 16, 173–8
Bruun, P. 1962: Sea-level rise as a cause for shore erosion. *Jl. Waterways and Harbors Division* 88, 117–30
Bruun, P. and Schwartz, M. L. 1985: Analytical prediction of beach profile change in response to a sea-level rise. *Z. Geomorph.* Suppl.-Bd. 57, 33–50
Colenbrander, H. J. (compiler) 1986: *Water in The Netherlands*. The Hague, TNO Committee on Hydrological Research
Giffen, A. E. van 1940: Die Würtenforschung in Holland. In, Haarnagel, W. (ed.) *Probleme der Küstenforschung in Südlichen Nordseegebiet*, Hildesheim, August Lax

Glopper, R. J. de 1969: Shrinkage of subaqueous sediments of Lake IJssel (the Netherlands) after reclamation. *Proc. Intl. Symp. Land Subsidence*, Tokyo. IAHS Publication No. 151, 487–96

Glopper, R. J. de 1986: Subsidence in the recently-reclaimed IJsselmeer polder 'Flevoland'. *Proc. Intl. Symp. Land Subsidence*, Venice. IAHS Publication No. 151, 487–96

Goemans, T. 1986: *The Sea Also Rises: The Ongoing Dialogue of the Dutch with the Sea*. Workshop on Sea-level Rise. UNEP/EPA Conference on the Health and the Environmental Effects of Ozone Modification and Climate Change. Crystal City, Va., USA

Health Council, 1987: CO_2 *Problem, Scientific Opinions and Impacts on Society*. The Hague, Committee of the Health Council, 140pp

Jelgersma, S. 1966: Sea-level changes during the last 10,000 years. In, Sawyer, J. S. (ed.) *World Climate 8000 to O BC*. (Proc. Intl. Symp.). London, Royal Meterological Society, 54–69

Jelgersma, S. 1979: Sea-level changes in the North Sea basin. In, Oele. E., Schüttenhelm, R. T. E. and Wiggers, A. J. (eds), *The Quaternary History of the North Sea*. Uppsala, Acta Univ. Ups. Symp. Univ. Ups. Annum Quingentesimum Celebrantis 2, 233–48

Jelgersma, S., Oele, E. and Wiggers, A. J. 1979: Depositional history and coastal development in the Netherlands and the adjacent North Sea since the Eemian. In, Oele, E., Schüttenhelm, R. T. E. and Wiggers A. J. (eds) *The Quaternary History of the North Sea*. Uppsala, Acta Univ. Ups. Symp. Univ. Ups. Annum Quingentesimum Celebrantis, 2, 115–42

Jelgersma, S., de Jong, J., Zagwijn, W. H. and van Regteren Altena, J. F. 1970: The coastal dunes of the western Netherlands: geology, vegetational history and archaeology. *Meded. Rijks. Geol. Dienst* NS 21, 93–167

Klijn, A.J. 1981: *Nederlandse Kustduinen Geomorfologie en bodems*. Wageningen, Centrum voor Landbouwpublicaties en Landbouw Documentatie

Leatherman, S. P. 1984: Coastal geomorphic responses to sea-level rise: Galveston Bay, Texas. In, Barth, M. C. and Titus, J. G. (eds) *Greenhouse Effect and Sea Level Rise: A Challenge for This Generation*. New York, van Nostrand Reinhold, 151–78

Linden, H. van der 1982: History of the reclamation of the western Fenlands and of the organizations to keep them drained. In, Bakker, H. de and Berg, M. W. van den (eds) *Proceedings of the Symposium on Peat Lands below Sea-level*. Wageningen, International Institute for Land Reclamation and Improvement (ILRI), Publication No. 30, 42–73

Ministry of Transport and Public Works 1989: Derde Nota Waterhuishouding (Policy Note on Water Management: 3rd Policy Document). 's Gravenhage, SDU Uitgeverij

Postma, H. 1961: Transport and accumulation of suspended matter in the Dutch Wadden Sea. *Netherlands Jl. Sea Research* 1(1/2), 148–90

Postma, H. 1967: Sediment transport and sedimentation in the estuarine environment. In, Lauff, G. H. (ed.) *Estuaries*. Washington, D. C., American Association for the Advancement of Science, 158–79

Postma, H. 1981: Exchange of materials between the North Sea and the Wadden Sea. *Marine Geology* 40(1/2), 199–213

Robin, G. de Q. 1986: Changing the sea-level: projecting the rise in sea-level caused by warming of the atmosphere. In, Bolin, B., Döös, B., Jäger, J. and Warrick, R. A., (eds) *The Greenhouse Effect, Climatic Change and Ecosystems* (SCOPE 29). Chichester, John Wiley and Sons, 323–59

Schoorl, H. 1973: Zeshonderd jaar water en land, bijdrage tot de historiche Geo- en Hydrografie van de kop van Noord-Holland in de periode circa 1150–1750. *Verh. Kon. Ned. Aardr. Gen.* 2, 534pp.

Schothorst, C. J. 1982: Drainage and behaviour of peat-soils. In, Bakker, H. de and Berg. M. W. van den (eds). *Proceedings of the Symposium on Peat lands below Sea-level*. Wageningen International Institute for Land Reclamation and Improvement (ILRI), Publication No. 30, 130–63

Straaten, L. M. J. U. van 1961: Directional effects of wind, waves and currents along the Dutch North Sea coast. *Geol. Mijn.* 40, 333–46

Straaten, L. M. J. U. van 1965: Coastal barrier deposits in south and north Holland, in particular in the areas around Scheveningen and IJmuiden. *Meded. Geol. Sticht.* NS 17, 41–75

Titus, J. G. 1986: *The Causes and Effects of Sea-level Rise*. Washington, D.C., Environmental Protection Agency

Veen, C. J. van der 1986: *Ice Sheets, Atmospheric CO_2 and Sea-level* (Utrecht, PhD Thesis)

West, R. G. 1968: *Pleistocene Geology and Biology*. London, Longmans.

Zagwijn, W.H. 1983: Sea-level changes in the Netherlands during the Eemian. *Geol. Mijn.* 62(3), 437–50

Zagwijn, W.H. 1986: *Nederland in het Holoceen*. Geologie van Nederland, Deel 1. Haarlem, Rijks Geologische Dienst

Ziegler, P. A. and Louwerens, C. J. 1977: Tectonics of the North Sea. In, Oele, E., Schüttenhelm, R. T. E. and Wiggers, A. J. (eds) *The Quaternary History of the North Sea*. Uppsala, Acta. Univ. Ups. Symp. Univ. Ups. Annum Quingentesimum Celebrantis 2, 7–22

6

Carbon Dioxide Increase, Sea-level Rise and Impacts on the Western Mediterranean: The Ebro Delta Case

M. G. Marino

6.1 BACKGROUND

There is a common consensus among the scientific community on the continuous increase of atmospheric CO_2 by 0.5 percent per year. A doubling or even a quadrupling of CO_2 and of other trace gases emitted, mainly due to human activities, will produce an increase on global average temperature in the range of 2°C to 4°C, with an increase towards the poles of 4°C to 8°C to a new climatic equilibrium (Hekstra 1986).

Impacts of this CO_2 build-up on climate and other ecological parameters of the planet were discussed at the WMO Villach conference, where increases of 1.5°C to 4.5°C in the average temperature for the next century were predicted, together with a sea-level rise of the order of 20 to 140 cm (Bolin et al. 1986: see also Preface and chapter 1).

There is a great deal of uncertainty in these predictions as the roles of the oceans in the CO_2 build-up are not completely understood (Crane and Liss 1983), climatic models need extra developments, temperature equilibrium and stratification of the oceans is a guess, CO_2 uptake by vegetation estimates vary greatly and the rate of melting or accumulation of polar icecaps depends on the assumptions made for its calculation. Nevertheless, the tremendous possible negative impacts of the climatic changes associated justify the acceptance of an assumption based on the agreement of the

Villach conference, i.e. increased temperature of 1.5°C and 20 cm sea-level rise by the year AD 2025, and 4.5°C and 140 cm by the end of the next century.

The main effects of this change in Europe will be the extension of dry summer periods, slight changes in the vegetation and crops, increased erosion and coastal dynamics, deeper saline water intrusion in aquifers and modification of general hydrological regimes.

6.2 MEDITERRANEAN SCENERY

Mediterranean scenery is rather different from that of the rest of Europe. Apart from climatic considerations, in the Mediterranean region the erosion processes and the loss of agricultural land are already a common problem, and forest fires are, unfortunately, not infrequent during the summer season. Water of adequate quality available is scarce, and salinization together with sea water intrusion are normal problems in many aquifers, especially on the coast, where agricultural, residential and industrial developments tend to concentrate. Superimposed on these activities, the tourist flood every summer exerts an extra demand that exceeds available resources in many of the new coastal resorts.

In many of the Mediterranean countries the littoral strip of land holds most, if not all, of the main economic activities, lacking the inland development common in the other European countries. Also, some of the more interesting natural areas are located on the coast, as is the case of the wetlands where many of the migratory bird species find a sanctuary.

Because of this already strained environment, it is expected that the predicted increase of the temperature and the related changes in the climatic conditions will accelerate the actual erosion and fire hazard of the Mediterranean and will make the problems of quality and quantity of available water more acute.

The rise in temperatures will also extend the ecological range of some pathogenic organisms or their hosts and vectors, so that the reintroduction of malaria and other illnesses, such as schistosomiasis in the southern European countries, cannot be ruled out. An increase in the use of pesticides and other palliative methods against these illnesses will then be necessary in the near future, together with a reassessment of the main health objectives for these areas.

Sea-level rise will have a special significance on the western Mediterranean coast owing to three main factors:

1   The most productive areas are located in low-lying lands such as deltas and lagoons where a rise in the sea-level will have definite effects, either by the flooding of these areas or by the alteration of the quality of the water used for irrigation.
2   There is less tradition of coastal defence in the western Mediterranean coasts than in northern Europe. The existing defences are not like those in countries, for example Holland, where even an increase of 50 cm in the sea level can be accommodated by the basic hydrological infrastructure, with relatively easy adaption in the next four to five decades (Hekstra 1986), and the economic capacity to undertake them is not equal.
3   Apart from the agricultural value, the lowlands and the coastline of the Mediterranean have an ever-increasing use for recreational activities based on the development of beaches, dunes and coastal lagoons where residential complexes are being built, on the assumption of the stability of the coast.

The recent history of the Mediterranean reveals many examples of changes in coastline that brought the decline in areas dedicated to agriculture or commerce (e.g. Rosas off the Catalonian coast). People adapted to these changes, moving their activities inland or to other more advantageous places, and natural areas were supplanted by new ones formed elsewhere in the process.

This flexible response is not as easy nowadays. The pressure on the coast has increased dramatically, reducing the available land and raising the value of the property on the littoral which, together with the lack of fertile areas inland, imposes an inertia to move and an emphasis on defence, now technically more attainable.

The new situation will especially affect the natural reserves, as the defence infrastructures will be concentrated on those areas of economic value, not allowing for the transformation of actual lowlands into wetlands, while the actual wetlands disappear after the sea rises and the changing dynamics of the coast resulting from it act on them.

Deltas and coastal lagoons will also be among the areas more affected by the sea rise because it will accelerate the erosion and coastal regression most of them are suffering today, mainly through human actions such as damming and dredging of rivers, and gas extraction.

On the other hand, it is expected that the main infrastructure of the Mediterranean will be adapted as the sea rises, in a gradual way, with no disruption for its ability to function. Although the total investment to adapt the ports and other engineering works to the rise in the sea will be

Impacts on the Western Mediterranean 127

tremendous in the long run, as it will be applied gradually, it is not expected that this transformation will be seen as a disaster.

Under these assumptions, the number of spots where the impact of the sea-level rise will be more acute are not as large as has been predicted in some cases. An examination of the map of the western Mediterranean (figure 6.1) shows among these possible 'hot spots' the following:

- Rhône, Ebro, Tiber and Medjerda deltas;
- Sebkha bou Areg, Ichkuel, the Etangs of the Gulf of Lions, Albufera de Valencia and Mar Menor among the coastal lakes and lagoons;
- Alcudia, Tunez, Arno, Oristano and Cagliari lowlands.

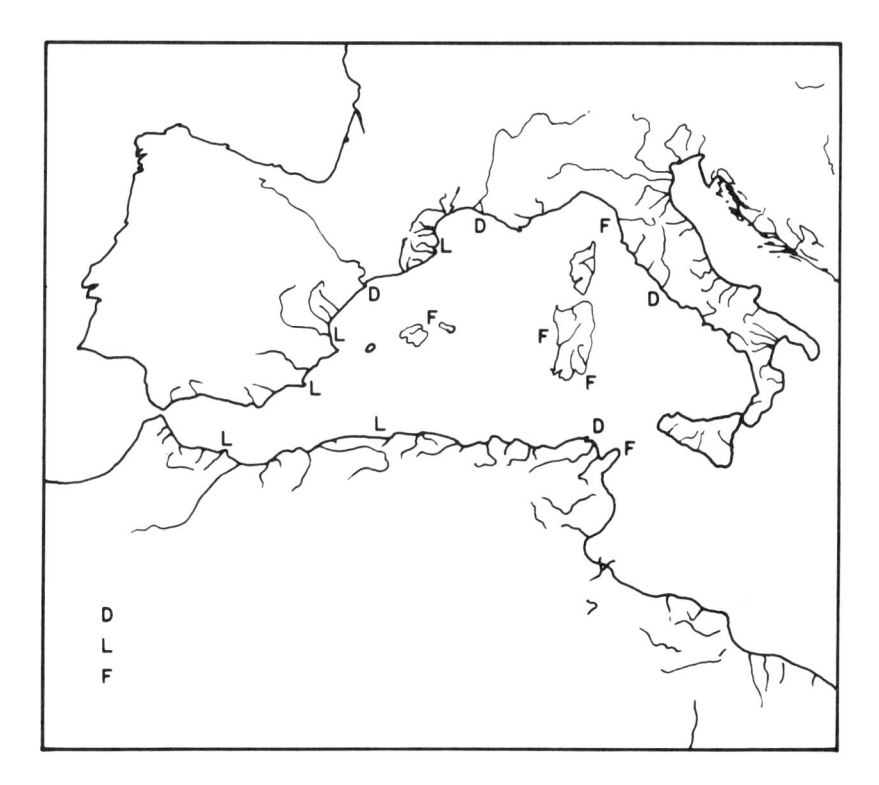

Figure 6.1 'Hot Spots' in the western Mediterranean, where sea-level rise will impact: D, deltas; L, lakes and lagoons; F, lowlands

6.3 THE EBRO DELTA

The Ebro delta is located on the northern Spanish Mediterranean coast. It covers about 350 square km and is the fourth largest delta in the Mediterranean.

The Ebro deltaic plain consists of three pronounced delta lobes extending 26 km seaward. The development of two of these lobes has notably increased the delta plain during the past four centuries. A major flood in 1937 resulted in a fourth lobe when the river opened a new mouth, abandoning the older lobe in less than twenty years (Maldonado 1977).

During historic times the Ebro delta has undergone remarkable modifications (figure 6.2). From 1970 onwards the growth of the delta has almost stopped, while the front and south lobes are subjected to quick erosion due to the damming of the Ebro and the man-made modifications on the discharge mouth which resulted in the loss of around 20,000 tonnes/year of sediments (Maldonado 1983). Approximately 96 per cent of the river load is retained by the Mequinenza dam situated upstream of the delta (Varela et al. 1983). Irrigation canals were constructed in 1857 and 1912 and made possible the cultivation of rice (Sorribes and Grau 1987).

The population of the two municipalities of the delta is 50,000. The principal activity is agriculture, with wheat and rice as the main products. Fisheries around the delta and in the lagoons are also extremely productive. In the bays protected by the lobes an expanding industry of mussel and oyster farms has grown in recent years.

Two of the lagoons in the southern part of the delta (Tancada and Encanizada) are classified as protected areas because of their value for many migratory birds which either stop here en route or come to winter (see figure 6.3). Some 260 species of birds have been recorded, and these represent 60 per cent of all European species (Sorribes and Grau 1987).

Two tourist resorts have recently been opened on the delta, one on the northern shore and another on the south, just in front of the white sand beaches that surround the whole area.

Present situation

The present situation in the Ebro delta is determined both by the coastal dynamics and by the actual or planned economic activities it is supporting. A schematic representation of the main processes of coastal dynamics acting on the Ebro delta is shown in figure 6.4 (Maldonado 1983).

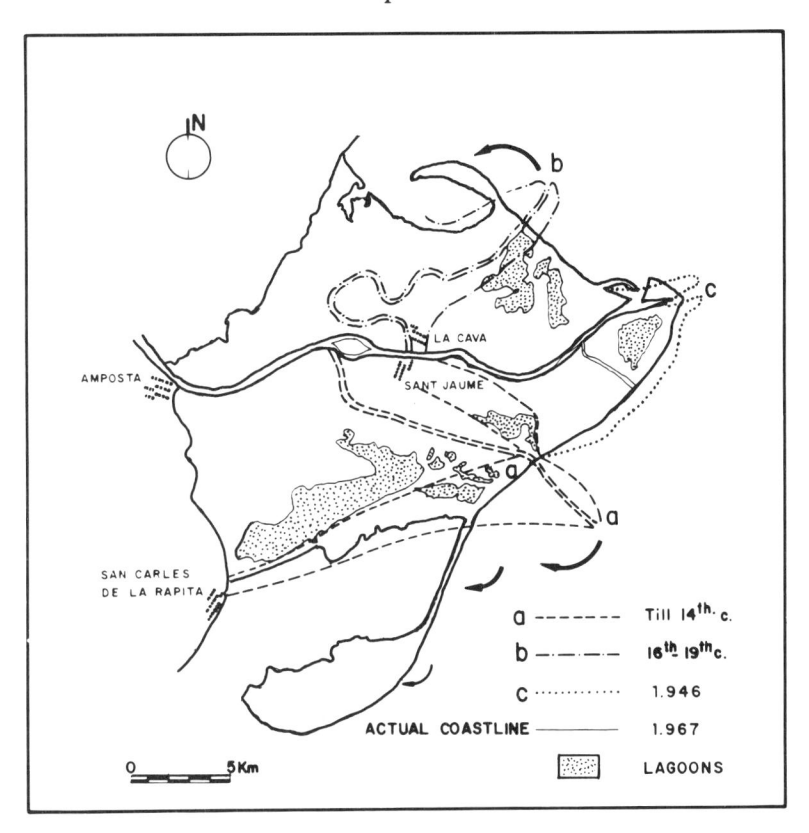

Figure 6.2 Lobe development in the deltaic plain of the Elbro delta from the fourteenth century to AD 1967.
Source: (From Maldonado 1983)

As is pointed out in this figure, the loss of sediments is not uniformly distributed in the delta, but is concentrated in those points where the dynamic equilibrium is more precarious, as in Cabo Tortosa, which is being eroded at a great speed. The Alfaques lobe, in the south, is also eroding, as is the Trabucador isthmus, now in danger of being disrupted, while, on the other hand, the northern lobe (El Fangar) is growing, although at a much lower rate.

The main economical activities in the delta will presumably continue to be agriculture and the fisheries for the near future. Tourist development is expected to remain restricted to the two areas already urbanized, especially that on the coast above the Trabucador isthmus.

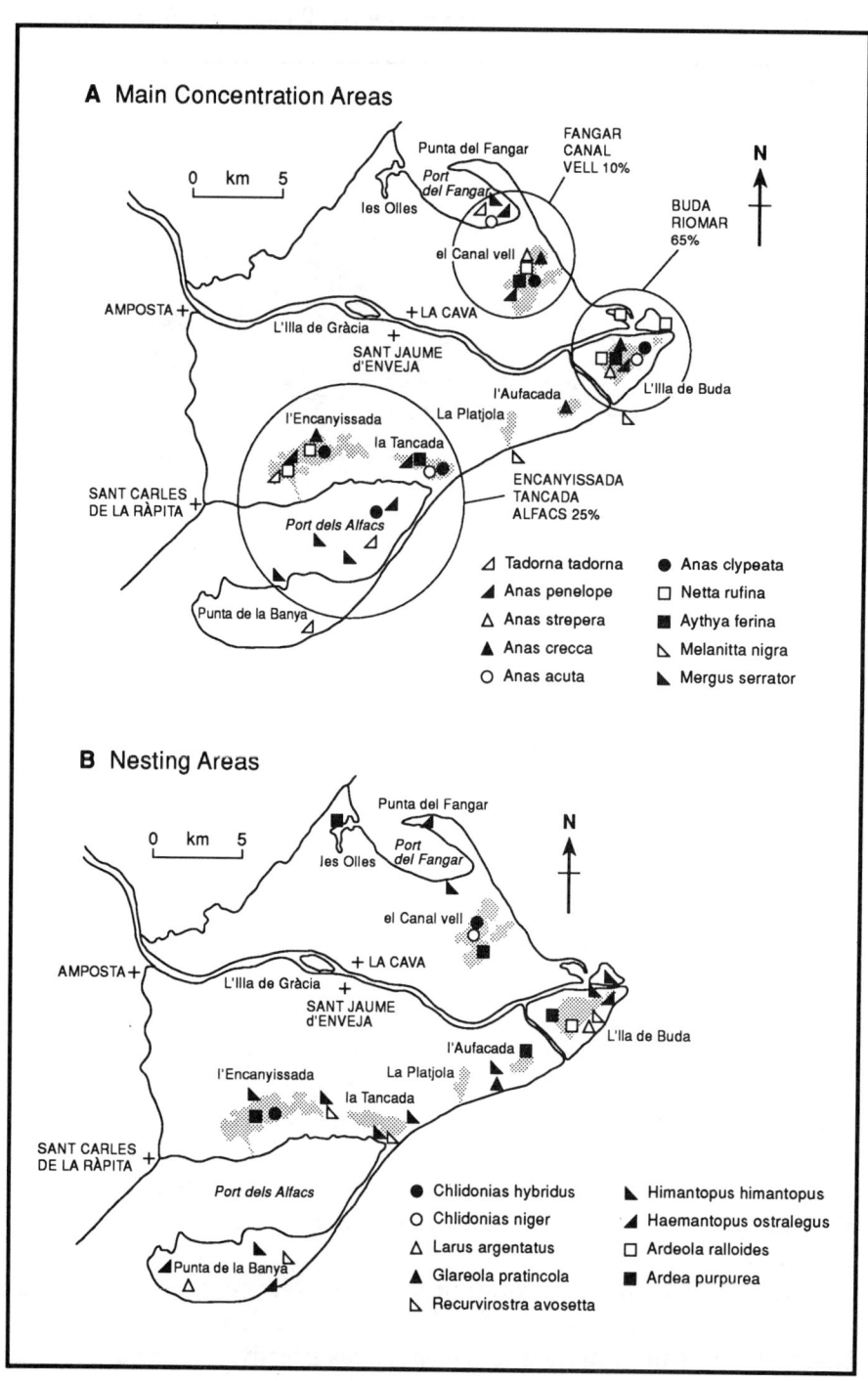

Figure 6.3 Migratory birds in the Ebro delta.
Source: (From Sero and Maymo 1972)

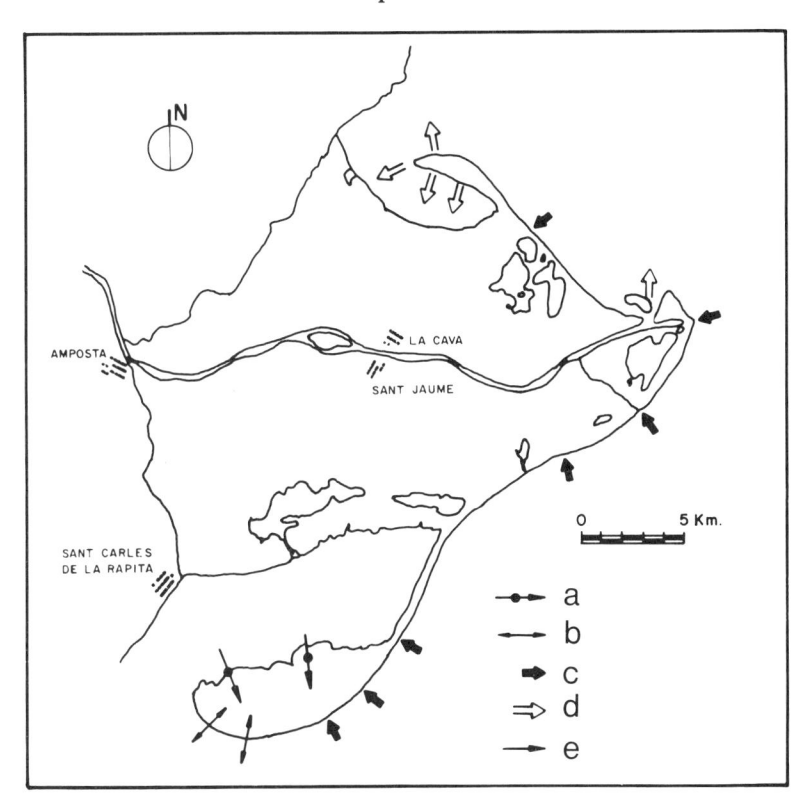

Figure 6.4 Main coastal dynamic processes in the delta: a subsidence; b stable; c erosion; d accretion; e sediment
Source: (From Maldonado 1983)

The only expected change in the economy of the delta is the further expansion of the shellfish cultivation in the bay behind the Alfaques lobe, where high water productivity and natural defence against the sea allow it. This area will thus become one of the most productive and valuable in the delta. The lagoons of La Encanizada and La Tancada situated on the inner coast of this bay and the one in the Isla de Buda are, on the other hand, the main areas for the aquatic bird communities in the delta, and for this reason they are proposed for conservation as a Natural Park.

Possible impact of sea-level rise

The possible effects of the actual dynamics of the delta are the subject of different research projects now under way in Spain. Also a UNEP case study is to be prepared in the framework of MED-POL using the methodology agreed at the meeting on Implications of Climatic Changes in the Mediterranean (Geneva, 11–13 May 1987).

Although these studies are far from being finished, a preliminary assessment of the possible problems can be made from the description of the actual situation. It seems clear that the main impact of the expected sea-level rise will come through the increase in the coastal dynamic processes already acting on the delta.

This increase will probably lead to the erosion of the Trabucador isthmus which is only 50–100 m wide and 0.5 m above the average sea level. Should this happen the Alfaques peninsula to the south would certainly be washed out by the sea in a short period of time, exposing the Alfaques Bay and the southern lowland of the delta to the sea. In the long run it is not pessimistic to assume that most of the delta will be eroded, returning to the shape it had in early days, as it is a very flat and a recent geological formation (see figure 6.5, where the evolution of the delta in the last few millennia is shown).

The main effects of these changes are:

Residential Most of the towns in the delta are located in its central part, which is also the highest (3–4 m above sea-level). Therefore, the main effect of the expected sea-level rise will be suffered by the tourist developments on the coast which, nevertheless, represent a minor proportion of that existing on the Mediterranean coast and only a marginal aspect of the economic value of the delta.

Agriculture and Fisheries The main activity in the delta is agriculture. In the short run the effect on agriculture will consist of the loss of valuable land and an increase in salinization (this can be offset by better irrigation practices using the Ebro water, if its quality remains as it is now, and the actual salinization trend ceases). As the delta represents one of the main agricultural areas of Catalonia, with an important production of rice, wheat and vegetables, this loss will have great significance for the region.

Fisheries are not expected to change as the Ebro discharge, which is the main source of nutrients for the area, will remain. The possible effects on stocks could arrive from the loss of nursery grounds in the northern and, especially, in the southern bays. Also, the actual aquaculture development in the southern bay will be in jeopardy if the bay is opened to the sea.

Nature Reserves Actual erosion processes are already reshaping the delta coastline, with a tendency for Cabo Tortosa to retreat and break down Isla Buda behind it. If the sea-level rises, most of the southern part of the delta

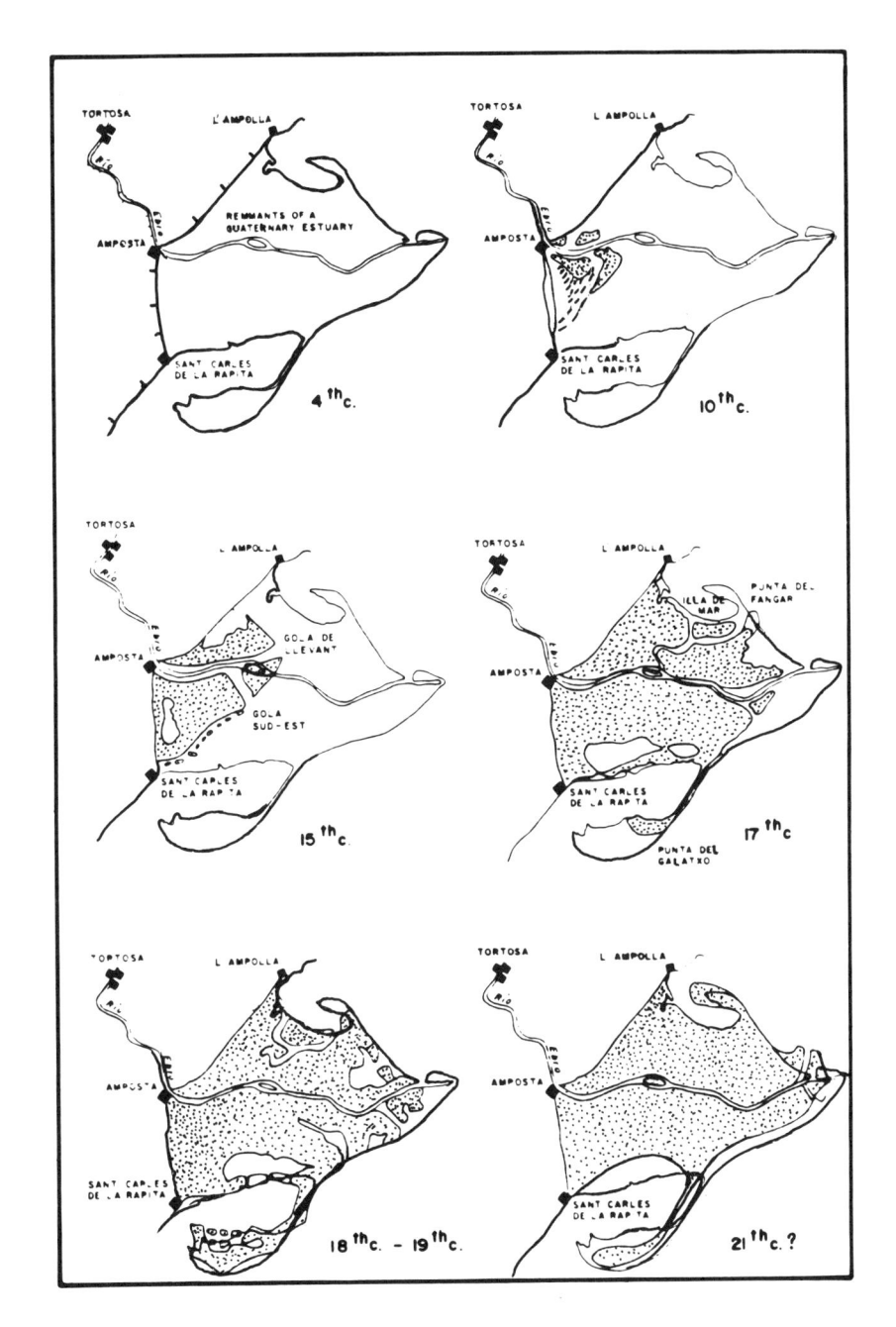

Figure 6.5 Evolution of the Ebro delta from Roman times.
Source: (From *Ministerio de Obras Publicas* 1976)

will be covered by water, opening the lagoons to the sea. Conservation of these natural areas will depend greatly on the defence actions undertaken and on the stability of the Trabucador isthmus, but its future is undoubtedly compromised.

Infrastructure There are no main infrastructures in the delta at the moment and the only important engineering work is the Port of Sant Carles de la Rapita, on the south coast. As the sea rise will be gradual and as this port is located just off the delta, no major structural problems are expected there, as defences will certainly be adapted. The problems for this port could come from the movement of the sediments that form the present Alfaques peninsula, so that, depending on it, the port could be silted up or need important extra dredging to keep it open.

6.4 CONCLUSIONS

The characteristics and problems of the Ebro delta have been presented. The main points are the following:

- The Ebro delta represents an important agricultural and natural resource in the western Mediterranean.
- Actual coastal dynamics are reshaping the delta's coastline, eroding the lobe of Cabo Tortosa and the southern peninsula and isthmus.
- A sea-level rise in the predicted range would accelerate the actual trend, even endangering the existence of the delta.
- The southern part of the delta is the area in greatest danger, mainly from the possible breach of the Trabucador isthmus.
- A proper understanding of the processes and dynamics that are acting on the delta is necessary to assess the problem better, and to decide on the most effective ways to reduce the negative effects of this process on society.

REFERENCES

Bolin, B., Döös, B., Jäger, J. and Warrick, R. A. (eds) 1986: *The Greenhouse Effect, Climatic Change and Ecosystems* (Scope 29). Chichester, John Wiley and Sons

Crane, A. and Liss, P., 1983: Roles of the Ocean in the CO_2 Question. *Marine Pollution Bulletin* 14, 441–7

Hekstra, G., 1986: Will climate changes flood the Netherlands? Effects on agriculture, land use and well-being. *Ambio*, 15, 316–26

Maldonado, A., 1977: Introducción geologica al delta del Ebro. In, Barcelona, Institució Catalana d'Historia Natural, *Els sistemes naturals del delta del Ebre*, 7–45

Maldonado, A., 1983: Dinamica sedimentaria y evolucion litoral reciente del delta del Ebro. In, Marino, M. G. (ed.) *Sistema Integrado del Ebro*. Madrid, 33–60

Ministerio de Obras Públicas (MOP) 1976: *Planes Indicativos de Usos del Dominio Público Littoral. Provincia de Tarragona*. Dirección General de Puertos y Costas, Madrid

Seró, R. and Maymó, J. 1972: *La Transformación Economique al Delta de l'Ebre*. Barcelona, Banca Catalana, Servei de Estudis

Sorribes, J. and Grau, J.–J. 1987: El delta del Ebro: una visión de conjunto. In, Bethemont, J. and Villain-Gandossi, C. (eds) *Les Deltas méditerraneens*. Vienna, Centre Européen de Coordination de Recherche et de Documentation en Sciences Sociales, 179–210

Varela, J., Gallardo, A. and Lopez de Velasco, A. 1983: Rentención de solidos por los embalses de Mequinenza y Ribarroja. Efectos sobre los aportes al delta del Ebro. In, Marino, M. G. (ed.) *Sistema Integrado del Ebro*. Madrid, 203–19

7

Sea-level Changes and Impacts on the Rhône Delta Coastal Lowlands

A. L'Homer

7.1 INTRODUCTION

A predictive study of the impacts of sea-level rise due to the so-called greenhouse effect applied to the Rhône delta coastal lowlands requires a precise assessment of the present evolutionary trends. Using the data, we may afterwards extrapolate the future shoreline in order to assess the impact of the sea-level rise on this deltaic environment located in the western part of the Mediterranean Sea. Slight subsidence has affected the delta during the historical period, and this factor must also be taken into account.

A wide offshore margin of 70 km fronting the Rhône delta which gradually slopes to a depth of about 100 m is associated with several fluvial mouths along the edge of a sea that is practically tideless. This physiography has generated sandy coastal barriers and lowlands along the Gulf of Lions coast (see also discussion by J.-J. Corre, chapter 8). The Rhône delta is well known in sedimentological literature as a model of a 'high destructive wave-dominated delta' (Fisher et al. 1969).

As the main distributaries have often shifted laterally in the past, the deltaic area protrudes southward and covers about 50 km of the shore of the Gulf of Lions. Its southern part consists mainly of an extensive pattern of former coastal barriers associated with abandoned river mouths. Elongated lagoons occur frequently between these old ridges. The area between the two distributaries of the River Rhône – the Grand Rhône and the Petit Rhône – is named the *Camargue*, part of which forms a National Reserve and Regional Park (figure 7.1).

Apart from being located at the outlet to the sea of a rather industrialized valley, the delta has several economic functions:

- *Agriculture* Cattle breeding, rice-growing, vineyards and orchards are linked to a vast drainage and irrigation network;
- *Salt extraction* The Salin de Giraud and Aigues-Mortes (plate 7) have saltpans, the overall output of which meets 25 per cent of French requirements;
- *Industry* The eastern end of the deltaic coast where the Fos Harbour was set up as well as the neighbouring port installations of the Berre lagoon serves as an industrial and port extension for Marseille (see chapter 11, section 11.1);
- *Tourism* Within the protected areas of the Camargue is a large nature reserve particularly important for migrating birds. For this reason and because of its wild beaches, the Camargue has been described by the media as one of the last Edens in Europe. This part of the delta accommodates more than 1 million people/year (Richez 1981).

Some coastal resorts and yachting harbours have also been developed, such as Saintes-Maries-de-la-Mer and Port-Camargue.

7.2 PHYSICAL ENVIRONMENT

Climatic data

During winter, the average minimum temperature seldom falls below 5°C (Heurteaux 1969). In summer, the mean maximum temperature usually exceeds 30°C for twenty-two days.

From September to April more than 50 per cent of annual precipitation is recorded. The dry season occurs from May to August (15 to 30 per cent). The annual average precipitation amounts to 539 mm at Salin de Giraud.

Winds blow throughout the year. There is a system of seasonally opposing winds. The prevailing winds are offshore winds (59 per cent of the total duration of wind) and blow from a north-east to north-west sector. The wind blowing strongly from the north-west is known locally as the Mistral (Kruit 1955). The onshore wind blowing from the south-east is called the Marin. It is usually accompanied by rainfall.

The high temperatures in summer associated with the effects of strong winds stimulate evapo-transpiration which ranges from 900 to 1,000 mm/year with a maximum monthly value of 280 mm in July. These special climatic conditions explain the development of salt pans within the coastal area.

As a consequence of the seasonal climatic changes, salinity of nearby sea lagoons, partly supplied by drainage waters, may vary from a minimum value of 10 g l^{-1} in March to a maximum of 30 g l^{-1} in August.

Data on the River Rhône

This 812 km long French–Swiss river (522 km of which is in France) has a complicated regime since its tributaries drain various mountainous regions. Low water discharge occurs between the beginning of August and the end of September.

At the head of the delta, just north of Arles, the Rhône bifurcates. The eastern branch, the Grand Rhône, is 50 km long and 10 m deep, shallowing to 4 m near the mouth, where the width of the river changes quickly from 400 m to 1125 m. The mean maximum water discharge is 1.400m³ s^{-1} (Astier et al. 1970). The Petit Rhône is 60 km long. There are no precise data for water discharge but it is calculated at 10 to 15 per cent of the Grand Rhône. According to Chamley (1971), the annual water discharge of the Rhône, which may amount to 55,000,000m³, exceeds those of the Po and the Nile.

Alluvium carried to the sea

The Rhône essentially brings sands and silts to the Mediterranean. The sediment load carried by the Grand Rhône was calculated last century at 17×10^6 m³ s^{-1} and at 4×10^6 m³ s^{-1} for the Petit Rhône. These data should be respectively equivalent to 40×10^6 t yr^{-1} and 9.5×10^6 t yr^{-1} (Vernier 1972).

Since 1972 the bulk of sediment carried to the sea by the two branches has been strongly reduced as a result of alteration work for the regulation of the course and hydroelectric facilities (sixteen dams) carried out by the *Compagnie Nationale du Rhône* and the French Electricity Board. In addition, we should mention the many harnessing effects and the consequences of the Durance River canal.

Thus the sediment load still evaluated in 1957 at 5.5×10^6 t yr^{-1} at the Grand Rhône mouth (van Straaten 1959) should amount nowadays to no more than 2.2×10^6 t yr^{-1} (Blanc 1977).

The subject of this chapter does not allow us to ignore the serious pollution linked to the Rhône. Indeed 90 per cent of the fresh flow into the Gulf of Lions is provided by the River Rhône.

Excessive levels of chemical elements (Pb, Cd, Cu, As) have been detected in the Rhône water. Sediments are notably contaminated by pesticides and organochlorides.

Moreover, the Rhône provides the strongest organic matter burden within the Mediterranean, estimated to be 433,000 t yr^{-1}. We know that organic matter works like a trap for heavy metals and synthetic organic compounds. Therefore the Rhône appears to be the main cause of environmental damage of the western Mediterranean.

Dynamic processes governing changes in shoreline

Sedimentation and erosion
Fine beach sands gradually move towards the open sea into muddy sands and silts then into marine muds. The rate of deposition evaluated using radio-nuclides from nuclear waste shows a rapid deposition for the first few miles in following the river flume (mean value of 2 cm yr^{-1}, Got and Pauc 1970).

In steady meteorological conditions, a marine current runs offshore in a westerly direction at a speed of 0.4 to 0.5 knots. By way of compensation a counter-current running from west to east may form along the seashore. In the shoreline vicinity, a complicated regime of currents frequently occurs depending on the prevailing swells and winds.

This local current associated with breaking waves is responsible for beach erosion and sand drift along the shore. Subsequently, wave-reworked sediment is redeposited in places offering better protection from marine processes (Kruit 1955, Oomkens 1970).

Three types of coastal barrier are distinguished in relation to their evolution:

1 Receding shorelines where there is not enough sand supply to compensate the washing effect by erosion;
2 Regularly prograding shorelines;
3 Steady shorelines, zones of relative equilibrium (= neutral point) located at the junction of the two previous types.

Receding shorelines These are the most common type. Nowadays the reduction in the bulk of sediment discharged at the river mouths is reflected by a lessening of the sediment carried along the shore which contributes to nourishment of the beaches. Consequently a great part of the sands carried along the shore is necessarily reworked either from the near-shore sea bottom or from former river mouths exposed to submarine abrasion. Furthermore although the protection structures defend the coastal barrier where they are set up, on the other hand they usually produce laterally an intensification of erosion of the unprotected shoreline.

Prograding shorelines Kruit (1955) showed that, at present, beach accretion is restricted to the immediate vicinity of river mouths and to stretches of the coastline that are protected from waves from the south-east.

Mouth bars A part of the sands and organic debris accumulated at the mouth of the Grand Rhône is reworked by the strong swells and redistributed along the shore to make up new mouth bars.

Long spits and coastal barriers oriented west–east The La Gracieuse spit illustrates this type of shoreline. When such sandy spits grow by extension they may isolate a new lagoon inland. This type of shoreline is the most common in the delta.

Convex spits oriented south–north When the longshore currents carrying sands reach a gulf environment, as their strength decreases, sands are deposited which form westward prograding ridges (mean rate of accretion of 11 m yr^{-1} for Beauduc spit and 18 m yr^{-1} for L'Espiguette, in front of the camp site).

Effect of the wind on the coastal barrier
The wind action leading to the formation of dunes (see chapter 8) is most noticeable where the shoreline is rapidly prograding. Thus the survey of ancient maps shows that the major dunes along the seashore of Beauduc and L'Espiguette spits were formed between AD 1850 and AD 1920. That process continues today but to a lesser extent. In particular, recent resort amenities set up within L'Espiguette spit by modifying the wind dynamics have reduced the possibility of dune formation due to the construction of long groynes and windbreaks and planting.

Furthermore, new dunes are regularly formed on each side of the mouth of the Grand Rhône as they advance. Also, just at the east of this mouth, there is a limited aeolian ridge which is regularly reconstructed where the beach is receding along the La Gracieuse spit.

For the rest of the shoreline, particularly at the east of Saintes-Maries, where the dunes are located at the back of the beach, they are not eroded by the sea. This is contrary to what happened for many decades along the Petite Camargue seashore.

Factors of variation of the water level in relation to the land surface

It seems necessary to remember these factors linked to the Rhône delta environment before studying the potential impact of a sea-level rise of some decimeters.

If we leave aside the barometric variations, four main factors may influence the variations of the water level in the littoral strip:

Effects of the tide
These are negligible since the average difference between the normal high and low tides amounts to 21 cm at Marseille with a maximum range of 30 cm.

Effects of the wind on the water level of the sea and the lagoons
The sea level usually rises under the effect of strong south easterly winds and falls during north-north-west winds.

At the time of storms caused by southeasterly winds, the sea level may attain extreme levels for several days of +1.10 m (information from Cie des Salins du Midi, Salins de Giraud). At that time, the coastal barrier is flooded and the sea water invades the backshore and the lagoon at the back of the shoreline. For instance, in January 1978, a sea-level rise of 1.80 m at the far end of the Golfe de Fos was recorded. At the same time, the La Gracieuse spit was submerged as well as the lower quarter of Fos (cf. Blanc 1979).

On 6 August 1985, a sudden storm surge flooded a beach of the Camargue causing damage and loss of life. The wind may also exert a pressure on the lagoons essentially during the high water season. Offshore winds may cause the water level to rise by 0.5 m against the northern bank of the Vaccares lagoon, cut in a former levee of the abandoned course of the Rhône de St Ferréol. The retreat of this bank since AD 1837 ranges from 100 to 150 m.

The fringes of the dunes within the lagoons are also eroded during the high water season (Petite Camargue but mainly in the backshore of the Beauduc spit.

Effects of evapo-transpiration

At the time of high water, the depth of the lagoons seldom exceeds 1 m. In summer, the combined effects of winds, heat, little rainfall and a reduction of drainage causes a distinct fall of the water level within the lagoons, which may fall to minus 40 cm. Then extensive areas around the lagoons dry out and are covered by a thin white salt crust.

Effects of subsidence

The damaging risk to the coastal barrier of the projected sea-level rise would of course increase in any zone affected by subsidence.

Evidence of a subsidence process since 6,000 BP The analysis of satellite imagery shows obviously that in the eastern part of the delta geomorphological features are less visible than those in the western part. This is due to the fact that eastward from a median line following approximately the Rhône de St Ferréol in its final course, the delta has been affected by slight subsidence (L'Homer 1975a), whereas the western part seems to have been affected in compensation by a slight uplift effect. For that reason, the well-preserved geomorphological features record the main sea-level positive variations from 6500 to 2000 BP. The history of these old ridges has been reconstructed (L'Homer et al. 1981).

This subsidence in the eastern part is testified by Roman ruins, the base of which lies 1 m below sea-level. In the same way, we have encountered by boring the top of a ridge 1.50 m below sea-level at the eastern end of the delta dated to 5860 ± 200 BP, while contemporaneous old ridges have accumulated 2 m above sea-level. The wharf of the Roman Fos harbour which is reached by diving 4 m below the sea surface is testimony to maximal subsidence in the eastern delta.

One of the results of this subsidence was the eastward shifting of the Rhône with the formation of the Grand Rhône branch that probably happened between the eighth and tenth centuries (L'Homer 1975b). The large extension of the Vaccares lagoon and the value of its depth slightly higher than otherwise are due to the associated effects of subsidence and recession by erosion of its northern bank.

If we take into account the sea-level rise that has occurred since the Roman period we may roughly estimate that the rate of subsidence in the eastern deltaic fringe ranges from 0.5 to 4.5 m, i.e., 5 cm to 45 cm per century, the medium rate certainly not exceeding 10 to 15 cm per century. These figures are not completely negligible in comparison to those predicted for the sea-level rise due to the greenhouse effect.

Is subsidence still acting nowadays? A comparative analysis of the successive geographical maps and aerial photos over forty years shows an anomaly in the area of the Beauduc spit. Indeed a clear extension of surfaces occupied by the lagoonal water is systematically noticed with, at the same time, a perceptible erosion of the surrounding dunes shown on the 1947 IGN map. This anomaly is very similar to what we would observe in the case of a slight rise of water level (roughly 30 to 40 cm) linked to a mechanism of subsidence. At once we think of the importance of this area as a 'model' for the survey of a sea-level rise in the future.

In fact, it appears that the landscape changes observed, are basically due to new sea water regularly introduced into this area for saltpan extraction. However, we cannot rule out a slight compaction of sediments since all the Beauduc area was built up by a progradation of sandy ridges on a former front delta made of silty and other muds.

7.3 SHORELINE EVOLUTION OVER THE LAST TEN YEARS

Before forecasting the potential impact of a sea-level rise by AD 2025, it is necessary first to sketch out the possible evolution of the coastal strip by this date. To this end we carried out an analysis of historical shoreline movements. The survey primarily considered the period from AD 1947 to AD 1984 well covered by maps and aerial photographs. Indeed the passing of forty years should have been ideal to predict the shoreline evolution in the next forty years.

However, the different modifications that occurred in the environment, and explained previously, led us to consider only the last ten years as a representative sequence for forecasting.

Figures given on the map (figure 7.1) represent the average annual value of recession ($-$) and progradation ($+$) for the shoreline. On the whole the general tendency is a mean recession of nearly 4 m yr^{-1}, the vicinity of the river mouth being the most exposed to erosion. As less sediment is supplied to the sea, there is a strong erosion process on each side of the Petit Rhône mouth (up to 10 m yr^{-1}).

Eastwards the shoreline recession still threatens Saintes-Maries-de-la-Mer in spite of the many defence structures built since AD 1914. Whereas the Grand Rhône mouth rates of recession of 12 to 30 m yr^{-1} are observed for the bank exposed to the south-western winds.

Among the most threatened zones are the shoreline on each side of the outlet de la Dent (point D) and the semi-natural beaches along the Petite Camargue. These two zones border land belonging to the Cie des Salins du

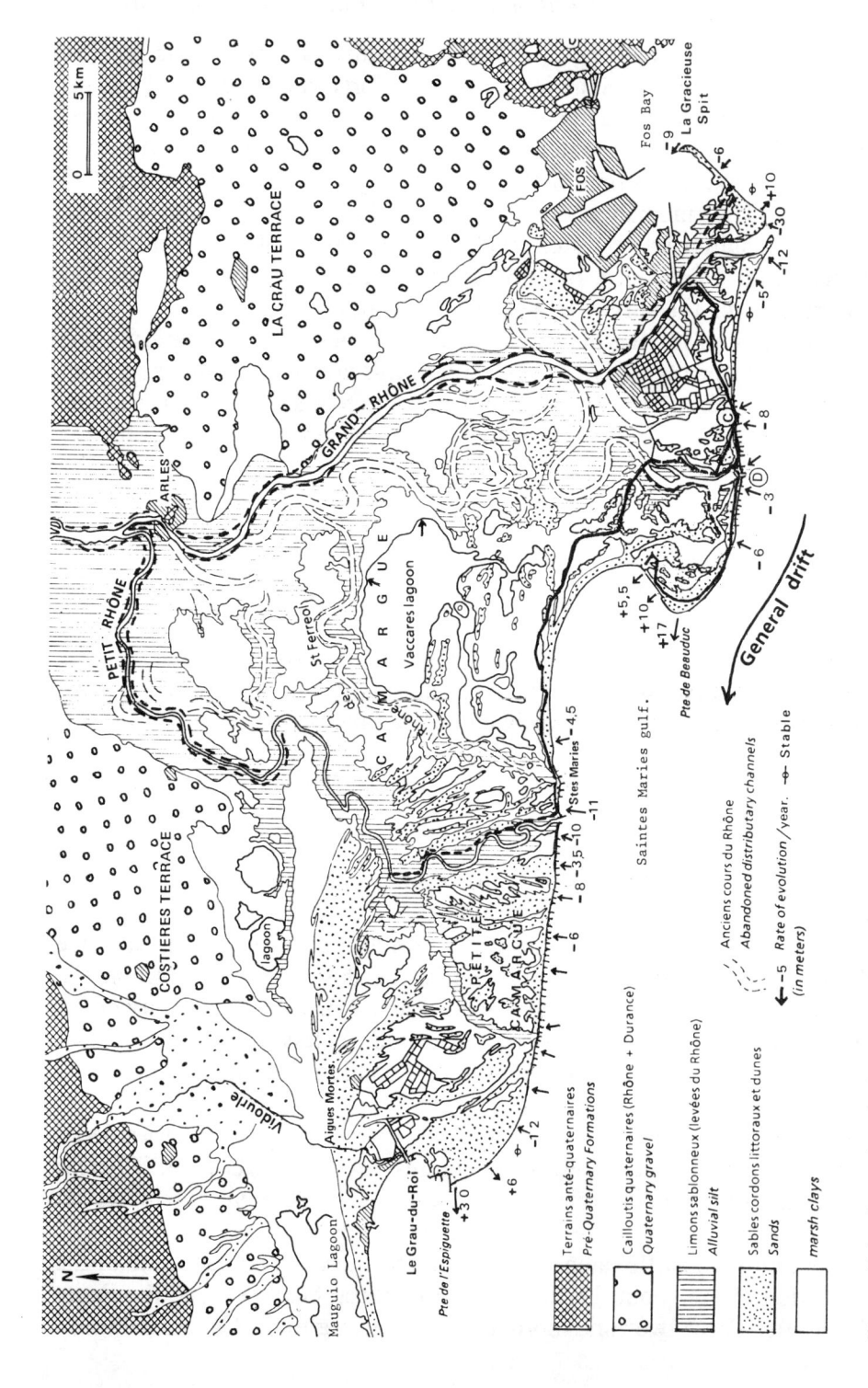

Figure 7.1 Annual rate of recession and progradation of the shoreline over the last ten years

Midi et de l'Est (CSME). The outlet de la Dent is located in the vicinity of a former Rhône outlet, which was abandoned in AD 1711 and has retreated since this date. The rate of recession was 9.5 m yr^{-1} from AD 1870 to AD 1906, then 5 m yr^{-1} from AD 1906 to AD 1931 (François 1937). During the last few decades the mean rate was 3.5 m yr^{-1} west of La Courbe-à-la-Mer (point C) and almost 10 m west of the Grau de la Dent. Today as a consequence of shoreline retreat, the sea dike protrudes at La Courbe. A 40 m retreat occurred there, as a result of the October 1987 storm (cf. Briand in Caillaud et al. 1990), which caused at the same time a deepening of the sea in front of this headland defence.

In order to address this critical situation, the Salins de Giraud (CSME) have set up new defences based on twenty-one groynes lying on 'geotextiles', and strengthening of the sea wall behind the coastal dunes (Caillaud et al. 1990). These protective measures proved to be effective. Not only has the erosion been stopped but for the past three years progradation of the shoreline has been observed (1 to 5 m yr^{-1} according to Briand).

Along the Petite Camargue seashore, the average rate of recession may be evaluated at between 4 and 5.5 m yr^{-1} over the last century. As a consequence of the reduction of the Petit Rhône sediment load, this rate increased to 6 m yr^{-1} over the last few decades with a maximum value of 18 m yr^{-1} locally registered by the CSME.

The 6 km-long headland defence set up in 1962 by the Salins d'Aigues-Mortes (CSME) and built by the means of sheet piles did not resist the storm surges for a long time, nor did a new sea dike, 400 m long, built in 1972. The infamous storm of 6–8 November 1982 caused significant damage to the defences and consequently to the Aigues-Mortes saltpans (damage estimated at 1.5 million French francs, Caillaud et al. 1990).

A new defence strategy was then drawn up by the CSME and based on the construction of thirty-eight groynes, 75 m long and 200 m apart. Today the formerly retreating coastal barrier (16.5 km of shoreline) between the Petit Rhône and the beginning of L'Espiguette spit is protected by such groynes. The recession has been notably reduced although, in places, the sea has succeeded in bypassing them.

7.4 FORECASTING OF SHORELINE EVOLUTION BY AD 2025

The figures given on the second map (figure 7.2) for retreat (−) or progradation (+) of the shoreline are only indicative. Indeed they are extrapolated from the present tendencies, i.e., they do not take into account either the predicted acceleration of the sea-level rise, or the many

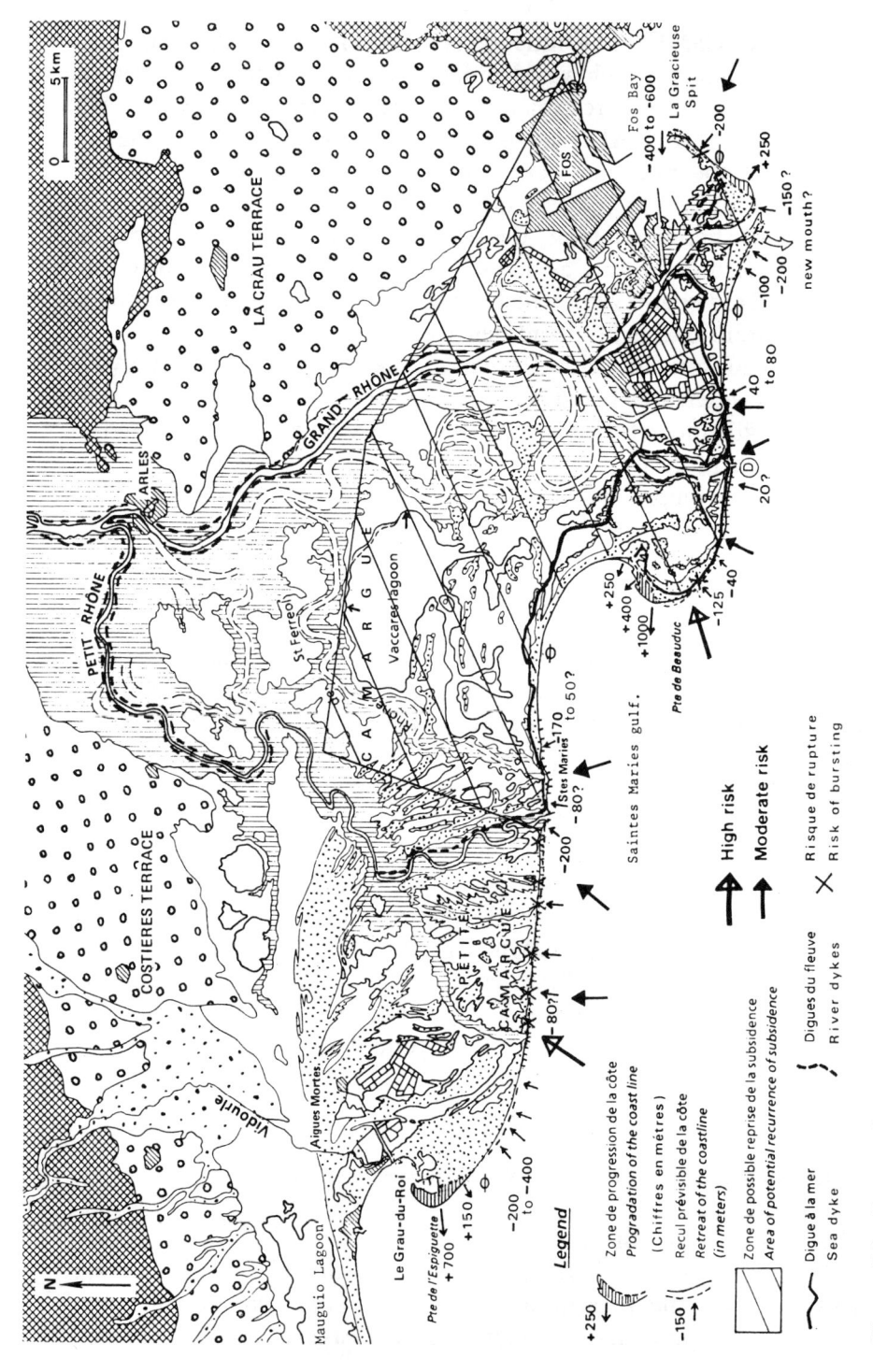

Figure 7.2 Forecasting shoreline evolution by AD 2025

Impacts on the Rhône Delta Coastal Lowlands 147

developments which could be carried out from now to AD 2025 all along the coast.

Nevertheless, those stretches of the shore that are now already unstable or threatened, will be even more so by AD 2025. These are, from east to west:

- the La Gracieuse spit, which may become detached;
- the inlet of La Dent (however, the groynes recently built seem to be efficient to consolidate the beaches);
- the south-western fringe of the Beauduc spit;
- the mouth of the Petit Rhône and the beaches situated on both sides of the promontory that defends Saintes-Maries-de-la-Mer;
- the shore of the Petite Camargue, particularly in the stretches where the residual coastal barrier is very narrow;
- the embayment between the point of Espiguette and Le Grau-de-Roi, which risks shoaling and silting up.

However, several questions remain:

- How will the offshore coastal defences resist storm surges, e.g. the breakwaters of the Courbe-à-la-Mer and Saintes-Maries-de-la-Mer or the groynes along the Grau de la Dent and the Petite Camargue?
- Will the mouth of the Grand Rhône continue to shift eastwards? Or, which is most likely, will a new mouth develop towards the SSW?
- Will subsidence continue at Beauduc, creating a real threat of inundation for the lagoons but also for the saltpans of that sector of the delta?
- Will the promontory of Saintes-Maries-de-la-Mer continue to protrude in the sea?
- Lastly, the coastal evolution at the point of Espiguette should lead to the building of a new extension of Port-Camargue just west of the present harbour.

7.5 IMPACT OF THE PROJECTED SEA-LEVEL RISE

On the Rhône mouth and fluvial sandy supplies

The reduction of the bed slope linked with the sea-level rise as well as the climatic scenarios (drier and warmer summers) should have the following consequences: an increased tendency to meandering for the final course of the river, silting up of the estuaries, and a lack of sands supplied in the summer.

On the shoreline

If we consider that the beaches of this highly destructive wave-dominated delta are built commonly to resist sudden 50 cm sea-level rises at the time of southern storm surges and that, within the lagoons, the water table may vary between 30 cm and even 40 cm depending on the season, it seems obvious that the Rhône deltaic environments should easily adapt to a low and very gradual rise during the next forty years.

However, we have to bear in mind that in many places the coastal strip is very narrow now and lacks sandy resources provided by previous dunes that have now disappeared. In these stretches during storms, the risks of breaching and of breaking over the coastal barrier, or the sea dike, will increase.

So in the future the major question will be whether the shore will or will not be capable of resisting the violent storms and healing up during periods of calm. The recent protective measures taken along the coast and the attempts to reconstruct dunes (to be developed), as well as the planting of trees in the Camargue carried out under the initiative of the Regional Park, might bear fruit. But would these measures be sufficient?

On the lagoonal environments

More salty water might enter the estuaries and the lagoons fringing the sea with harmful consequences for vegetation. This effect will be similar to what has happened in the Petite Camargue where the pine trees have been killed as the coastal dunes have been destroyed and the lenses of fresh water tapped.

The sea-level rise will produce a general deepening of the lagoons without any possibility of compensation as the result of sedimentation. Indeed today there are neither sediment supplies from the sea since the building of the sea wall of the Camargue in AD 1855, nor alluvium brought by the Rhône floods since the complete embankment at the end of the nineteenth century.

We may presume an inundation of the lagoonal banks and of the flat islets during extreme water levels and an increase of erosion as a consequence of water deepening will occur. Some of these lagoonal islands provide breeding sites for birds, especially for pink greater flamingoes. In the 1960s several islands were being rapidly eroded by waves stirred up by the Mistral, interfering with the birds' breeding. With the assistance of the CSME, new artificial islands were constructed within the shallow lagoons where flamingoes used to nest. Since that period the salt company has tried to maintain a constant water level within these lagoons (PNUE 1989).

Consequences for agriculture

More fresh water taken from the two Rhône branches will be needed to compensate the salty water intrusion, and some adaptations of the large irrigation and drainage network will have to be made.

Fortunately the variations of water level and their relations with the fauna and the flora are well known thanks to the many studies carried out since the 1960s, in particular those of the researchers of the Tour du Valat laboratories. Presently a vast programme of hydraulic studies is in progress with the help of all the partners concerned with the Camargue.

Impact on industrial activities

Of course, the expected sea-level rise will present problems for some industrial facilities located on or near the seashore, e.g. for the Fos Harbour docks and buildings for which it seems to be possible to prevent these disadvantages by minor dikes or by slight elevation of the ground level in some places.

On the other hand, the situation of the saltpans will become worrying and even disastrous in the Salins de Giraud in the event of additional subsidence. Indeed we have to remember that water depth in the salt-marshes is only 30 cm and even 15 cm in the saltpans where crystallization is achieved. Therefore it is of great consequence to set up gauges and indicators in order to survey a possible subsidence of the Beauduc area.

It must be borne in mind that the western part of the delta also has been affected by a light subsidence since the Roman epoch. We cannot eliminate the assumption of a return to subsidence. All the area around the Grand Rhône mouth represents, with the Beauduc area, the zones where some process of subsidence is most likely in the future. (cf. map in figure 7.2).

Impact on coastal resorts

Important investments have been made along the western end of the delta (Carnon, La Grande Motte, Le Grau-du-Roi). There are important stocks of sands in this area to compensate coastal erosion. It is not so for Les Saintes-Maries yacht harbour.

7.6 CONCLUSIONS

The preceding reflection shows how prospective surveys made for the long term are of considerable interest. By anticipating the coastal evolution in the future (including the effects of a sea-level rise), it might become possible to take preventative measures in time. Such a strategy should certainly be better and less costly than that of attempting to take curative measures afterwards.

The protective measures and testing carried out during the last decade seem to be in the right direction. They must be reinforced and coordinated all along the deltaic shoreline. In addition a better control of the water discharge and bulk of sediment carried to the sea should be promoted by a hydraulic management of the Rhône more conformable to environmental requirements with the help of the National Company for the Rhône and the French Electricity Board.

As the additional effect of a possible subsidence might locally increase the risks, in particular for the saltpans of Salin de Giraud, the setting up of gauges in areas well protected from the southern storm surges and, in addition, water table indicators within the lagoon are strongly recommended.

The very succinct survey of the potential repercussions of the sea-level rise predicted on the deltaic environments, calls for a deeper reflection on this theme by the different specialists of the Rhône delta.

ACKNOWLEDGEMENTS

We would like to thank the following who kindly supplied information: Mme Crombé (Regional Park of Camargue), MM Febvre and Froidevaux (CSME), M. Briand (Geoconcept, Marseille), M. Pont (Université Lyon I).

REFERENCES

Aloisi, J. C. 1986: *Sur un modèle de sédimentation deltaïque. Contribution à la connaissance des marges passives* (Thèse Un. Perpignan), 162 pp., 1 vol. with annexe

Astier, A. (dir. publ.) 1970: *Camargue. Etude hydrogéologique, pédologique et de salinitè. Rapport général.* Dir. départ. agric., 352 pp

Bethemont, J. 1987: Sur l'organisation de l'espace en Camargue. In, *Les Deltas méditerranéens*. Vienna, Centr. Eur. Coord. Rech. Doc. Sc. Sociales, 211–64

Bertrand, J. P. and L'Homer, A. 1975: *Le Delta du Rhône. Guide d'excursion.* IXème Congr. Int. Sédim., Nice, 65 pp

Blanc, J. J. 1977: *Protection des littoraux sableux. Méthodes d'étude. L'exemple de la Camargue.* CNEXO

Caillaud, A., Boudet, G., Gieulles, D. and Briand, O. 1990: Le littoral de Salin de Giraud. Evolution et programme de travaux de stabilisation. In, *Littoral 1990*, (CR ler Symp. Int. EUROCOAST), 729–33

Chamley, H. 1971: *Recherches sur la sédimentation argileuse en Méditerranée* (Thèse d'état, Univ. Aix–Marseillie), 401 pp

Clairefond, P. 1977: *Le Golfe des Saintes-Maries-de-la-Mer (Camargue). Etude sédimentologique. Aménagement et protection* (Thèse, Univ. Aix–Marseille II), 141 pp

Corre, J. J. 1988: Implication des changements climatiques. Etude de cas: Le Golfe du Lion (France). In, *Rapport collectif PNUE (Méditerranée)*, 131 pp.; condensed translation by G. Sestini 16 pp

Fisher, W. L., Brown, L. F., Scott, A.J., McGowwen, J. H. 1969: Delta systems in the exploration for oil and gas. *Colloq. Presented by Bur. of Ec. Geol.*, Un. of Texas, 77 pp., 108 pl

François, L. 1937: Etude sur l'évolution actuelle des côtes de Camargue. *Et. Rhodan.* 12 (2–3), 92

Got, H., and Pauc, H. 1970: Etude de l'évolution dynamique récente au large de l'embouchure du Grand Rhône par l'utilisation des rejets du Centre de Marcoule. *C.R. Acad. Sci. Paris*, 271: 1956–9

Greslou, M. 1984: Catalogue sédimentologique des côtes françaises. Côtes de la Méditerranée de la frontière espagnole à la frontière italienne. Partie B, de Sète à Marseille. Eyrolles, *Coll. de la Dir. des Et. et Rech. d'E.D.F.*, 106–87

Guy, M. 1975: Changements dans les voies d'eau naturelles, variations climatiques et variations du niveau moyen des mers. Actes Colloq. 'DuLéman à l'océan – Les eaux en Gaule'. Univ. Tours, *Caesarodunum* 10, 95–101

Heurteaux, P. 1969: *Recherches sur les rapports des eaux souterrainnes avec les eaux de surface (étangs, marais, rizières). Les sols halomorphes et la végétation en Camargue* (Thèse sc. Montpellier), 400 pp.

Jelgersma, S. and Sestini, G. 1988: *Impact of a Future Rise in Sea-level on the Coastal Lowlands of the Mediterranean.* First draft, UNEP Meeting, Split, October 1988, 36 pp., 17 figures

Kruit, C. 1955: Sediments of the Rhône delta. *Verhand. Konink. Neder. Geol. Mijn.* Deel XV, BL2, 357–514

L'Homer, A. 1975a: *Notice explicative carte géologique à 1/50,000 des Saintes Maries-de-la-Mer.* Orléans BRGM, 35 pp

L'Homer, A. 1975b: *Notice explicative carte géologique à 1/50,000 d'Arles,* Orléans, BRGM, 72 pp

L'Homer, A., Bazile, F., Thommeret, J. and Thommeret, Y. 1981: Principales étapes de l'édification du delta du Rhône de 7000 BP à nos jours; variation du niveau marin. *Oceanis,* 7, (4) 389–408

Oldham, R. D. 1929: Historic changes of level in the delta of the Rhône. *Quat. Jl. Geol. Soc. London* 86, 64–93

Oomkens, E. 1970: Depositional sequences and sand distribution in the post-glacial Rhône delta complex. In, *Deltaic Sedimentation* (Soc. of Econ. Pal. and Min. Spec. Pub. No. 15), 198–212

Parc Naturel Régional de Camargue 1988: *Protection de la nature. Sauvegarde des rivages de Camargue.* Courrier du PRC No. 37

PNUE 1989: *High and dry. Mediterranean Climate in the Twenty-first Century,* issued by Medit. Co-ord. Unit and Progr. Activ. Centre Oceans and Coastal areas of the UNEP, 48 pp.

Richez, G. 1981: Développement touristique, consommation d'espace et protection de la nature en Camargue. Aix-en-Provence, *Et. et Mém. Centre des Hautes Et. Tourist,* 23 pp

Russell, R. J. 1942: Geomorphology of the Rhône delta. *Ann. Assoc. Amer. Geography* 32(2), 149–254

Straaten, L. M. J. U. van 1959: Littoral and submarine morphology of the Rhône delta. *Second Coastal Geogr. Conf. Baton Rouge Proceed., Nat. Acad. Sci., Nat. Res. Council,* 233–64

Surell, E. 1847: *Mémoire sur l'amélioration des embouchures du Rhône.* Nîmes, Imprimerie Cévenole, 141 pp., 4 plates

Vernier, E. 1972: *Recherches sur la dynamique sédimentaire du golfe de Fos* (Thèse 3è cycle Marseille), 72 pp

8

The Coastline of the Gulf of Lions: Impact of a Warming of the Atmosphere in the Next Few Decades

J.-J. Corre

8.1 INTRODUCTION

Most scholars agree today that there is a warming of the climate on a world scale. From AD 1900 to 1940, the increase in mean temperatures was 0.5°C in tropical regions and 2°C in northern regions. A decreasing trend lasted until 1960 and was followed by a renewed increase.

The specialists believe that there are many reasons for this development. Some of them have to do with the origin of natural oscillations in temperature of greater or smaller amplitude and frequency, the trace of which was detected by the palaeoclimatologists studying the Quaternary. However, the increase of carbon dioxide in the atmosphere also contributes to the rise in air temperature through the increase of the 'greenhouse effect' that it induces.

This phenomenon would have several foreseeable consequences:

- displacement of the heat thresholds to which all beings and especially plants are currently submitted (assumptions are for colder winters and warmer autumns along the French Mediterranean coast);
- more active cyclonic circulations;

- modification of pluvial rhythms: drier hot seasons and wetter cold seasons;
- rise of the mean sea level, through an accelerated melting of glaciers and especially of the polar ice caps and through thermal expansion of surface oceanic water masses.

In order to establish a long-term strategy for the development policy of its coastline, we must integrate these predictions into the Mediterranean context. None of the foreseeable consequences of the climatic warming can be neglected, but they must be evaluated within a regional framework, since certain independent climatological parameters (hydrographic, tectonic, biological, historical) may support, and others may weaken, the conclusions drawn from observations carried out in other regions, even those not too far removed.

8.2 THE GULF OF LIONS: PHYSICAL ENVIRONMENT

Geography

The Gulf of Lions (figure 8.1 and figures 8.2–8.4) from Perpignan to Toulon describes a large arc of a circle of approximately 270 km, open on the south-east located between 3° and 6°E in longitude, and 42.5° and 43.5°N in latitude. Its borders are on the south-west crystalline massif of the Albères mountains, which forms the demarcation line of a rocky, very jagged coastline, and on the north-east the calcareous chain of the Estaque to the east of Fos. Most of the littoral is bordered by a low coast often dotted with lagoons, with occasional rocky outcrops: Cap Leucate, La Clape mountains, Cap d'Agde, Sète.

The delta of the River Rhône is located at the eastern end of the Gulf. It covers 173,640 ha. The alluvial deposits of the river have contributed greatly to the geomorphology of about 40 per cent of the coastline. Within the context of the western Mediterranean basin, the whole gulf area forms a structural unit thus meriting a separate study.

Geology and geomorphology

From north of the Albères to the Canet (figure 8.2) there is a major subsidence area filled with Neogene sediments (Monaco 1971). The Flandrian transgression was considerable in the area and led to the formation of a coastal strip accompanied by deltaic and lagoonal sedimentation. The present-day ponds are the remnants of lagoons filled by river

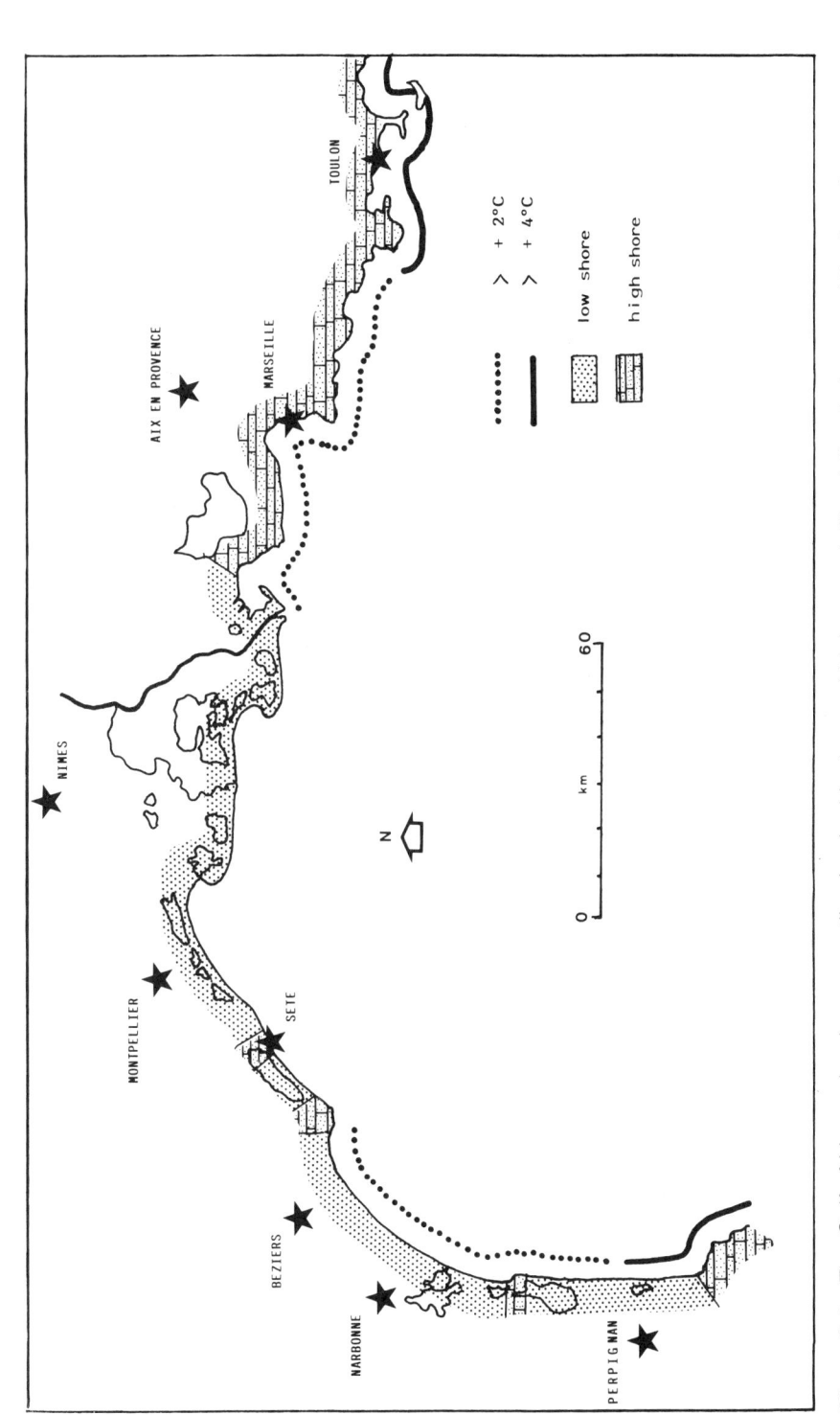

Figure 8.1 The Gulf of Lions, showing a coastline backed by low and high ground. The delineated areas of the low shore are lagoons. Average minimum temperatures (T̄ min) along the coast are shown by continuous and dotted lines.

inputs. More to the north, under the Leucate sand bar, one can pick out a regressive marine facies, then sands.

It is known that c. 2510 BP there was a sand bar between Albères and the Leucate lake. Further north up to the borders of the department of Hérault (lake of Vendres), the coast seems to have been much more indented with gulfs like those of Leucate and of Narbonne (occupied by the Bages-Sigean and Vendres lakes) and bordered by islands (Leucate, now enclosed behind the shore, La Clape mountain, etc.), although recent borings do not provide the proof of marine sedimentation (Duboul-Razavet and Martin 1980). This part of the coast developed rapidly because of the supply of alluvial deposits: 500 years later there is mention of the Leucate sand bar. Through progressive closing of the bays with alluvial deposits of fluvial and lagoonal origin, the coastline became regularized and the lagoons took on their present form. The dune system is only a few hundred metres in width and at the back of the beach it is mostly limited to a simple ridge. Further north, the Thau lake is an old gulf occupying a syncline closed by a narrow, very extensively levelled sand bar.

From Sête to the Grande Motte, the coast does not seem to have developed very much during the Flandrian Age. Bazile (1974) showed that in the Carnon region there are sedimentary series, corresponding to it and forming deposits of a rather limited width (<1 km), which follow one another. The system of dunes is relatively wide but of low altitude (<10 m) and greatly altered by the winds from the land.

From the Grande Motte onwards, the sand bar has an easterly direction and divides into several sandy strips spreading out like a fan; lagoons are integrated amongst them. This system extends throughout the Petite Camargue and gives a coastal formation which is several kilometres wide. It is abruptly cut off near the Petit Rhône, at right angles to the Saintes-Maries-de-la-Mer branch by a geological irregularity due possibly to faulting or folding, which lies perpendicular to the coast. Each one of the branches seems to mark out various stages of a considerable progradation of the coast. These surface units would have begun forming during the Flandrian transgression, as shown by Bazile (1974) and L'Homer et al. (1981). They continued forming until modern times, as witnessed by the formation of Cap Espiguette. The system of dunes is very diversified. Its altitude, even though it is the highest on all the coast, does not exceed 12 m.

Beyond Saintes-Maries-de-la-Mer, towards the east, the landscape consists of a network of lakes, earth banks and remnants of dune strips partially filled in by alluvial clay deposits. This landscape is the result of the combined action of fluvio-palustral deposits preponderant in the north and of lagoonal-marine deposits especially important in the southern areas

(Astier 1970). It is characteristic of the Camargue and a major element of the Rhône delta.

The Camargue substratum consists of gravel, the top plunging in a NE–SW direction from −5 m to −42 m. Its topography is rather confused, probably consisting of terraces cut by gullies and with closed depressions. The Flandrian transgression has left traces of five successive phases separated by still-stands, if not slight regressions of sea-level. At its maximum extension the transgression does not appear to have gone beyond the northern limit of the Vaccarès lake (see figures 2 and 5 in L'Homer et al. 1981). During the first four phases, the alluvial deposits were primarily lagoonal or marine. During the period of the fourth phase, which can probably be attributed to the Boreal-Atlantic transition (Pons et al. 1979) (*c.* 7500 BP), a big barrier was formed which blocked any further forward movement. It seems to have been linked with large terrigenous inputs, too early to be considered as the consequence of neolithic farming activity. From then on a phase of general advancement begins, due to river inputs. It continued until quite recently, since the coast towards the end of the fourth century was roughly 4 to 13 km to the north of where it is today (Greslou 1984).

The whole recent history of the Camargue, up to the mid-nineteenth century, is dominated by the shifting of the distributaries of the River Rhône, the successive mouths of which have slowly moved eastward (see figure 5 in L'Homer et al. 1981).

On the basis of surface deposits we can distinguish three ecological sectors from north to south: the Haute Camargue, where the altitude varies between 4.5 m and 1 m NGF with marshes which are near zero m NGF (*Nivellement géneral de la France*) altitude; the Moyenne Camargue including Lake Vaccarès with lower general altitude and marshes in spots lower than the surface of the sea; and the Basse Camargue which is the third ecological sector. It is located to the south of Lake Vaccarès and is characterized by more or less distinct rows of dunes separated by lagoons; altitudes vary between +7 m NGF for certain dunes and −0.3 m NGF in the lake depressions. The Haute and Moyenne Camargue were formed from a network of fluvial meanders encircling the marshes, the Basse Camargue through lagoonal-marine sedimentation (sands and brackish clayey silts).

Today the coast of the delta is advancing towards the sea through two large blunted capes veering toward the west: the Beauduc headland, at right angles to the Camargue, corresponds to the inputs of the Grand Rhône; the Espiguette headland, at right angles to the Petite Camargue, corresponds to those of the Petit Rhône (shown on figures 7.1 and 8.4).

The climate today

The climate of the Gulf of Lions is of the Mediterranean subhumid type with cool to temperate winters according to the Emberger classification, or of the IV 3 type according to Walter's classification. It has also been called 'transitional', because some years it can display oceanic climate characteristics, or more rarely continental climate characteristics (Baudière and Emberger 1959).

Generally speaking, the rise of the Azores anticylone in the summer protects the region from the cyclonic disturbances coming from the Atlantic, but there are years when it settles at a lower latitude, and this allows the penetration of circulations from the west (Estienne and Godard 1970).

In the winter, the inland relief and the continental thermal high pressures displace the perturbated currents northwards; however, the weakening of these high pressures may allow the perturbations of the western circulation to pass through. In any event, two-thirds of the perturbations affecting the region are of local origin caused by meridional circulations of cold polar air reaching the Mediterranean. In particular, they give rise to the southeasterly winds which can be violent and bring with them rain and storms. Annual rainfall ranges between 400 and 750 mm (540 mm in the Camargue) with fifty to ninety-five rainy days a year. Mean annual temperatures range between 14 and 15°C.

The average of the minimum temperatures ($\bar{t}$ min) is in this connection especially important for vegetation because it localizes the distribution threshold of the various species; from the SW to the NE, $\bar{t}$ min decreases (figure 8.1), by +4°C on the Albères coast to +0.9°C in Montpellier and it increases up to +1.7°C in the Camargue and +8°C in Monaco, near the Italian border. The median part of the gulf is thus relatively cold in the winter and this causes a gap from a biogeographical point of view.

The prevailing winds have a NW–SE orientation with a strong preponderance in the north-west sector. Winds are frequent and at times violent; they blow 208 days year^{-1} at more than 38 km h^{-1} (10.8 m s^{-1}) and on eleven days/year at more than 74 km h^{-1} (20.5 m s^{-1}) with peaks at 135–65 km h^{-1} (37.5–45.8 m s^{-1}).

The sea

Tides are weak, the amplitude not exceeding 30 cm at spring-tide average. On the other hand, the variations due to the oscillations of atmospheric pressure and especially to the wind effect are much more important. The sea is caused to rise by south-east winds and caused to fall by north-west winds. Greslou (1984) reports record values of +1.80 m east of the Camargue (toward the Gulf of Fos) and −0.5 m NGF at Port-La-Nouvelle (South of the Gulf of Lions). In the area of the Camargue coastline over a seven-year period (1967–73) the oscillations ranged between the following extremes: −0.39 m NGF and +0.90 m NGF, i.e. a level of difference of 1.29 m (Greslou 1984).

In the northern part of the gulf (Sète–Camargue), the prevailing orientation of the waves is from the south-west or the south-east. Maximum wave height, with a probability of a 100-year period, is calculated at 5–10 m depending on the site, the annual probability from 3 to 6 m.

The general currents circulate too far out in the sea to affect the coastline dynamics. On the other hand, the currents caused by the waves which are oblique to the beach play an important role in the tangential migration of sediments by longshore drift.

According to French National Geographic Institute (I.G.N., *Institut Geographique National*) data there was a net 10 cm rise from 1885 to 1979 with an acceleration between 1944 and 1955 followed by a decrease (Pirazzoli 1987).

8.3 COASTAL ACTIVITIES

The past

The coastal zone of the Gulf of Lions had for a long time been considered economically marginal, because it lies outside the principal currents of economic activity of the region. Thus traditional activities and especially handicrafts developed there: fishing and shellfish in the lakes, exploitation of salt, hunting, agriculture, extensive breeding of horses and bulls (*manades*).

Summer tourism, severely limited because of the mosquito problem, was confined to the urban centres of the interior. The harbours of Sète and Port-La-Nouvelle remained the principal centres of commercial and industrial activity situated on the coast (see figures 8.2 and 8.3).

The environment had reached a certain level of equilibrium. When the agricultural or fishing activities made it necessary, the landowners took various measures of protection against the sea. They aimed primarily at preserving the dune strip unbroken at right angles to the exploited areas and at managing the *graus* (channels between the sea and the lakes). In 1867, Régy, a civil engineer working for the government, published an important report in which he set out a management plan for the lagoons of the Hérault department coastline, including a plan for the protection of the coast (a planting programme and the creation of artificial dunes), all of which gives us a clear picture of the concern of the authorities as long ago as the nineteenth century to secure the area against the aggression of the sea.

In the Camargue the first attempts to control the environment go back to the twelfth century (Picon 1979), but true concern for the protection of the coastline did not begin until the nineteenth century, when those involved in salt production reclaimed stretches of the Basse Camargue after an unsuccessful attempt to put them under cultivation. In 1859 the sea embankment was completed. In 1869 the distributaries of the Rhône were finally embanked between training walls, and this put an end to the migration of the river and to the risk of flooding.

As early as 1929, most of the Basse Camargue was made into a reserve. This was done because of the conflict between salt makers and farmers for the management of the water resources and because of the overriding importance of the area as a biotope. The status of reserve will give the delta its natural image, and this will foster both international tourism and large-scale scientific activities (Biology Station of the Tour du Valat, CNRS laboratory, until 1986, the Camargue National Reserve, Regional Park).

The situation today

Currently, the interest of the coast of the Gulf of Lions and the financial stakes involved have greatly developed under the influence of several large scale development programmes in:

- Agriculture: development of irrigation and drainage networks in the Camargue for rice growing and management of ponds and water flushes for water fowl. In the Languedoc, construction of the canal of the Bas-Rhône and large-scale infrastructure projects of irrigation and drainage for local diversification of crops;
- Tourism: for the whole coastline, except in the Camargue, a mosquito combating scheme was the prerequisite; when this had made the beaches more attractive, a concerted construction plan was launched (new towns, extension of existing urban areas) directly fronting the sea, making the whole of the Gulf of Lions one of the new tourist beaches in Europe. Currently more than 40 per cent of the coastline has been developed along these lines;
- Industry: mainly the creation of a new harbour, metal works and petroleum complex in Fos-sur-Mer;

The Gulf of Lions 161

- Environment: creation of the Regional Nature Park of the Camargue and classification as Reserve of the Biosphere of Camargue. These actions underscore the importance of this region both for its rôle for the preservation of the European avifauna and for the uniqueness of its biotopes (Riège forest of *Juniperus phoenicea*), as well as for the richness and diversity of its wetlands (506 natural stretches of water of >0.5 ha covering 40 per cent of the Camargue and ranging from fresh water to water with excessive salinity (Britton and Podlejski 1981).

8.4 DEVELOPMENT OF THE COAST TODAY

A systematic study of the development of the coast based on aerial photographs (IGN 1942–81) and on bibliographical data (Rueda 1985) for the years 1942–6 to 1970–80 gives the following results (figures 8.2, 8.3 and 8.4).

- Eastern Pyrenees and Aude (from Albères to Vendres lake), 90 km of coast, 27 km of which are bordered by dunes. In general, the coast has advanced, with some recession near the mouths of rivers (Tech and Têt) and around rocky headlands (Albères, Cap Leucate). The picture is as follows:

  53 per cent of the coastline advanced by 0.5 to 1.5 m year^{-1}, and in places up to more than 3 m year^{-1};
  27 per cent of the coastline receded by 0.4 to 1.2 m year^{-1}, and in places up to more than 3 m year^{-1};
  20 per cent of the coastline is stable (with variations of 0.3 m year^{-1} in one direction or the other, taking into account a margin of error in measurement).

- Hérault (from Vendres lake to the Grande Motte), 84 km of coastline, of which 62 km are bordered by dune ridges. The coast shows the same phenomenon of recession near the mouth of rivers and rocky headlands. It is however less systematic because of the existing sea defence works. In summary,

  9.5 per cent of the coastline advanced;
  24.4 per cent of the coastline receded;
  66.1 per cent of the coastline is stable.

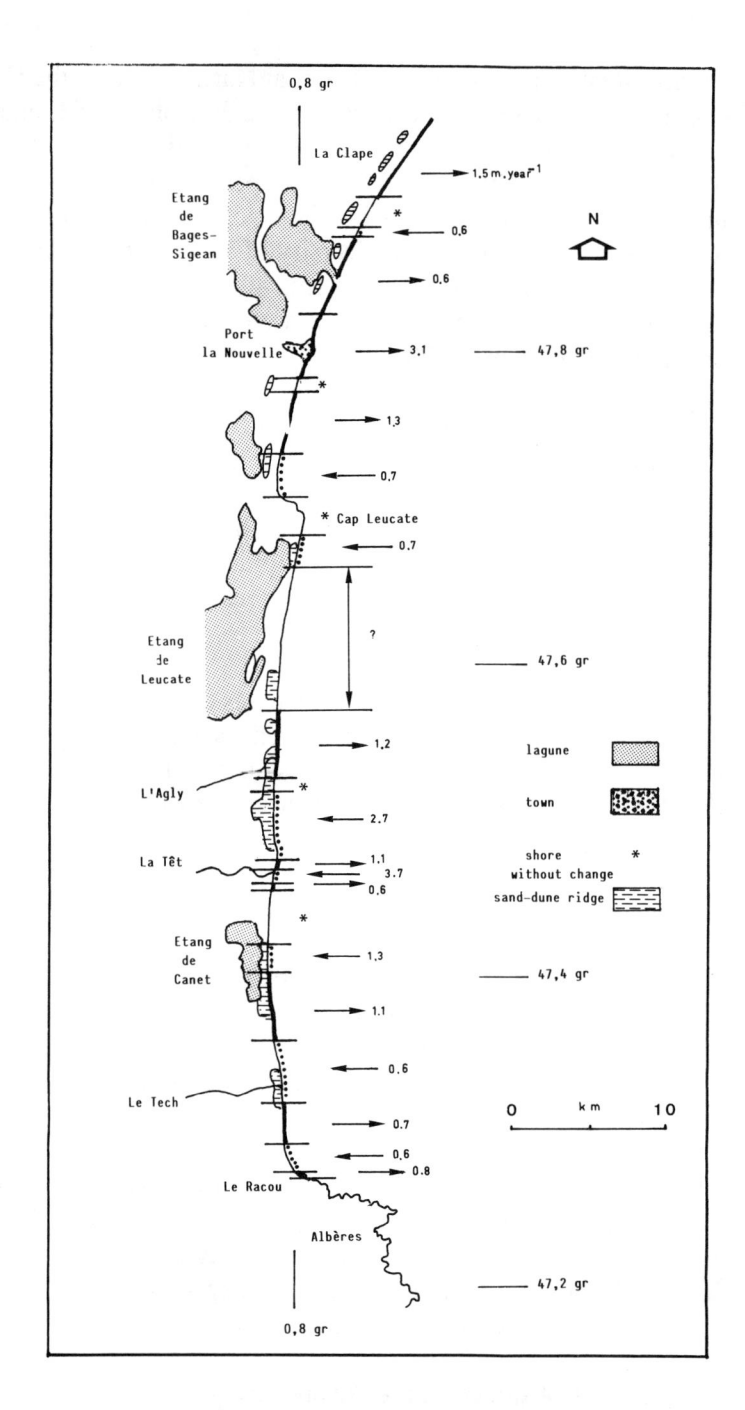

Figure 8.2 Coastline advance or retreat in the western part of the Gulf of Lions between Albères and La Clape, 1942–6 and 1970–80 in m year^{-1}

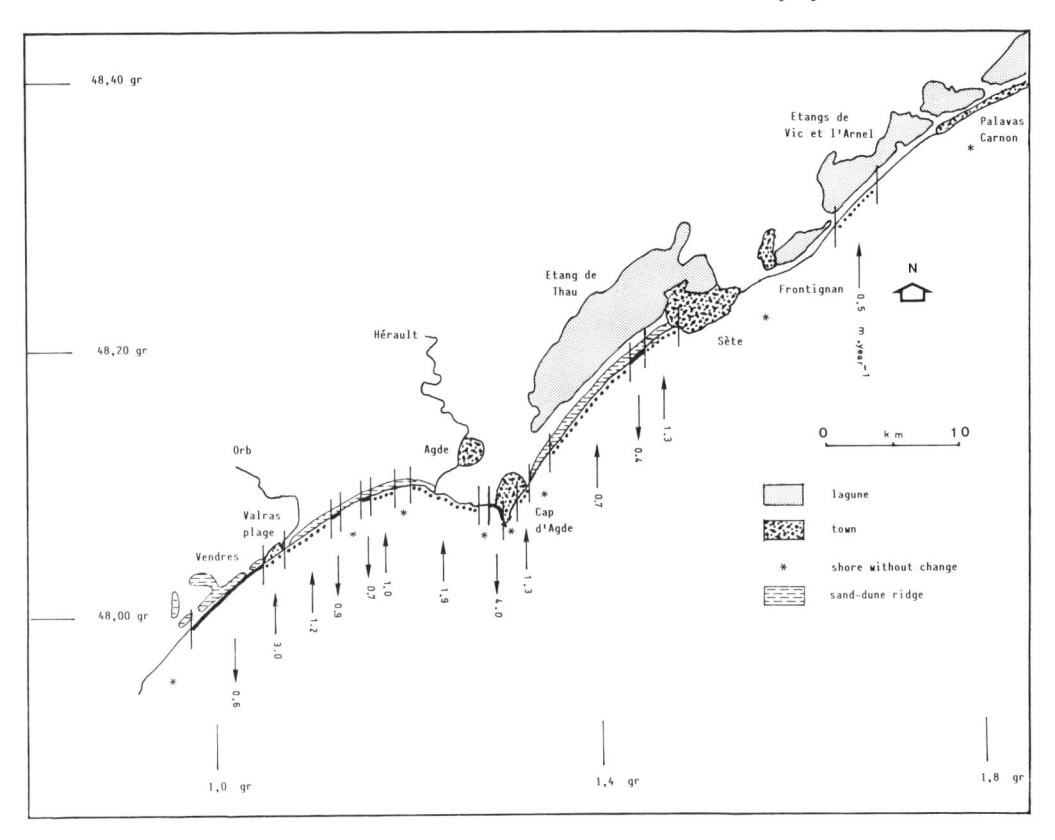

Figure 8.3 Coastline advance or retreat in the middle part of the Gulf of Lions from Vendres to Palavas Carnon since 1950 in m year^{-1}.
Source: (After Rueda 1985)

- Gard and Bouches-du-Rhône (from the Grande-Motte to the Gulf of Fos), 103 km of coastline of which for the Gard 17.6 km are dunes, out of 33 km for the department as a whole. The whole coastal dynamics are linked with the slow westward movement of the old and current deposits of the two mouths of the River Rhône. The Espiguette point advanced by 18 m year^{-1}; the construction of dikes at its tip has changed the direction of this forward movement and in places it has advanced 20–25 m year^{-1}. The Beauduc point advanced 11 m year^{-1}. Near the river mouths the recession of the coast is considerable, approximately more than 4 m year^{-1}. This latter figure should be correlated with the decrease in the solid inputs of the river, which was 40×10^6 t year^{-1} for the

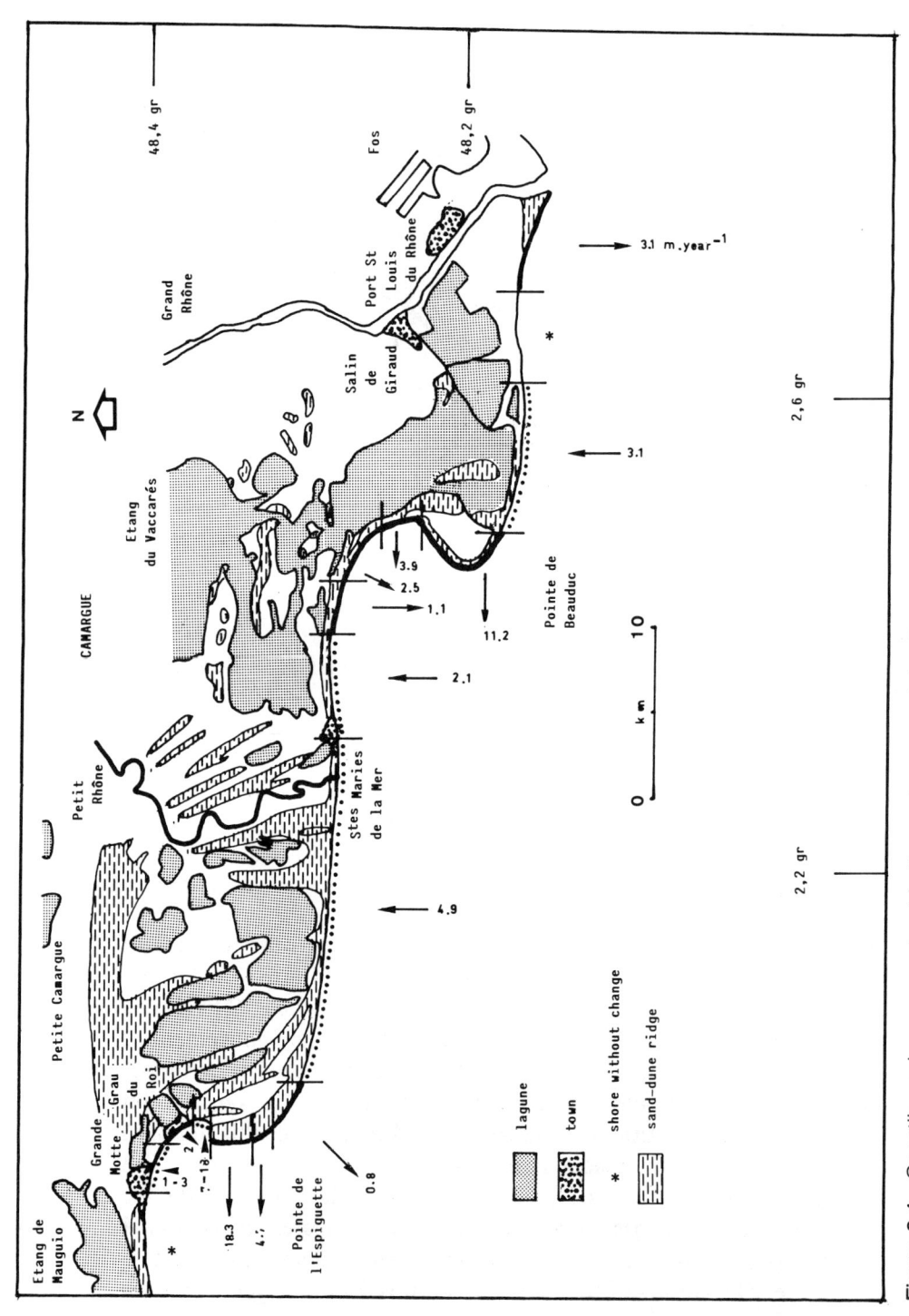

Figure 8.4 Coastline advance or retreat in the eastern part of the Gulf of Lions from Grande Motte to Fos, 1942–6 and 1970–80 in m year^{-1}

Grand Rhône and 9.5×10^6 t year^{-1} for the Petit Rhône in the nineteenth century (Surrel 1847). These values seem to be underestimated because important quantities of sand at this time were not taken into account. Recent output of sediments are mainly silts, and the values are estimated to be 5.5×10^6 t year^{-1} (van Straaten 1977) for the Grand Rhône, or less (2.2×10^6 t year^{-1}) according to other authors (e.g. Blanc 1977). The displacements of the beach can be summarized as follows:

28 per cent of the coastline advanced;
62 per cent of the coastline receded.

These statistics correspond to resultants; it seems on the basis of the first observations carried out on the Hérault coastline that the resultants are, at least in certain places, the outcome of cycles of advance and retreat, as has been observed on other coasts over periods of ten years or more (Bird 1986).

The great storms have formidable consequences: in three days the 1982 storm caused the coast to recede by several dozen metres in certain places and disrupted the dune strip.

It is noteworthy that the coast has generally become narrower at right angles to the mouths of the rivers. Bird (1986) showed that this is common throughout the Mediterranean basin. Many working hypotheses have been advanced in order to explain it. Amongst the most interesting, because it is the least fatalistic, is that which questions the method of managing the catchments (agriculture, forests, dams, sand pits).

Finally, changes in the state of the sea can be observed. In particular, Rueda (1985) drew attention to an increase in the wave height and hence its energy between the years 1880 and 1960.

8.5 IMPACT OF A POSSIBLE RISE OF SEA-LEVEL AND A WARMING OF THE CLIMATE

If we assume that there will be a warming of the climate we shall have to consider two series of consequences:

- those due to the rise of the sea level and to the activation of a cyclonic circulation;
- those affecting biogeography.

Rise in sea-level and activation of a cyclonic circulation

The 1982 storm can serve as a reference point for a projection for the impact of wave action and of a rise of the level of the sea.

The truncation of the dune front will accelerate and the sands will be remobilized. The vegetation cycles which are particular to these scarred areas will become shorter and the maturation stages which characterize them will be reached with even greater difficulty (Corre 1987). More intense wind erosion will follow and will bring about the displacement of the dune strip if no precaution is taken. The segmentation of the sand dune border will create passages through which sea water will infiltrate. This would have a double impact: the waves will carry behind the dunes sand in suspension, which means a loss of sediments from the cycle of the beach–dune interaction during coastal erosion; and secondly, the sediments transported in this way accumulate in the depressions, the floors of which are raised. As the topographical level rises, the environment becomes drier. That would be injurious because these depressions in the summer become true wet oases for part of the dune fauna. Their restoration will be very long or can be achieved more quickly through very considerable upheavals involving a phase of bare soil.

The erosion can quickly lead to break-up of the sand bar and increase the sea–lagoon water exchange, thus bringing about changes in the fauna and its possible exploitation.

The Camargue has greatly eroded sand-dunes and a relief which is very flat; it would thus be seriously affected by a considerable rise of the sea-level and this, despite the existence of dikes, would have an impact on the balance of the aquifer. Would the Camargue lose its interest as a biotope? It is difficult to give an opinion, but the history of the Reserve is full of upheavals caused by the various methods of water management of the delta. Therefore there is no reason for being pessimistic.

Concerning the economic activities on the coast as a whole, there is the danger that the consequences of a possible rise in sea level and an increase in the frequency of storms will be serious. Most of the construction on the coast has been built directly on the upper part of the beach itself after levelling of the sand dune barrier, and one may expect damage due to the erosion of the foundations. This is exemplified by the big storms which occur at the present time.

Biogeographical consequences of the rise in temperature

Assuming that there will be a rise in temperature of between 2°C and 4°C, it would be interesting to know if this would correspond to a general rise of seasonal temperatures or a more continental temperature pattern. If it is expected that winter will be colder, the area of the existing winter isotherm will enlarge and it will be interesting to follow the behaviour of some plant species sensitive to cold such as *Thymelaea hirsuta* and *Mesembryanthemum edule* which are at their thermal limits and so would be a good indication of this tendency. It will be of some interest if winter temperatures rise.

8.5 CONCLUSIONS

It is fortunate that the insalubrity of the coast in the past centuries caused most big urban centres of the Gulf of Lions (Perpignan, Narbonne, Béziers, Montpellier, Nîmes, Arles) to be built sufficiently far back from the beach (figure 8.1), so that they are protected from the serious impact of a foreseeable rise of the sea-level.

On the other hand, the coast with its tourist and harbour installations is particularly vulnerable owing to the narrowness of the strip of sand for most of its length, the low altitude of the dunes, their levelling for construction purposes and their decayed condition due to abandonment for the period which separated the farming phase from the tourist development phase.

At the present time, the major risk would be an increase in the frequency and the seriousness of the storms which is linked to the activation of cyclonic circulations. The great storm of 1982 fortunately acted as an alarm signal for public authorities. A number of actions and control measures were launched or extended, either experimental or definitive in order to reinforce an existing defence policy against the sea, to respond to the most critical situations. The aim of these actions and measures is: to control tangential transits (rockfill); to help to bring back and to hold the sand on the beach and on the dune (network of semi-permeable barriers, nets, planting schemes); and to create an obstacle to storms while at the same time maintaining a wide enough strip of usable beach by establishing artificial dunes, dikes and beach nourishment.

ACKNOWLEDGEMENTS

Assistance with this work was provided by CNRS (*Programme interdisciplinaire de recherches sur l'environnement*) and PNUE (*Programme des nations unies pour l'environnement*). I thank the *Entente interdépartementale pour la démoustication* for its cartographic data and dune statistics. I thank also Dr S. Jelgersma for her encouragement and useful comments on this chapter.

REFERENCES

Astier, A. 1970: *Camargue. Etude hydrogéologique, pédologique et de salinité*. Rapport général, Direct, départ. de l'agric. (B. de Rhône), 352 pp.
Baudière, A. and Emberger, L. 1959: Sur la notion de climat de transition, en particulier dans le domaine du climat méditerranéen. *Bull. Serv. Carte Phytogéogr.* sér. B., 4(2), 95–117
Bazile, F. 1974: Nouvelles données sur l'âge des cordons littoraux récents du Golfe d'Aigues-Mortes. *Bull. Soc. languedocienne Géogr.* 8(3–4), 199–206
Bird, E. C. F. 1986: *Coastline Changes. A Global Review*. Chichester, John Wiley and Sons, 219 pp.
Blanc, J. 1977: Recherches de sédimentalogie appliquée au delta du Rhône, de Fos au Grau du Roi. *CNEXO*, 75/1193, 69 pp.
Britton, R. H. and Podlejski, V. D. 1981: Inventory and classification of the wetlands of the Camargue (France). *Aquatic Botany* 10, 195–228
Corre, J.-J. 1987: Les peuplements végétaux et la gestion des côtes basses du golfe du Lion. *Bull. Ecol.* 18(2), 201–8
Duboul-Razavet, C. and Martin, R. 1980: La sédimentation holocène au niveau de trois étangs du littoral Languedoc-Roussillon. *Bull. Soc. Languedocienne Géogr.* 15(1–2), 69–86
Estienne, P. and Godard, A. 1970: *Climatologie*. Paris, Librairie Armand Colin (Collection U, Série Géographie), 367 pp
Greslou, M. 1984: Catalogue sédimentalogique des côtes françaises. Côtes de la Méditerranée de la frontière espagnole à la frontière italienne. Partie B, de Sète à Marseille. Eyrolles, *Coll. de la Dir. des Et. et rech. d'EDF*, 106–87
IGN 1942: Mission Nîmes–Camargue
IGN 1942: Mission Perpignan–Rivesaltes
IGN 1946: Mission Le Grau du Roi–Lunel–Sommières
IGN 1946: Mission St Chinian–Pézenas–Sète
IGN 1970: Mission Saintes-Maries-de-la-Mer
IGN 1978: Mission FR 9068/145
IGN 1980: Mission Limoux–Narbonne
IGN 1980–1: Mission Foix–Leucate

L'Homer, A., Bazile, F., Thommeret, J. and Thommeret, Y. 1981: Principales étapes de l'édification du delta du Rhône de 7000 BP à nos jours; variations du niveau marin. *Oceanis* 7(4), 389–408

Monaco, A. 1971: *Contribution à l'étude géologique et sédimentalogique du plateau continental du Roussillon (Golfe du Lion)* (Thèse d'Etat, USTL, Montpellier), 295 pp.

Picon, B. 1979: Aperçu de l'histoire socio–économique de la Camargue. *Terre et Vie, Rev. Ecol.*, Suppl. 2, 31–48

Pirazzoli, P. A. 1987: Sea level changes in the Mediterranean. In, Tooley, M. J. and Shennan, I. (eds) *Sea-level Changes*. Oxford, Basil Blackwell, 152–81

Pons, A., Toni, Cl. and Triat, H. 1979: Edification de la Camargue et histoire holocène de sa végétation. *Terre et Vie, Rev. Ecol.*, Suppl. 2, 13–38

Régy, P. 1867: *Assainissement du littoral méditerranéen du département de l'Hérault*. Rapport de l'Ingenieur en chef. Ponts et Chaussées, département de l'Hérault. Archives départ. Hérault, 213

Rueda, F. 1985: *Le littoral de l'Hérault et du Gard*. Serv. mar. et de navig. du Languedoc-Roussillon, 154 pp.

Straaten, L. M. J. U. van 1957: Dépots sableux récents du littoral des Pays-Bas et du Rhône. *Géol. Mijn.* 19, 196–213

Surell, E. 1847: *Mémoire sur l'amélioration des embouchures du Rhône*. Nîmes, Imprimerie cévenole

9

The Impact of Climatic Changes and Sea-level Rise on Two Deltaic Lowlands of the Eastern Mediterranean

G. Sestini

9.1 INTRODUCTION

As compared to the south margins of the North Sea, the extent of coastal low-lying plains on the Mediterranean shores is fairly limited (figure 9.1). The demographic, economic and ecological importance of these areas is nevertheless very considerable on a local, national and international scale. Besides settlements involving millions of people, vital agriculture and fishing resources, as well as industrial, commercial and communication centres, and the increased recreational use of beaches, most areas still contain little-modified natural ecosystems of irreplaceable value.

In the western Mediterranean the Ebro delta in Spain (Marino 1988, Sorribes and Grau 1987) is small (285 km^2), but is a rice-producing and fishing area of primary importance for the country; it is currently threatened by erosion, since dams across the River Ebro in the Catalanid coastal ranges hold back 96 per cent of the river sediment (see also chapter 6).

In France, the 235 km arc of the Gulf of Lions, from Narbonne to Marseille (pop. 1,100,000) comprises a series of coastal lagoons, low plains, and the delta of the River Rhône. Economic activities include agriculture (large reclaimed and irrigated areas with cattle breeding, rice and vegetable

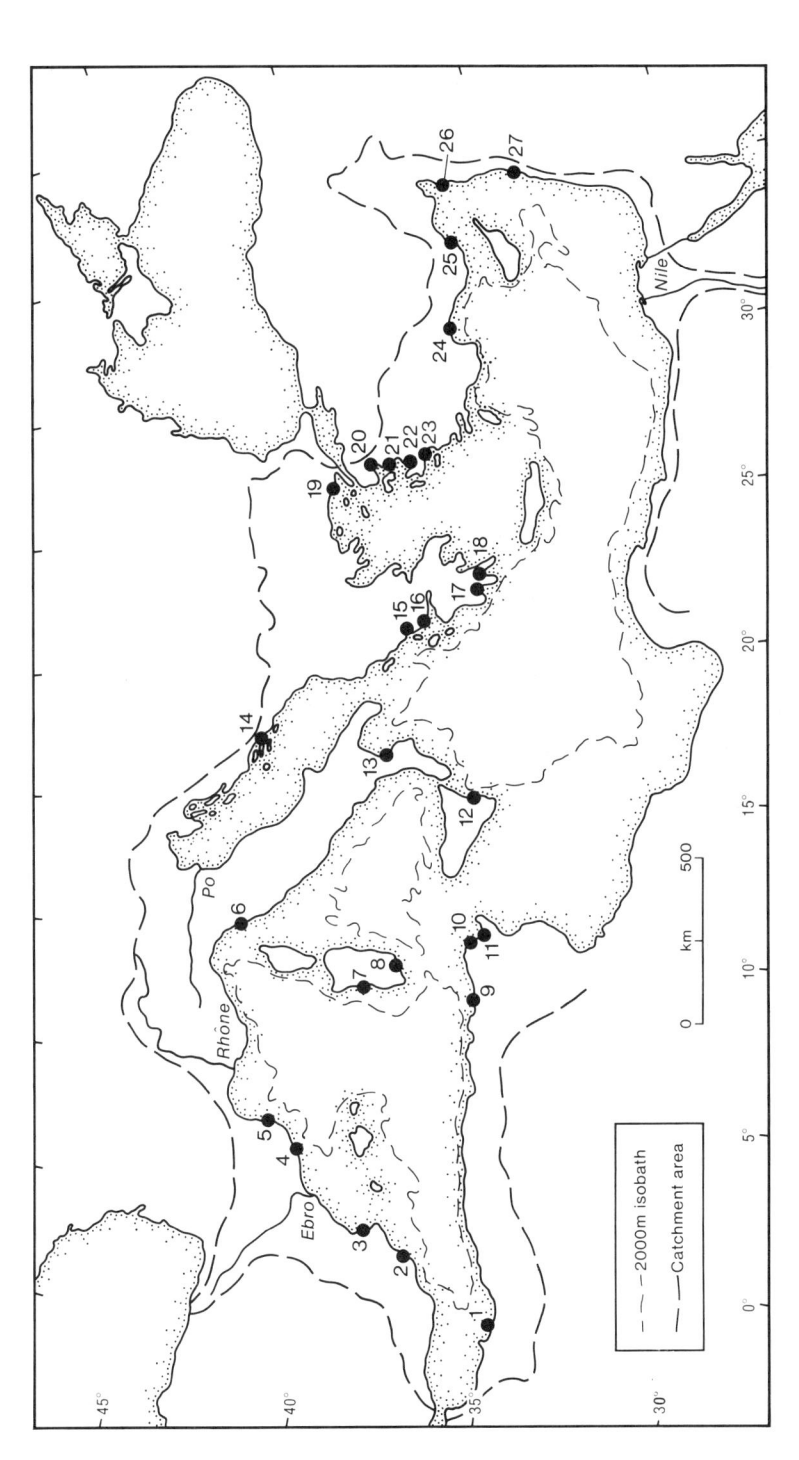

Figure 9.1 The main alluvial-deltaic areas of the Mediterranean coast. In addition to the named deltas, the following have been numbered: 1, Mar Chica and Mouluya; 2, Mar Menor; 3, Valencia; 4, Llobregat; 5, Rosas; 6, Arno; 7, Oristano; 8, Cagliari; 9, Gulf of Annaba; 10, Medjerda–Birzerta; 11, Tunis; 12, Simeto; 13, Crati, Gulf of Taranto; 14, Neretva; 15, Amvrakia; 16, Achelaos; 17, Messenia; 18 Marathon; 19, Evros; 20, Edremit; 21, Bakir Cay; 22, Gediz; 23, Kücük and Buyuk Menderes; 24, Aksu; 25, Goksu; 26, Latakia; 27, Akkar. The 2,000 m isobath is shown and is based on the map in Pirazzoli (1987).

growing, vineyards), important salt pans, industry (Sète, Fos de Mer) and a string of fast-developing beach resorts (Corre 1988; see also chapter 8).

In the northern Adriatic the coastal lowland of the Veneto and Emilia-Romagna regions of Italy, are best known for the ancient cities of Venice and Ravenna, for the lagoons and nearby Po delta (wetlands of international standing) and for a string of beach resorts that yearly attract millions of tourists. A population of 1.2 million is sustained by complex agricultural, commercial and industrial activities, of economic importance extending well beyond the region.

In the south-east Adriatic, the coast of Albania includes 190 km of flat beaches, marshes, lagoons and deltas (A. Sestini 1940; Paskoff 1985b) related to the rivers Drin, Buna, Matia, Skumbini, Semeni and Vojussa (figure 9.2). The lowlands (population around 700,000) constitute a large proportion of prime agricultural land in a country that is mostly mountainous and rugged. Vlone and Dürres are the country's ports and main industrial centres.

In Greece the most significant deltaic shorelines of ecological and agricultural value are in the bays of Arta and at the mouth of the Achelaos river, the Axios delta in the Thermaikos Gulf, near Thessaloniki (Georgas and Perissoratis 1989, Sivignon 1987), and the Nestos and Evros deltas.

In Turkey, small but economically important deltaic plains occur on the Aegean coast (Gediz near Izmir, the Büyük Menderes delta) and on the Mediterranean coast (the Aksu plain near Antalya, the Goksu delta near Silifke). The larger delta complex of the Rivers Ceyan and Seyhan, near Adana (figure 9.2) forms the Cukorova plain (Erinc 1953; Göney 1976; Evans 1971), in and around which live over 2,000,000 people.

In North Africa the only alluvial plains related to modern deltas are in Egypt and in north Tunisia. The lower Nile delta (to +5 m) not only contains 20 per cent of the country's population, but also is its main commercial outlet with the vital industrial and recreational centres of Alexandria and Port Said. Agriculture and fishing account respectively for 15 and 60 per cent of national production. One third of the Mediterranean migrating and wintering birds is estimated to use the deltaic lagoons (Meininger and Mullie 1981). The Nile delta coast is still not too extensively exploited, but population and economic pressures are urging development plans that involve new settlements, harbours and industries.

The lowlands of Tunisia are of more limited extent; they include the plains of the River Medjerda and the lagoons of Tunis, Garh el Melh and Bizerta (all fishing grounds) and Lake Ichkeul (a bird sanctuary) (figure 9.2). This region is the population and economic hub of Tunisia with over 1,500,000 inhabitants.

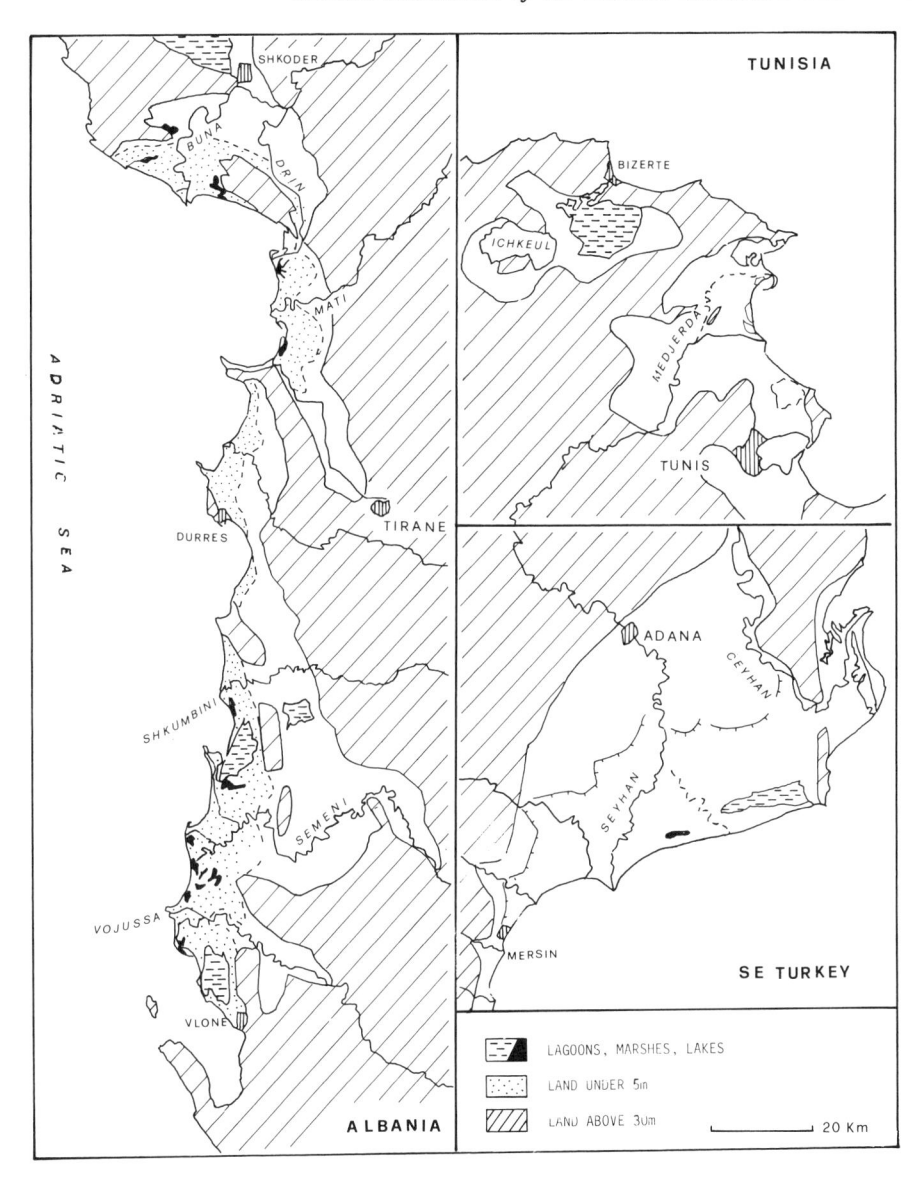

Figure 9.2 Deltaic areas in Albania, Tunisia, and south-east Turkey

Because of their ecological fragility, related to the land–sea transitions, these coastal lowlands are the most vulnerable to climatic changes involving hydrology, ecosystems and sea-levels.

There is now a consensus in the scientific community that a significant global warming will result from the greenhouse effect of the rising concentration of CO_2 and other radiative gases in the earth's atmosphere, due to industrial emissions; and that if their build-up is allowed to continue a doubled concentration will induce, sometime in the next century, average increases of temperature in the range of 1.5°C to 4°C (Bolin et al. 1986; Barth et al. 1984, UNEP/WMO/ICSU 1986). Climate scenarios for a doubled CO_2 predict a northward shift of climatic zones, a more continental climate in central Europe with higher winter temperatures and drier summers, and a warming of north Atlantic waters.

In the Mediterranean region the present areas of lesser and unreliable rainfall (North Africa, south Spain, Sicily, interior Turkey) could move northward and become more extensive. In the Alpine region river regimes would certainly be altered by the upward retreat of the snowline and the disappearance of glaciers, and especially by greater rain- versus snowfall in winter. Overall, rainfall would continue to be dependent on the cyclonic patterns that affect the western and central parts in winter, and on their relations with relief (e.g. more precipitation over the western Pyrenees, the eastern Alps, the Balkan mountains and north-western Greece). It is not yet clear how precipitations could change, but certainly rates of evaporation would increase (Wigley 1988).

Particularly important would be the effects of changed rainfall quantity and patterns and of increased evaporation on soils and hydrology. They could range from increased soil erosion, with consequent river alluviation, and increased soil salinity, to slower recharge of aquifers. Wetland ecology would be very sensitive to water temperature and salinity increases. The composition and patterns of natural vegetation would also gradually, but dramatically, change, owing to a northward and upward shift of vegetation zones.

Since temperature oscillations in the last 2,000 years have been within 1°–2°C at most, there is no doubt that such a substantial change of climate would have profound effects on agriculture and marine resources. There is evidence, for instance, that much smaller temperature variations in historical times, yet accompanied by deterioration of rainfall to drought conditions, have caused social and economic disruptions (Lamb 1982). In Egypt, the cyclic variability of Nile floods (Hassan 1981, Hamid 1984), related to climate and precipitation changes in East Africa, had resulted during Dynastic times in periods of economic stagnation, sometimes accompanied by political upheavals (Butzer 1976).

Another main consequence of a warmer atmosphere is an accelerated rise of sea-level, due to the melting of Alpine and polar glaciers and to the thermal expansion of oceanic waters. Sea-level has been rising, not only since the last glacial maximum (120 m rise in 17,000 years, at rates of 8 to

12 mm/year until 6,000–7,000 years ago), but in more recent historical times, and especially in the last 100 years. During the latter period the Mediterranean sea-level has risen 1.3 mm/year (Pirazzoli 1985). In Holland the rise from 1870 to 1980 has been 18–20 cm, accompanied by a 15–44 cm increment in the level of high tides (Hekstra 1986 and chapter 2). More globally, discounting local subsidence and uplift, the average rise has been calculated at 1.22 mm (0.9–1.4 mm) a year (Gornitz and Lebedeff 1987); between 1940 and 1975 sea-level has risen 1.5 ± 0.3 mm/year (Emery 1980), with a possible recent increment to 2.4 ± 0.9 mm (Peltier and Tushingham 1989).

Estimates of the increase of sea-level rise in consequence of the predicted global warming vary according to the CO_2 scenarios used, to models of oceanic thermal expansion, and especially to the behaviour of the polar ice caps, such as Greenland versus Antarctica, and of the western Antarctic ice shelf. Conservative to moderate estimates range 13–39 cm (by 2025), 24–52 cm (by 2050) and 38–91 cm (by 2075) (Hoffman 1984; Robin 1986). Hekstra (1986) has argued that a reasonable estimate for the last quarter of the twenty-first century would be an increase of 69 cm, while the Villach 1985 Conference (UNEP/WMO/ISCU 1986) concluded that a global warming of 1.5°–4.5°C would lead to a sea-level rise of 20–140 cm (see also discussion in chapter 1 and figure 1.1).

In terms of physical impacts, increases of <30 cm should be considered to be moderate, because they could be coped with by gradual adjustments to existing coastal defences and by acceptance of modest losses. Greater increases (>50 cm), however, would certainly have catastrophic consequences involving hard economic decisions about the cost of coastal protection and political decisions about what to protect and what to abandon.

A significant rise of sea-level, coupled with a possible increase in storminess and storm surges, and maybe of higher tidal ranges, would cause the retreat of beaches, and possibly the transformation of some lagoons into bays, the flooding of reclaimed lands, salt wedges to move further inland in rivers, as well as direct damage to harbours, towns and roads. The impact analysis of these effects is complicated, however, by the growing anthropic interference with natural environments and the enormously accrued economic value of the coastal regions. The parameters of the natural systems have been or will be altered by the superposition of such factors as urban and industrial pollution, the 'freezing' of the coast by tourist developments and by defence structures, changes to river water and sediment discharge, and (locally) accelerated land subsidence due to excessive groundwater extraction.

The aim of this chapter is to summarize the historical evolution and the present physical state of the two largest coastal lowlands of the Mediterranean that are of major socio-economic importance nationally and internationally: the northern Adriatic coast of Italy and the coast of the Nile delta in Egypt. The former is in a state of intensive land use, the latter is rapidly developing in the face of coastal retreat consequent to the sediment cut-off caused by the completion of the Aswan High Dam in 1964. Further, a preliminary evaluation is made of how sea-level rise could affect the two regions physically, in comparison with the other eastern Mediterranean deltas, and of the parameters to be considered for economic risk analysis, in the context of the postulated climatic changes.

9.2 THE NORTH-WEST ADRIATIC COAST

The 300 km-long coast from Monfalcone, near Trieste, to Rimini limits a lowland 15–25 km wide, generally under 2 m elevation. This belt is composed of lagoons, salt and fresh water marshes, and of reclaimed lands (the remnants of former lagoons and inter-distributary bays), separated by the more elevated channel systems of several rivers that flow from the Alps and the Apennines. Extensive areas, especially around the Po delta and north-east of the Venice lagoon, actually lie below sea-level (figure 9.3).

The coast is made of a series of beach-dune barriers the distribution of which is related to the littoral currents and longshore drift of sands derived from the cuspate mouths of the Isonzo, Tagliamento, Piave, Brenta, Adige, Reno and Fiumi Uniti rivers, and from the larger, lobate delta of the Po river. With the exception of a few stretches north of the Po (e.g. Tagliamento delta, Sile mouth to Lido outlet, Chioggia to Adige mouth), where the sand barrier is wider (1–4 km) and originally had dunes 3–5 m high, the Veneto–Friuli coast is characterized by narrow, gentle beaches interrupted by numerous waterways (estuaries, canals) and backed by lagoonal waters or reclaimed lands below sea-level. The shores of the Isonzo delta are very low, muddy with tidal flats (Brambati 1987), and the Grado-Marano and Caorle lagoons are limited by mobile low barrier islands. In the Po delta there is a discontinuous external margin comprising sandy islands and beach ridges with large lagoons and marshes, whereas most of the reclaimed lands of the inner delta lie 2–3 m below sea-level. In the west they are crossed by a band of north-south sandy ridges (former shorelines) that are partly wooded or built up and rise a few metres above sea level (figure 9.4). The river branches and the outer margins of agricultural land are defended by dikes generally 2–2.5 m high.

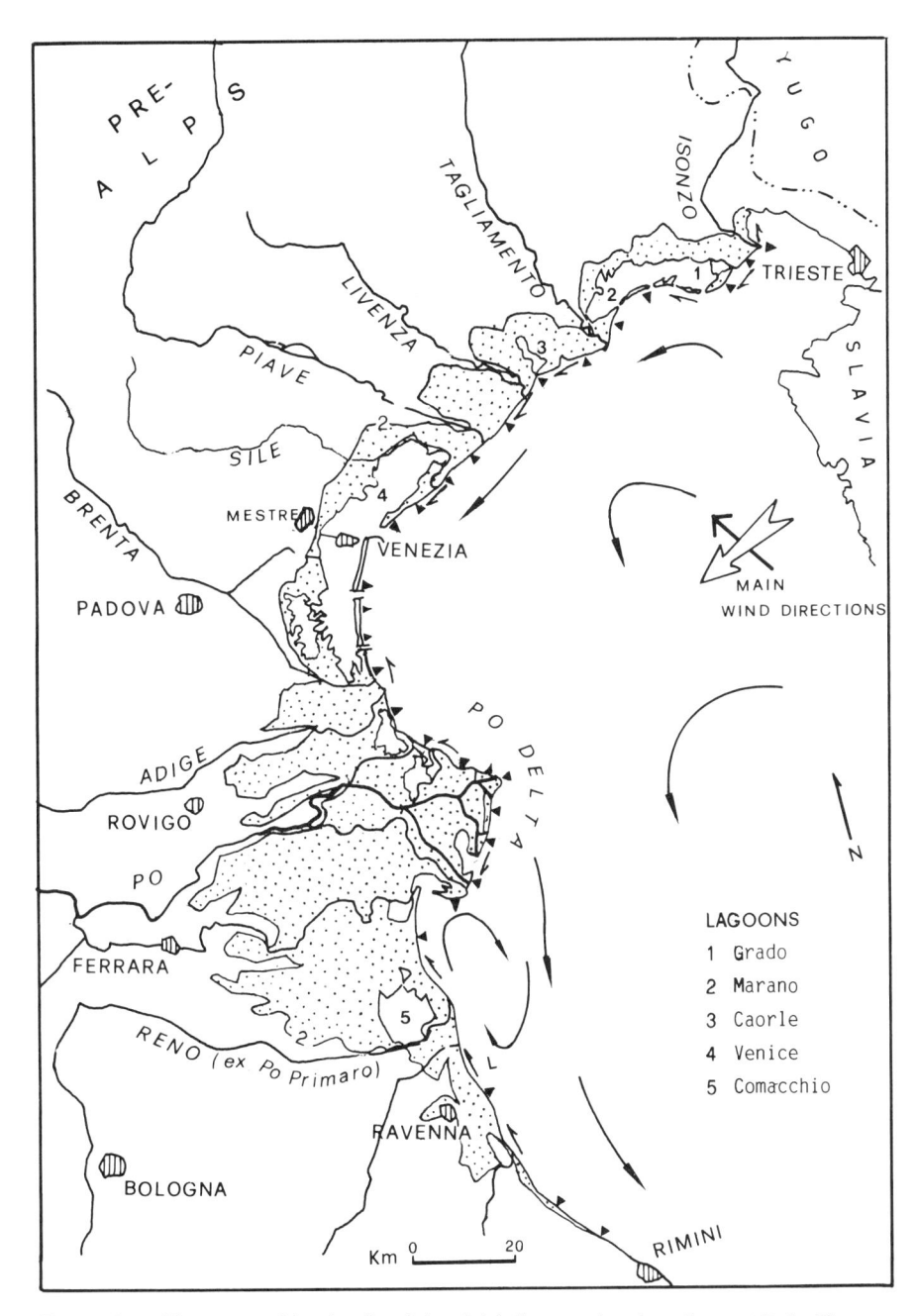

Figure 9.3 The coastal lowlands of the Adriatic margin of north-east Italy. The lowlands below 2m are shown by dot shading. Eroding and prograding shorelines are shown by solid arrow heads.

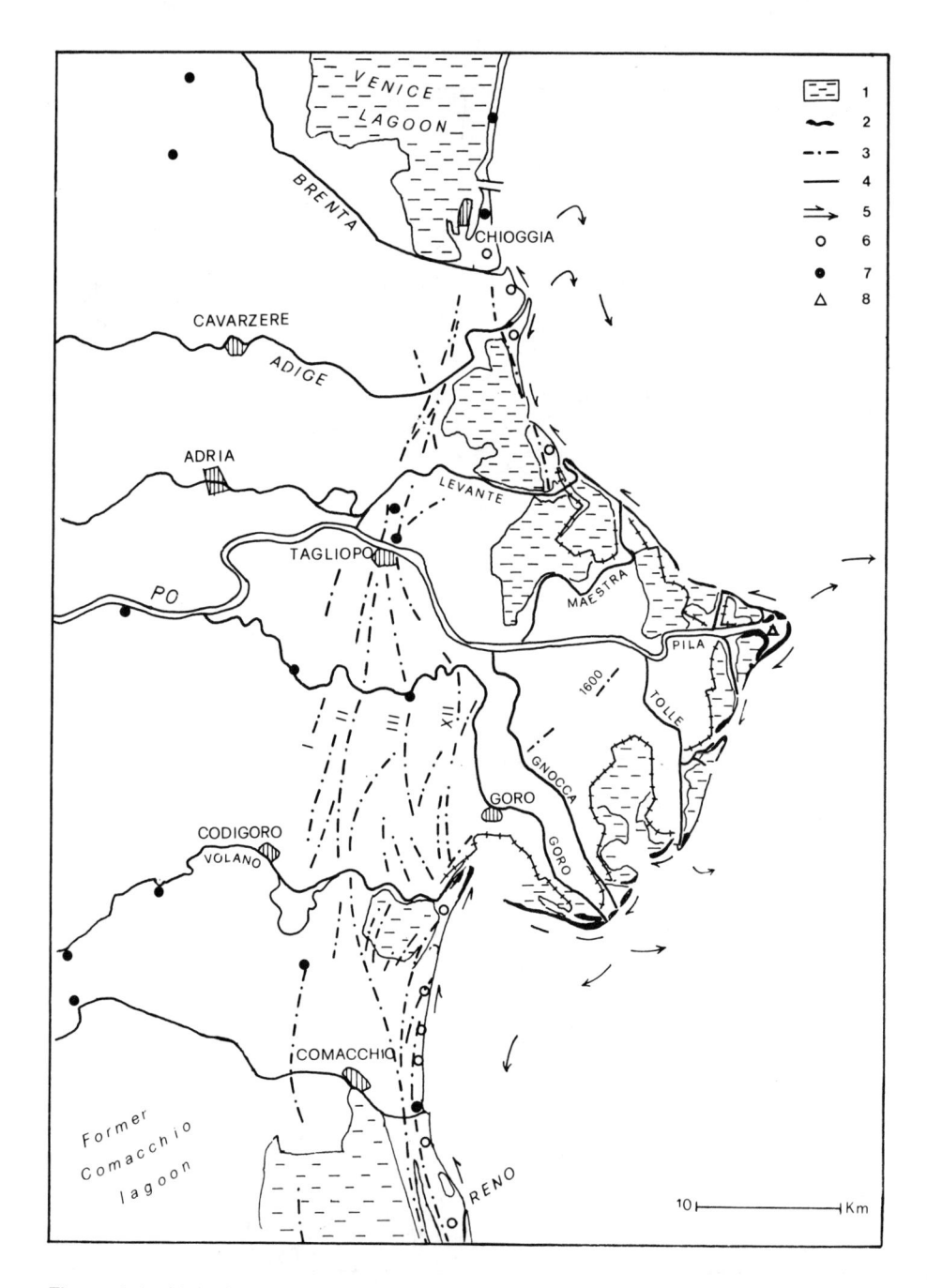

Figure 9.4 Main features of the Po Delta: 1, lagoons, marshes, tidal flats; 2, sand bars, beaches; 3, ancient beach-dune ridges (I, II Etruscan; III-XII Roman); 4, dikes, seawall; 5, directions of sediment dispersal; 6, beach resorts; 7, main towns; 8, power station

The coastal zone of Romagna, from the Po to Cervia, is a belt 3–10 km wide made of arcuate beach-dune ridges, with relics of lagoons and marshes. The dunes, 4–5 m high in their original state, had been extensively planted with pines in the first part of this century, but have been largely destroyed in the last decades (Cencini et al. 1979). The present Comacchio lagoon is one of the remnants of a vast wetland that occupied most of the inter-distributary low areas east of Ferrara and north of Ravenna.

The north-west Adriatic coastline bears a large number of fixed structures (seawalls, groynes, jetties, offshore breakwaters) built mainly in the last fifty years to protect the agricultural lands and the beaches and to regulate the entrances to lagoons and to several canalized harbours (e.g. Grado, Porto Buso, Lido, Malamocco, Chioggia, Porto Garibaldi, Porto Corsini, Cervia, Cesenatico, Rimini). As much as 70 per cent of the littoral from the Isonzo to the Adige, and 50 per cent of the Volano (Po delta) to Rimini coast is now thus armoured.

The Romagna and Veneto lowlands are considerably affected by land subsidence. This is derived from the continued tectonic subsidence of the Neogene Padan-Adriatic basin, and the sedimentary compaction of clays and peats. A rate of sinking of 1.3 mm/year has been estimated in the Venice area (Pirazzoli 1983) reaching at times, in the Pleistocene, 2.6–4.5 mm/year (Fontes and Bertolami 1973). In the Po delta and Ravenna regions natural subsidence has been greater, about 0.5 mm/year between 1900 and 1950 (Bondesan and Simeoni 1983; Ciabatti 1967; Carbognin et al. 1985a, b).

There has been, in addition, subsidence caused by the draining of lagoons and marshes (up to 1.3 m) and in some areas (e.g. Venezia-Mestre, Po delta, Ravenna) by the overextraction of ground water between 1955 and the late 1970s. In the Po delta the pumping of methane-bearing waters caused a sinking of up to 30 cm a year, with considerable negative impacts on drainage; in Venice sinking amounted to 12 cm between 1952 and 1969 (Carbognin et al. 1981, Gatto and Carbognin 1981).

In regard to the meteorological and marine factors that control coastal stability (Franco et al. 1982), air circulation over the northern Adriatic is dominated in winter by the strong predominant ENE to NE winds ('bora') and the less frequent south and south-east winds (Libeccio, Scirocco). The summer and spring westerlies are weaker and subordinate. The scirocco and bora winds build 2–3 m waves, the former even higher (Cavaleri et al. 1981), and there is considerable wave energy concentration on the coast, also because of wave refraction by delta promontories and offshore shoals. The relative effects of the NE versus SE waves vary as the overall orientation of the coast changes; the Friuli coast is relatively low-energy for instance, in contrast to Romagna, while the stretch from the Tagliamento to the Po delta is the most exposed to all storm surges. Many parts of the coast were severely inundated by storms in the 1960s.

Marine currents flow anti-clockwise, active longshore currents move beach sands along the coast, though not far from the river mouths (CNR 1985). In the Po delta they are dispersed north and south of the Pila branch, while in Romagna the drift is northwards (Gazzi et al. 1973). Natural sediment redistribution is interfered with by the fixed structures, especially groynes and jetties, which have instigated complex nearshore circulation patterns; sediment is trapped upstream of the jetties or deviated seaward, and generally substantial erosion is taking place 'downstream' (Brambati 1984; Cencini et al. 1979; Gatto 1984).

The northern Adriatic and the Gulf of Gabes have the highest tides in the Mediterranean Sea. From a nodal point in the central Adriatic the tidal wave rotates anti-clockwise, reaching maximum spring tide elevations of 80 cm at Trieste and 60 cm at the Po delta. Between October and April (mainly November–December) the combination of south-east winds, low pressure, low river discharge, high spring tide and seiches can raise the water level along the northern coast as much as 1–1.6 m. This event of *acque alte* has become a progressively more frequent phenomenon; in the Venice lagoon it has been aggravated by land subsidence, the reduction of lagoonal surface by reclamation and the deepening of lagoonal outlets (Pirazzoli 1983).

At the maximum of the Holocene transgression (6000–7000 BP) the sea had reached a line 5–20 km inland of the present coast (Favero and Serandrei Barbero 1978). A phase of active regression followed, with the formation of cuspate deltas, and the progressive enclosure of lagoons by beach ridges and the development of barrier islands. The Venice lagoon extended from the River Piave to the Brenta; the Gado Lagoon was formed after the gradual switching of the River Isonzo eastwards from the tenth century (Brambati 1987).

Between Venice and Ravenna in Etruscan and Roman times (3000–2000 years BP) the mouths of several branches of the Rivers Po, Adige and Brenta developed modest triangular deltas that had a rate of advance of 4 m/year. Coastal changes have been traced in detail from morphological and historical evidence (Ciabatti 1967; Fabbri 1985; Veggiani 1974). Until the early twelfth century the Po discharged south of the present delta (Primaro, now Reno, and Volano branches). A major flood-break west of Ferrara in the mid-twelfth century diverted it to the east, then in the fourteenth–seventeenth centuries there was a gradual build-up of discharge and bedload following greater upland rainfall, associated with the 'Little Ice Age', deforestation and increasing flood control (Zunica 1978). River flow became more confined owing to the channel deviations and dikes built by the Venetian Republic to keep the Brenta, Adige and Po from discharging into the Venice lagoon. Consequently, the present lobate delta started to form and to grow at a faster rate (100 m/year) in the sixteenth–seventeenth centuries, first towards the south (Gnocca, Goro, Tolle branches), then to the east (Pila branch, 65 m/year progradation with high floods), resulting in the gradual enclosure of elongated marine bays (figure 9.4 and plate 8).

North of Venice, modifications included the deviation and straightening of the lower reaches of the Sile, Piave and Livenza, both to protect the Venice lagoon and for flood control (the Veneto and Romagna rivers tended to flood easily because of elevated channels). South of the Po, until the early part of this century, the Reno and Fiumi Uniti had more prominent cuspate deltas.

Man's activities in exploiting and disrupting the coastal environments began a drastic acceleration in the early 1950s with the development of the harbours and industrial complexes of Porto Marghera and Ravenna involving the destruction of tidal flats, the filling of marshland, the digging of deep access canals and the extension of long jetties into the Adriatic; with overuse of ground water, and with the wide-ranging construction of tourist facilities (hotels, marinas, camp sites and related infrastructures) resulting in the almost total destruction of the wooded dune belt. With the exception of the Po delta, as much as 52 per cent of the Isonzo to Rimini shoreline is now built-up; many summer resorts are narrow urban strips with high-rise buildings.

With this has coincided a dramatic decrease of river discharge, due to reforestation and slope control, the retention of water and sediment in hydro-electric reservoirs, the diversion of water for irrigation and the dredging of sand from river beds (Bevilacqua and Mattana 1976; Dalcin 1983). The Veneto and Romagna beaches have become unstable and have been retreating along most of the coast since the early 1950s even at the mouths of the main rivers (Brambati et al. 1978; Cencini et al. 1979; CNR 1983; Zunica 1971, 1976).

Growth at the edges of the Po Delta has slackened, and continued subsidence has caused retreat (Carbognin et al. 1985). In the period 1965–73 the solid load of the river Po had decreased from 16.9 to 10.5 million tons/year. The bedload, a quarter of these figures, had become barely sufficient to satisfy the calculated sand budget of 3.1 million tons/year, necessary to keep the delta margin in equilibrium (Dalcin 1983).

The industrial dredging of sand from the river bed, officially given as 100 million tons for the period 1958–81, is estimated to have been up to six times greater. The sand still delivered to the sea came, in part, from the erosion of the river bed itself, which has already been lowered by 5 metres. Sand input will substantially decrease, however, when engineering measures will have to be undertaken to prevent the bed from reaching below sea-level. A sand deficit lasting at best thirty years, possibly even 100 years, is forecast, even if all dredging operations were to be stopped.

9.3 THE NILE DELTA

An extensive part of the Nile delta between Alexandria and Tineh lies below 2 m elevation with lagoons and agricultural land, mostly of recent reclamation, bordered by a coastal sand belt 1–10 km wide, made of beaches, backshore plains and dunes. The topographic depressions of the present and former wetlands alternate with ridges 1–3 m high formed by the Nile, canals and older stream channels (figure 9.5).

The coastal lakes and lagoons, almost twice as extensive as in the last century, have been reduced by siltation and especially by land reclamation. South and east of Alexandria a large agricultural area 1–3 m below sea-level, lies on the sites of former lakes Mareotis and Abuqir; it is drained by pumping at Mex (Alexandria) and El Taba (Abuqir), and marine flooding is prevented by the 10 km-long Mohamed Ali seawall (figure 9.6). The Idku, Burullus and Manzala lagoons are brackish (salinity 0.8–5‰) as they are fed by the Nile drains, but the Bardawil lagoon is entirely outside the Nile's influence and is hypersaline (30–40‰ to 100‰). The lagoons communicate with the sea through narrow outlets; the single outlets of the Idku and Burullus lagoons are now protected by short jetties, the Manzala lagoon has two canalized outlets; the two natural outlets of Bardawil are often blocked by sand drift. The lagoons are mostly shallower than 2 m and are dotted with islands, many of which represent ancient dunes, river banks and beach ridges (Frihy et al. in press). The irregular marshy south and south-west edges of the Idku, Burullus and Manzala lagoons have been straightened by post–1964 reclamation, with a surface reduction of about one third each.

In the west, the Alexandria coast from Agami to Abuqir is cut into a fossil dune ridge 5–15 m-high, there are 2–3 m-high cliffs and small pocket beaches. Besides the harbours of Alexandria-Dakheila and Abuqir, the coast is intensely built-up, with a major 20 km-long road by the sea (the Alexandria *corniche*) protected by a bulkhead wall. Recent and historical

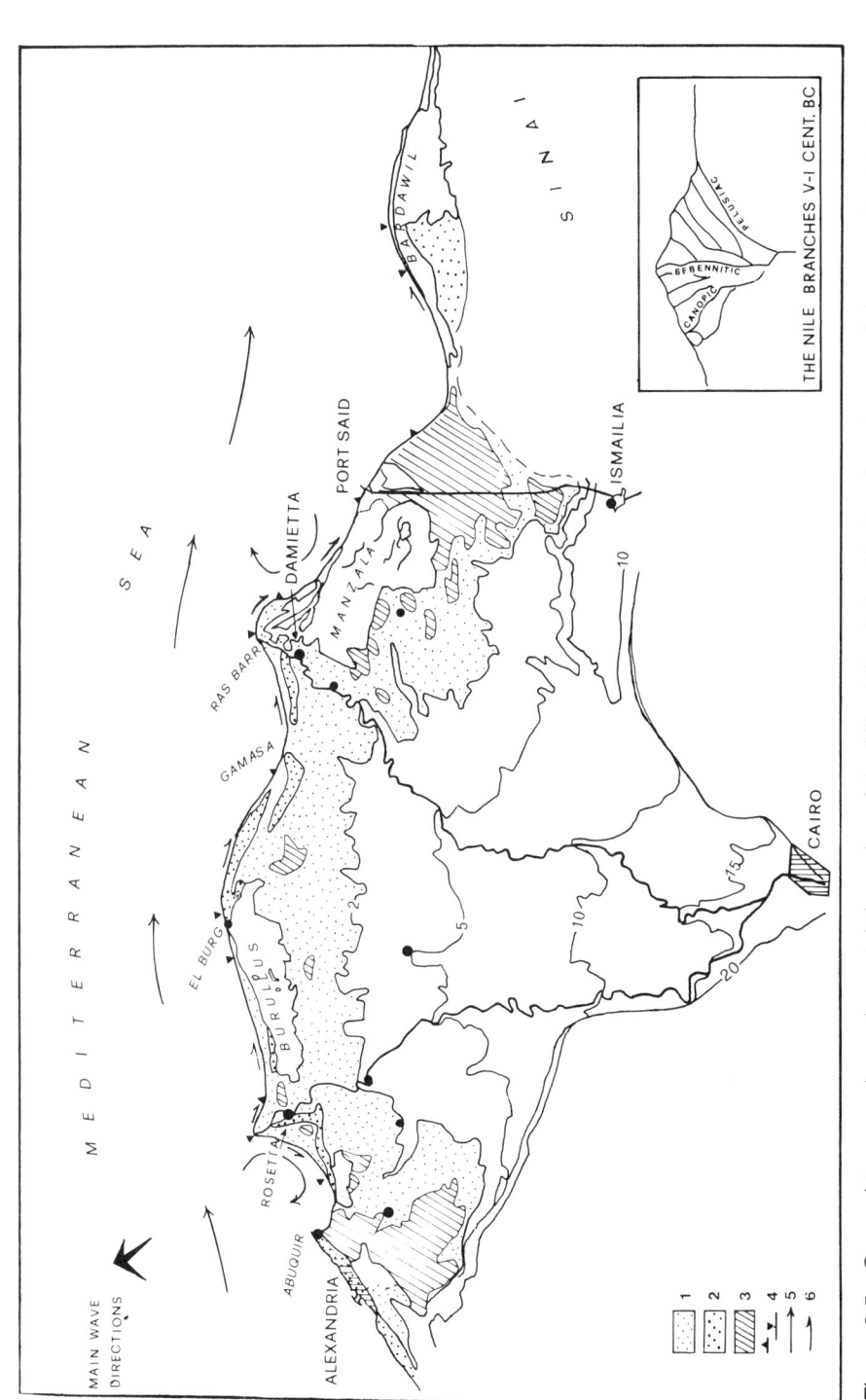

Figure 9.5 General topography and coastal dynamics of the Nile delta: 1, land below 2 m; 2, coastal dunes; 3, land below sea-level; 4, shoreline advance (upper) or retreat (lower); 5, surface currents; 6, longshore drift direction

beach accretion ridges are prominent east of the Rosetta, Burullus, Damietta and Port Said headlands. Aeolian dunes are common in the coastal sandbelt, except between Damietta and Tineh; the main dune fields (Idku to Rosetta, El Burg to Gamasa, and south and west of Bardawil) reach heights of 10–20 m; they are still in a natural state and not fixed by vegetation. The sand deposits of the coastal barriers are generally 5–10 m thick and rest on either lagoonal clays, silts and peat (at Idku, Burullus; G. Sestini 1989) (figure 9.6) or on marine sands (at Rosetta, Damietta promontories; El Askary and Frihy 1986).

Considerable subsidence of the coastal zone is indicated by the 10–50 m-thick layer of post-8000 BP nearshore marine, lagoonal and deltaic sediments, with rates of deposition as much as 5 mm a year, in the north-east part of the Nile Delta (Coutellier and Stanley 1987). At the margin of Abuqir Bay, 15 m of now compacted lagoonal/beach deposits represent sedimentation since 4800 ±150 radiocarbon years BP. Active subsidence in historical times is also evidenced by the drowning of Ptolemaic and Roman settlements in Abuqir Bay and in the Burullus and Manzala lagoons. In the Suez Canal region, north of Ismailia, a steady sinking of 1.2 mm/year since 1900 has been documented by Goby (1952), while at Port Said tidal gauge records for the period 1922–50 have suggested a subsidence of 4.8 mm/year (Emery et al. 1987), a figure closely comparable with that obtained by Stanley (1988) from geological studies.

Before 1964 Nile river discharge into the Mediterranean was concentrated in the summer months, the August–September flood peak; reworking and dispersal took place in winter. Marine surface currents north of the Nile delta flow eastwards (part of a general anti-clockwise gyre), but near the coast clockwise eddies have developed in response to coastal shape, a weak one in Abuqir Bay, a stronger one (60 cm/sec.; Coleman et al. 1980) off the Damietta headland (see figure 9.5). Constant north-west surface winds over the central and eastern Mediterranean in summer generate swells with periods 9–10 secs and heights of 0.8–1.5 m, and lengths of 100–200 m. Wave fetch can be as much as 1,000–1,500 km from the Ionian and Aegean seas. From November to April WNW- to ESE- moving depressions cause cyclonic storms that last four–seven days, a dozen of which strike the north-east Egyptian coast each winter. Storm waves have periods of 7–8 secs, are 1.5–2.5 m, sometimes 3 m, high and approach mainly from the north-west, secondarily from the north, seldom from north-north-east or north-east (Manohar 1981). Wave energy concentration is particularly high on all north-east-trending coastal stretches and promontories (Inman and Jenkins 1984; Manohar 1981). The north-west-oriented stretches (including western Abuqir Bay) are attacked more by north-west

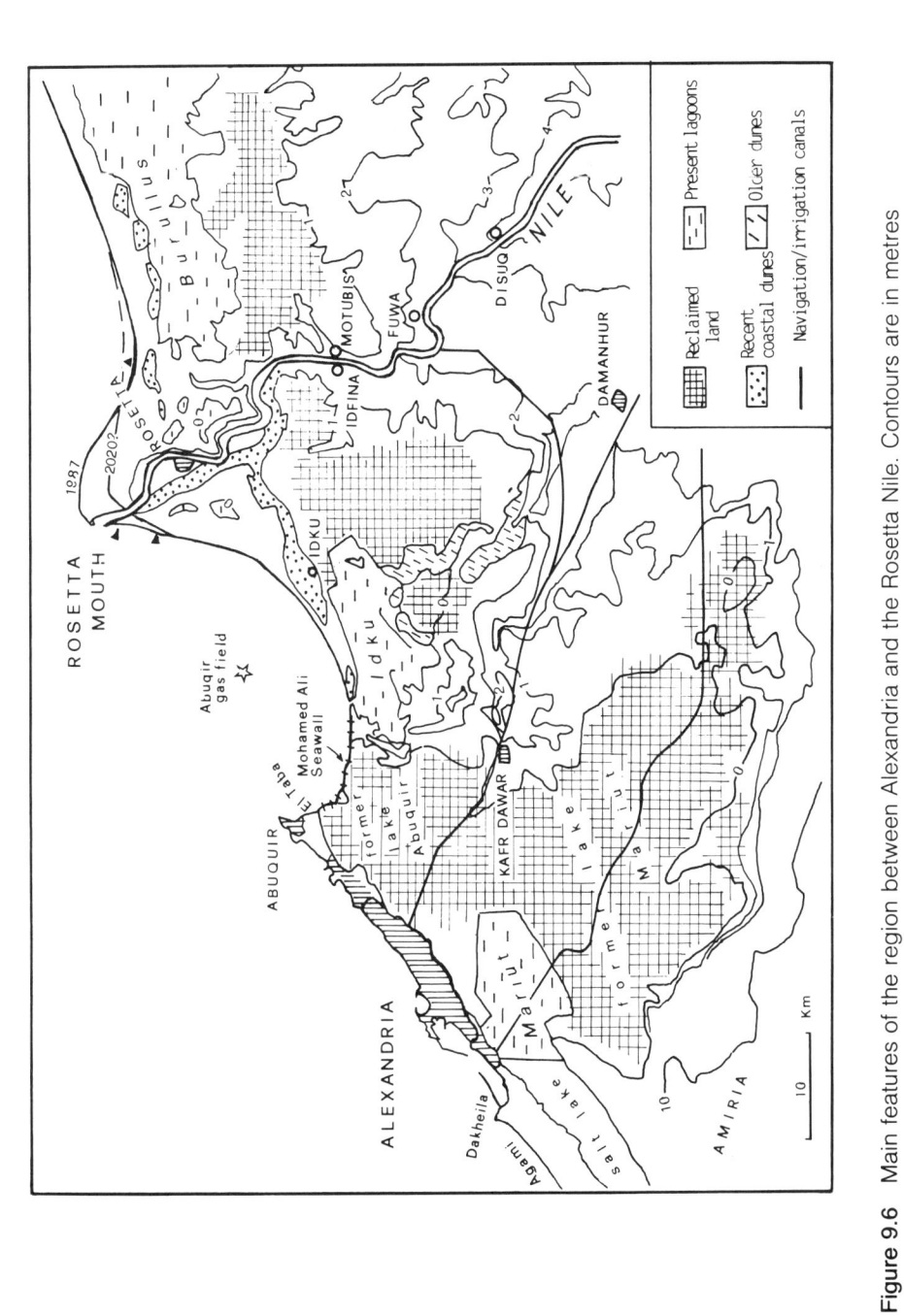

Figure 9.6 Main features of the region between Alexandria and the Rosetta Nile. Contours are in metres

and north waves in consequence of refraction, than by the less frequent north-east and north-north-east waves.

The oblique wave approach generates an active longshore current system, eastward flows (see figure 9.5) with mean velocities of 20–50 cm/sec. (max. 80–140 cm/sec.) that drive beach and near-shore sands along the entire shore of the south-east Mediterranean, as far as Israel. Several stretches of the coast are characterized by actively migrating large beach cusps. Erosion and intensity of sand transport are considerable: conservative estimates are that one million m^3 of sand are moved yearly (though being as low as 60,000 m^3 in the shallow zones and embayments east of the promontories). In addition, large amounts of sand have been and continue to be subtracted from the beaches by the offshore winds. During winter storms, fine-grained bottom sediments in areas deeper than 10 m, are stirred into suspension and moved seaward, mainly to the east (Summerhayes et al. 1978).

The tidal excursion is low, with a mean range of 12–30 cm, but extreme ranges of 60–130 cm have been recorded with a twenty-year cycle of recurrence (Sharaf El Din and Moursy 1977). Normally, the main effect of tidal oscillations is on the dynamics of the lagoon outlets. During storms, however, sea-level can rise up to 1.5 m (wave set-up in conjunction with high tides). Thus several stretches of the coast tend to be flooded each year, e.g. eastern Abuqir Bay, the barrier north of Burullus, the shores of the Gamasa embayment, and between Port Said and Tineh.

Subsurface stratigraphic evidence (Coutellier and Stanley 1987; Sneh et al. 1986; Sestini 1989) indicates that the delta coast started to prograde about 8000–7000 radiocarbon years BP (like the other Mediterranean deltas, e.g. the Ebro, the north Adriatic coast, the Rhône; L'Homer et al. 1981 and chapters 6–8), from a shoreline situated just south of the present lagoons. Several distributaries formed small delta lobes. From Herodotus' and Strabo's accounts there were in the fifth–third centuries BC at least six river mouths; the main flow was through the western (Canopic) and the central (Sebennytic) Nile branches. The Rosetta and Damietta branches were then no more than canals. The disappearance of the older distributaries occurred mainly in the second–fifth centuries AD, the eastern branches (Tanitic, Pelusiac), and perhaps the Sebennytic branch, persisting until the ninth century (Tousson 1934).

In Abuqir Bay the extent of the Canopic delta lobe is well documented by archeological and geological data (cf. references in G. Sestini 1988). The growth of the Rosetta and Damietta cuspate heads can be estimated by reference to the position by the sea of the respective towns at the time of their foundation in the ninth century AD. Off Rosetta advance from AD

900 to AD 1800 was 10 km, at Damietta 15 km; south-east of Port Said, progradation between AD 850 and AD 1800, dated by historical landmarks, was about 9 km.

The lagoons and lakes may have developed not only out of flood basins (especially during the cycles of higher Nile floods), but also from the closure of inter-distributary bays by the coast-wise extension of spits. The expansion of lagoons and marshlands during the last 2,000 years appears to have been irregular, in relation to subsidence versus sediment supply, from time to time intensified by major earthquakes (Tousson 1934; Ben Menachem 1979).

Coastal changes can be followed more precisely from topographic surveys since early AD 1800. Throughout the last century there was a steady advance of the Rosetta and Damietta promontories, respectively averaging 30 and 10 m/year, and accretion in all the embayments towards the east. An exception, however, was a slow retreat at Burullus (El Burg to Baltim), a tendency probably dating back to the disappearance of the Sebennytic mouth. About 1910, an overall 25 per cent decrease of Nile discharge, due to the reduced monsoonal rainfall over eastern Africa (Rossignol-Strick 1983) initiated a period of coastal instability and of headland recession.

Before 1964 the river discharge at Aswan was on average 84×10^9 m^3 per year, a sediment load of 160×10^6 tons/year, 75 per cent of which was silt and clay carried in suspension. Between Aswan and the sea, as much as 60 per cent of the water was lost because of use for irrigation, evaporation and seepage, while 30–35 per cent of the Aswan sediment load was deposited in the Nile valley and delta. Actual mean water discharge into the sea may have averaged $50–60 \times 10^9$ m^3/year, largely (65 per cent) through the Rosetta branch, and sediment discharge $100–15 \times 10^6$ tons/year (Said 1981), only a quarter of which was of sand-coarse silt grade. While clay and fine silt were dispersed by surface plumes tens of kilometres to the north and to the east, the sand supply became insufficient to maintain the promontory's shape. The rate of retreat at Rosetta, from 1915 to 1964, amounted to 30 m a year.

Since 1965 there has been no flood in Egypt and all water flow has been controlled; near the sea a meagre flow of $2–4 \times 10^9$ m^3/year has to be maintained to ensure navigation and to prevent salt intrusion. The headland recession has drastically increased to 80–130 m/year (Smith and Abdelkader 1988; Frihy 1988), with the progressive cutting of the former beach ridges. Serious erosional problems have occurred near the Burullus outlet, where the dunes are retreating by 6 m/year (total retreat here since AD 1800 has been 800 m). A state of recession with greater lower beach-face gradients has been affecting the western shores of Abuqir Bay

(Khafagy and Manohar 1979), where the Mohamed Ali seawall has recently needed substantial repairs. At Alexandria, overtopping of the *corniche* bulkhead wall is frequent in winter and the pocket beaches are shrinking.

Accretion, however, has continued north-east of Rosetta, at Gamasa and east of Ras el Barr, where a 15 km-long spit has grown in twenty-five years. The south-east migration of this spit and of coastal bulges (1 km wide by 5 km long) will eventually threaten the two Manzala outlets. At the northwest side of the 4,800 m-long Port Said jetty there has been strong accretion for years, but a state of sand deficit now occurs both to west (8 to 15 km from Port Said) and to the east, where the coast can no longer advance as in the past; the Bardawil spit may become unstable. Similarly, it is likely that accretion west of the 2,000 m-long New Dumyat jetty (figure 9.7) might further compromise the existence of the Ras el Barr resort and the maintenance of the Damietta Nile mouth, which is now almost blocked by sand bars and a spit. In conclusion, as the processes of coastal dynamics continue, the only source of sand at present is the shore itself, and the shallow marine offshore to 30 m depth, where areas of relict sands have been mapped (Summerhayes et al. 1978).

9.4 DISCUSSION

The physical impact of the rise of sea-level on the lowland coasts can be predicted, even modelled quantitatively on the basis of the present parameters of morphology, hydrodynamics, sediment budgets, land subsidence and the effects of artificial structures. Equally, the impacts of altered rainfall distribution on surface and groundwater could be modelled quantitatively, and the effects of increased air temperatures and of changed soil-water parameters on biosystems could be estimated, at least qualitatively, which could give some idea of impacts on agriculture and fisheries. What is much more difficult to estimate however, is the impact of these physical and biological changes on the future socio-economic framework of the threatened lowlands.

Even without an acceleration of sea-level rise the Adriatic and Nile delta coasts are now in a state of retreat, owing to reduced or ceased sediment supply, in conjunction with exposure to high wave energy and active longshore currents. Damage from storm surges, with associated flooding, has been and is occurring. In contrast, the other deltas of the eastern Mediterranean (Albania, Greece, Turkey, north Tunisia) are in lower-energy situations with considerable sediment inputs from rivers, at least until the recent erection of dams and under the present climatic conditions.

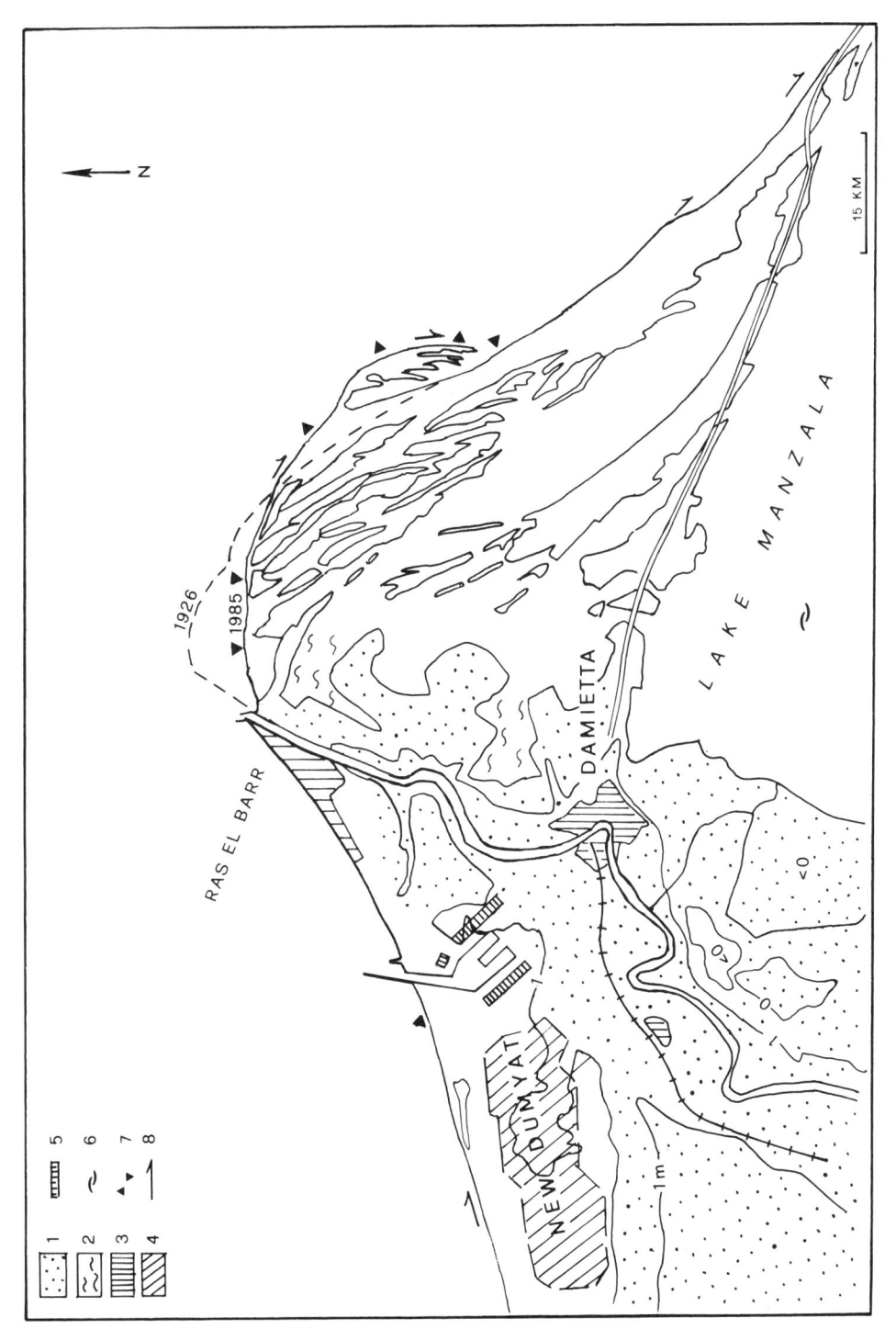

Figure 9.7 Main features of the region of Damietta: 1, cultivated land; 2, aquaculture; 3, present urban areas; 4, planned urban areas; 5, planned industrial-commercial estates; 6, open lagoonal fishing; 7, shoreline erosion versus accretion; 8, littoral sand transport

The shorelines of these deltas have been advancing (Erol 1983; Paskoff 1985a, b; Stournaras and Marcoupoulu-Diacontini 1985, Shuisky 1985). So long as the rivers are not irreversibly dammed, coastal advance might balance the effect of sea-level rise, though problems could arise in regard to lagoon and river outlets.

As the level of the sea rises, a normal beach and barrier island would be expected to migrate gradually inland (Bruun and Schwartz 1985). Actual examples of this recession are indicated by the response of the Caorle lagoon barrier island to a storm in 1977 (Catani et al. 1978); by the 126 m retreat that occurred at Lido Adriano (Ravenna) from 1957 to 1977 in consequence of a 45 cm subsidence (Carbognin et al. 1985b); and by the recession of the Burullus beach barrier, west of El Burg, in the last centuries, which has shifted southwards over lagoonal deposits (G. Sestini 1988). The overtopping of the lagoonal barriers, where low and devoid of dunes, is however a definite possibility with flooding behind them, eventually magnified by the deepening of existing or newly created passages.

Initially (next two–three decades?) a 25–30 cm rise of sea level would not cause *per se* any flooding of the north Adriatic lowlands, but it would worsen the present situation of beach and lagoonal instability. Higher water levels in the lagoons and the flooding of estuaries and canals, especially in association with land subsidence (e.g. Romagna) would continue. The beaches and the edges of the Po delta will continue to retreat, in spite and because of defence structures (unless they are planned more rationally). All protective structures (breakwaters, seawalls, dikes) will have to be raised periodically, and beach nourishment schemes intensified. The cost of beach protection and maintenance will thus increase, and the viability of some beach resorts will become questionable. The same considerations apply to the Nile delta coast, with repeated need to adjust existing or forthcoming protective structures, especially those from Alexandria to western Abuqir Bay, at the Burullus outlet, at New Dumyat-Ras El Barr and Port Said. The Alexandria beaches will be further reduced and the Ras El Barr resort will have to be shifted to the southwest.

An additional sea-level rise above 50 cm, and up to 100 cm, would certainly be disastrous, at least for parts of the two coastal regions, in view of the present-day storm surges of 1–1.5 m above mean sea level, unless measures of artificial protection have been undertaken previously, such as raising and stabilizing dunes, erecting suitable seawalls and blocking canals and estuaries with systems of sluices. Beaches for recreation use will continue to exist, but contemporary sea-front installations of tourist resorts and of exposed towns (e.g. Port Said, Alexandria, and those of the Veneto and Romagna coasts in Italy) will suffer extensive damage. In Italy, higher water levels in the lagoons would certainly accelerate the physical and social deterioration of several towns (e.g. Venice, Chioggia, Comacchio) unless the access of tidal waters is entirely regulated as proposed in the second stage of the 'Progetto Insulae' scheduled to start in 1994. The maintenance of the adjacent agricultural lowlands would require the elevation of all marginal dikes and an increasingly polder-type management.

In Egypt the main concerns would be a greater wave attack on harbour structures, the retreat of the headlands, and the management of the lagoons. At Rosetta a normal extrapolated retreat on a 2 mm/year sea-level rise could bring the shore to within 6 km of the town by AD 2020; with a 6–7 mm/year rise the shore would be 4 km nearer (figure 9.6). The high seawall that is being constructed in the area will prevent this; though, it could, at that time, make a negative impact on the supply of sand to the Burullus lagoon barrier. In the Damietta area the main danger is from the counter-effects of the New Dumyat harbour installations and the flooding of the residential and industrial quarters that are being developed on land that is below 1 m elevation (figure 9.7). The maintenance of the status quo in the lagoons (water circulation and salinity) is essential to fishing and aquaculture, which relies on the efficiency of the outlets. The latter are vital to fish migrations, and tidal flushing is necessary to prevent choking by aquatic plants and to control pollution from fertilizers and pesticides (Beltagy 1985). Wider and additional outlets would in fact be beneficial, but the higher water and salinity levels would damage agriculture and increase soil salinity, which is already high (Khashef 1983).

The above simplified scenarios of physical impacts suggest that there would be an increasing need to 'protect' present and future coastal uses, or else local economies will gradually deteriorate. The next stage would be either a drastic, and very expensive, construction of seawalls along many shores and lagoons, or acceptance of substantial disruptions and losses. The immediate task should be therefore the identification of all high risk areas, and a re-examination of the present factors of coastal dynamics in the context of increasing air/water temperatures and sea-levels. Storm penetration maps should provide a scientific basis for proper coastal zone planning and protection.

Coastal zone management, however must be based on cost-effectiveness, which means an assessment of the 'value' of the threatened land uses, not only in terms of their present functions, in the context of the local needs and of the importance of the lowland concerned to its hinterland and farther, but especially of those of decades ahead. The primary needs are determined by the present level of population and its trends of growth or decline, and the wider economic role of the region by external market forces. For instance, the future relevance of local industries, agriculture and ports, will largely be conditioned by world-wide commodity prices and trade trends, such as those of mineral and energy raw materials, with their effects on heavy and chemical industries; by those of cereals and industrial crops; and by the demand for consumer goods in a competitive international society. The role of individual ports may change in response to altered trends of maritime trade (e.g. Italy, Belgium) and cannot be physically maintained.

The two regions considered present different demographic and economic situations on which to base an evaluation of the economic impacts of climatic changes. The north Adriatic lowlands have reached a stage of advanced land use intensity, with complexly interrelated agricultural, industrial and tertiary activities, that are tied to national and supranational (EC, overseas) markets, with the support of a well-developed communications network, including motorways, railways, the commercial ports of Venice, Marghera, Ravenna, Chioggia, and Venice airport (figure 9.8a). Resident population within the +5 m contour is not high (1.8 million), in parts with low densities (<100/km^2 figure 9.8b); it is stationary, if not declining owing to a zero growth rate and to migration. Present economic expansion is focused in several areas, mostly situated above 2 m (exceptions are Chioggia, the central Po delta and Ravenna–Porto Corsini).

Agriculture in reclaimed lands produces mainly cereals (wheat, soft corn), soya beans, fodder; in parts sugar beet, rice, fruit, grapes, vegetables. There is active fishing and aquaculture in the lagoons, mainly serving a wide urban market (including the summer resorts). There are two levels of industrial activity: (1) three major centres of heavy/metallurgical and chemical industries (Monfalcone, Porto Marghera, Ravenna); (2) a large number of scattered, smaller-scale and diversified industries that range from food processing to consumer goods and are essentially geared to export markets. The initial development of the Ravenna industrial centre was associated with the exploitation of several natural gas fields. Tourism, both cultural (Venice, Ravenna) and recreational (beaches) makes a substantial contribution to the regional GDP. In 1987 the stretch from Grado to Chioggia attracted in summer 7 million tourists, with 37.7 million day visitors, against its resident population of about 120,000 inhabitants.

The main threats to be foreseen, therefore, are to the physical existence of Venice (and other towns of artistic-historical importance), to the tourist industry, to the activities of important harbours, and to specialized agricultural production, such as the vegetable orchards on the sandy soils of coastal barriers and relict beach ridges. As regards extensive agriculture in the reclaimed lands below sea-level, it might be more economical to turn at least parts of them (e.g. in the Po delta, Venice and Caorle lagoons) back to their original lagoonal state, in favour of fishing, which is at the present a more efficient and remunerative activity than cereal-growing on soils prone to salinization. Lagoons and marshes could act as buffer zones between the open sea and higher land, and in parts as nature reserves. Industrial and other activities in the areas <1 m above sea-level would probably move gradually inland without excessive disruption; losses of agricultural land might be compensated by higher productivity in a CO_2−richer atmosphere UNEP/WMO/ICSU 1986), provided water resources are still adequate. The lowlands are bound to lose population, in consequence also of the deterioration of summer resorts.

By contrast, the socio-economic setting of the Nile delta coastal zone is dominated by the basic needs of a growing population: food production, employment, housing. The present population in the belt ±<3 m contour, is about 10 million; on a 2.7 per cent rate of growth (the average for 1961–81, United Nations 1982) it will increase to 12.5 million by AD 2000, though there might be a decrease in growth rate (to 2 per cent) as a result of official policies and changes in social attitudes. Population is not evenly distributed; apart from the heavy concentrations of Alexandria, Damietta and Port Said, it is relatively low in the agricultural lowlands.

During the next decade, demographic and economic development will be concentrated in the following pole areas and axes (figure 9.9): Alexandria to Idku, Rosetta-Idfina, Damietta and Port Said areas; additionally, along the high-ground alignments Kafr El Dawar–Damanhour, Kafr El Sheikh–Siudi Selim, Mansoura–El Manzala–El Mataria. Development includes the expansion of present industries, which are essentially geared to the primary needs of the country (food processing, textiles and clothing; chemicals, fertilizers, cement; small-scale metallurgical, mechanical and manufacturing). These industries are in part dependent on imports of raw materials and on energy sources, either foreign or local (Alexandria oil refineries, Abuqir Bay gas fields, Abu Madi gas field) which could underpin the development of the Damietta area. Future developments will probably include an increasing production of consumer goods. Recent urban and agricultural expansion, however, has created serious pollution problems, especially in the sea, lakes and canals near Alexandria.

Agriculture and fishing are basic to local consumption and their continued and improved efficiency will be vital to the country's survival; few crops (e.g. cotton) are grown for export. The multi-harvest, seasonal crop-rotation system could be favoured by higher temperatures, but water supply could become critical (also in relation to salinization and pollution from fertilizers). The Nile delta coastal zone is almost totally dependent on the Nile for its surface and underground water needs. Variations of rainfall

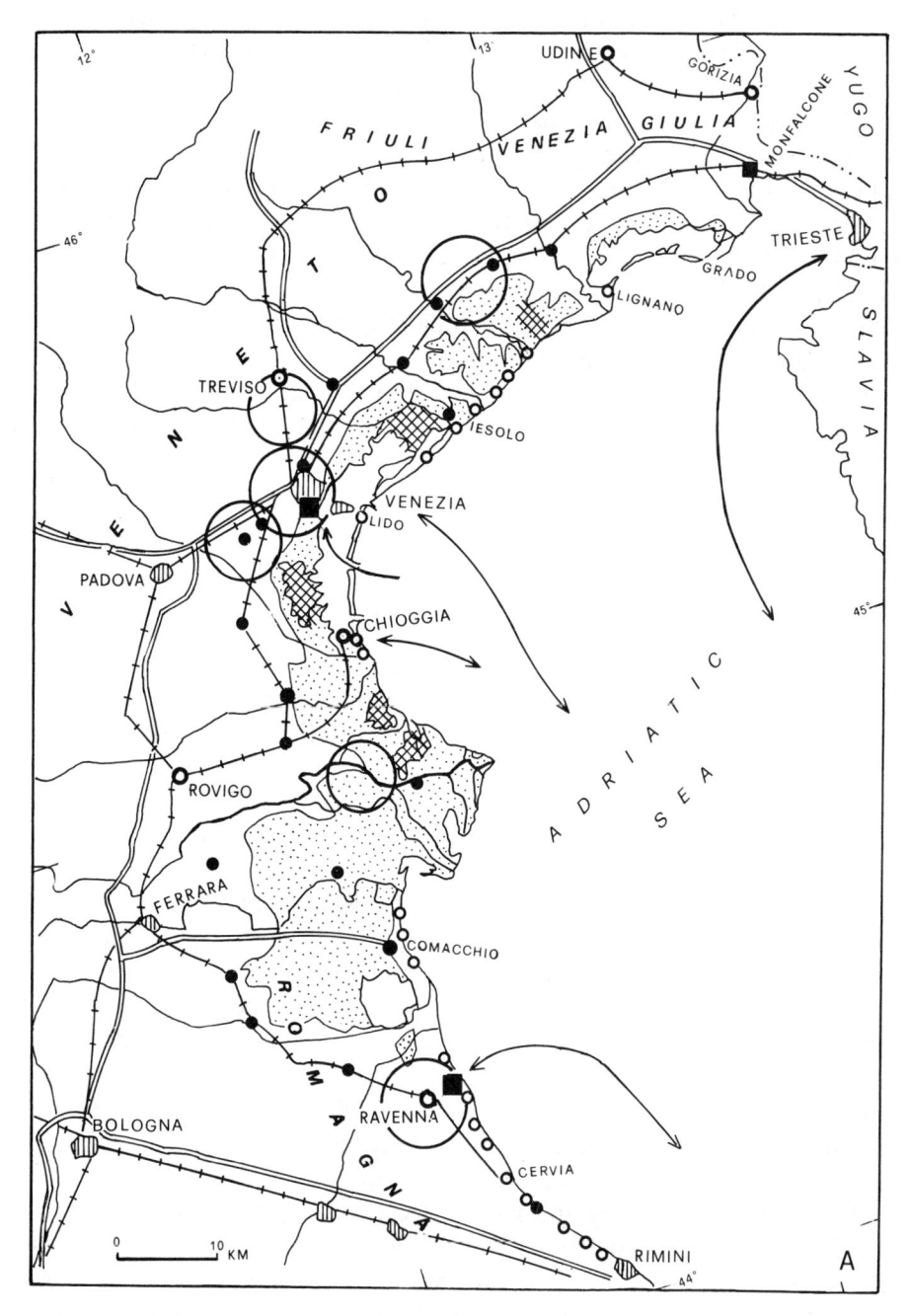

Figures 9.8 The coastal zone of Italy along the north-west Adriatic coast: A, Economic activity; B, Population density

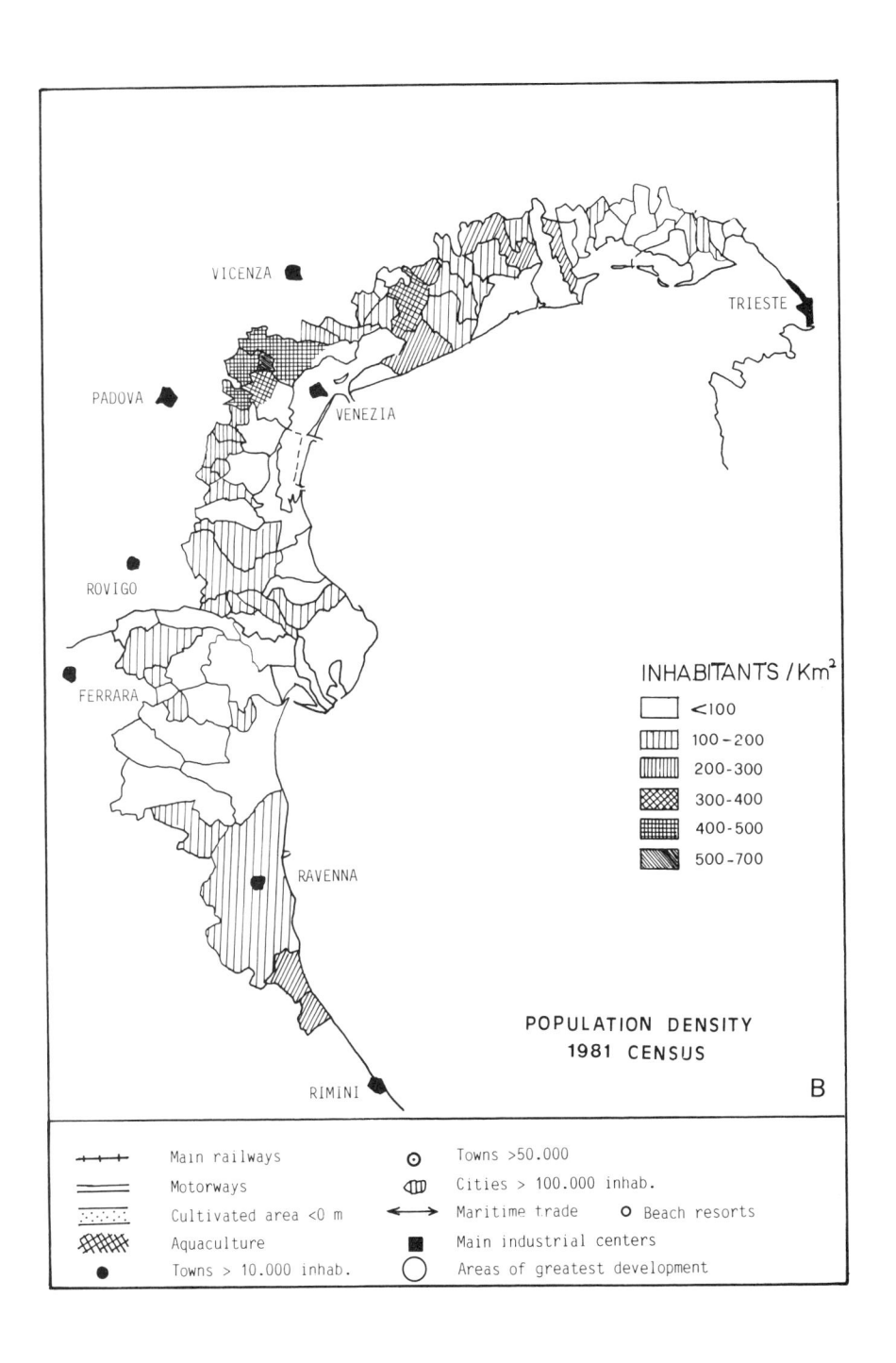

POPULATION DENSITY
1981 CENSUS

INHABITANTS / Km²
- <100
- 100 - 200
- 200 - 300
- 300 - 400
- 400 - 500
- 500 - 700

B

Legend:
- Main railways
- Motorways
- Cultivated area <0 m
- Aquaculture
- Towns > 10.000 inhab.
- Towns >50.000
- Cities > 100.000 inhab.
- Maritime trade
- Main industrial centers
- Areas of greatest development
- Beach resorts

Cities labeled: VICENZA, TRIESTE, PADOVA, VENEZIA, ROVIGO, FERRARA, RAVENNA, RIMINI

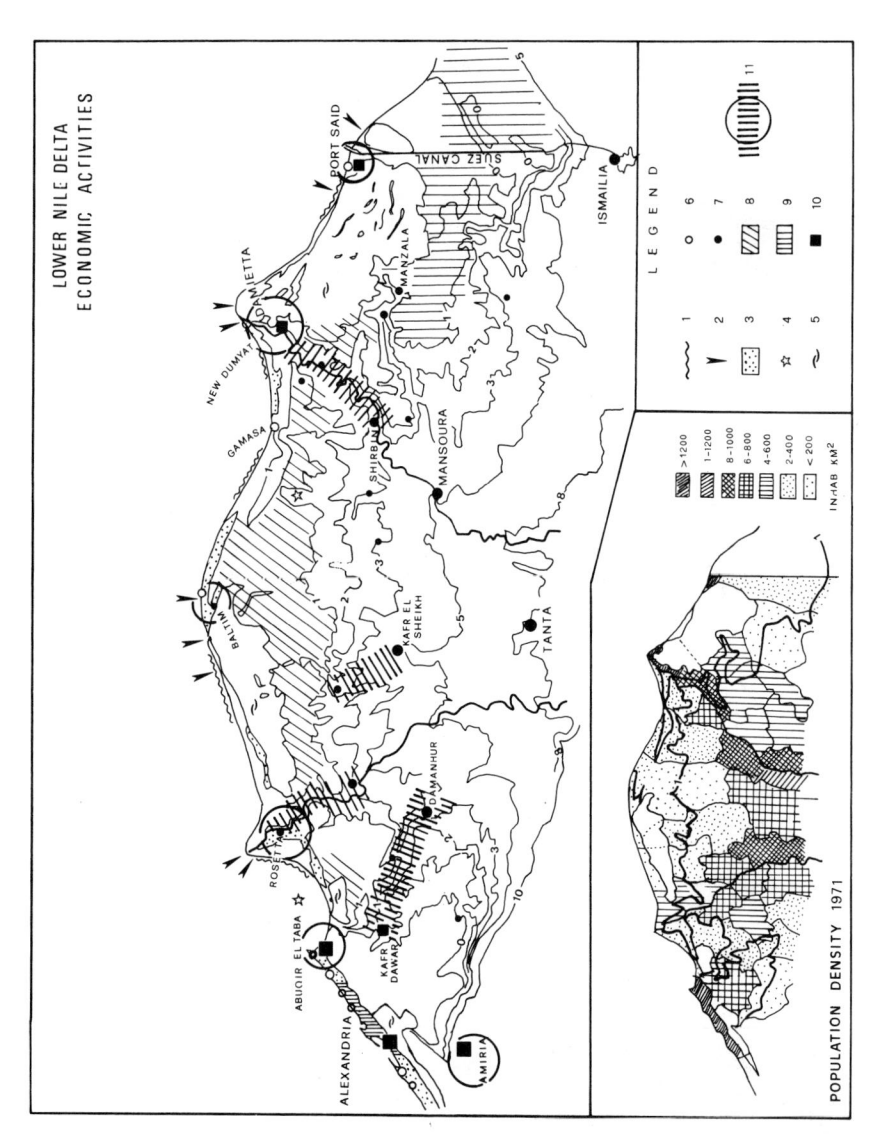

Figure 9.9 Summary of population, present and planned economic activities in the lower Nile delta region: 1, stretches liable to surge flooding; 2, shoreline retreat; 3, dune protection; 4, natural gas fields; 5, fishing (including aquaculture); 6, beach resorts; 7, district capitals; 8, recent land reclamation; 9, present and planned reclamations; 10, industrial centres; 11, poles and axes of future demographic and economic development

in the Nile catchment areas have a profound effect on Egypt's water resources, witness the low water-level situation of Lake Nasser reservoir, in consequence of the 1979–86 period of drought, which has imposed consideration of ways to replace water-intensive crops, like rice and sugar cane, with others like corn and sugar beet.

Lagoonal fishing, especially aquaculture, is a growing resource. If gradually adjusted to physical changes, it should not be unduly affected. Beach tourism in the delta is important at a national level, an essential social commodity, given the overcrowded state of Egyptian cities. The waning of Alexandria's beaches has already resulted in the fast development of the coast west of Agami, as well as of north Sinai.

The most serious negative impacts of temperature and sea-level rise on the Egyptian coast would be on the efficiency of ports and on the management of lagoonal fishing and of lowland agriculture; thus, indirectly, on population centres, which are tied to ports and agriculture-related industries.

Alexandria will lose its attraction as a summer resort city, but the recreational use of beaches is not threatened elsewhere. Coastal developments in the next decades should therefore be carefully controlled and directed to locations that could be most economically and effectively maintained in order to avoid the mistakes made elsewhere, which would eventually lead to considerable disruption and protection expenditure.

8.5 CONCLUSIONS

1 The behaviour of the north Adriatic and Egyptian deltaic coasts during past stages of sea-level rise is no guide to prediction of future coastal behaviour, as generally shorelines have advanced (except where deltaic lobes had been abandoned by channel switching). River sediment inputs were sufficient to keep the coasts in equilibrium with subsidence. This situation has continued, so far, in the deltas of Albania, Greece and Turkey.

2 Most of the deltaic lowlands of the Mediterranean Sea are experiencing more or less serious environmental problems derived from agricultural, industrial, urban and tourist developments during the last two decades, that have been carried out with little regard to the responses of natural systems. Problems range from water pollution and salinization, to land subsidence, shoreline erosion, restriction and deterioration of wildlife habitats.

3 Regarding the effects of temperature rise and changed precipitation patterns, especially on water resources, soils and biosystems, little can be said specifically until a reliable scenario for the doubling of carbon dioxide is available for the central and eastern Mediterranean (and East Africa) that is based on actual topography, has a high spatial and temporal resolution, simulates realistically the patterns of observed climate, and gives information on how the regional weather will be affected by the altered, larger-scale circulation patterns. There is still no definite indication as to whether and how precipitation over the Mediterranean would change. The effects of a temperature rise on agriculture and fishing might generally be positive (though negative for some species). Changes, overall, could be gradual, which would allow time for research and technical adaptations, but the impact of exceptional events and of inter-annual variability could in fact, be more relevant.

4 The effects of sea-level rise are more predictable. They are the direct wave impact on exposed coasts (e.g. the Venice lagoon coastal barrier, beach resorts) and on harbour installations (Alexandria, Port Said, La Golette-Tunis, amongst others), and the consequences of flooding through estuaries, canals, lagoons, more serious for agriculture than for the increasingly more valuable lagoonal fishing.

5 Perspective actions to be taken would be either to protect entire coasts and lagoon margins by special walls or to accept that choices must be made, between irreplaceable coastal uses, such as national and military harbours, towns of historical-artistic value, lagoonal resources or specialized agriculture; and adaptations, such as (1) land uses that can be shifted elsewhere (e.g. industries, roads, airports), and (2) a different approach to beach recreation (i.e. less urbanized), the replacement of extensive, uneconomical crops in lands below sea-level, with lagoons destined to aquaculture and nature reserves. The lagoons would act as a buffer belt, since their inner margins can be more easily protected than the exposed coast.

6 There is above all the necessity to initiate research on all climatically induced changes and to plan and control coastal development well in advance of the postulated sea-level rise, in order to avoid the man-made disequilibria already experienced in many parts, and to take into account the cost-effectiveness of future protection.

REFERENCES

Barth, C. and Titus, J. C. (eds) 1984: *Greenhouse Effect and Sea-level Rise: A Challenge for This Generation*. New York, van Nostrand Reinhold

Beltagy, A. I. 1985: Sequences and consequences of pollution in the northern Egyptian lakes. I Burullus. *Bull. Inst. Oceanography and Fisheries*, ARE 11, 79–97

Ben Menachem, A. 1979: Earthquake catalogue for the Middle East (92 BC–1980 AD). *Boll. Geofisica Teor. Applicata* 21, 245–313

Bevilacqua, E. and Mattana, O. 1976: The Op River basin: water utilization for hydroelectric power and irrigation. In, Italian Contributions to the *23rd International Geographical Congress*, Moscow, USSR. Rome, Consiglio Nazionale delle Ricerche, 181–9

Bolin, B., Döös, B. R., Jäger, J. and Warrick, R. A. (eds) 1986: *The Greenhouse Effect, Climate Change and Ecosystems* (SCOPE 29). Chichester, John Wiley and Sons

Bondesan, M. and Simeoni, C. 1983: Dinamica ed analisi morfologico-statistica dei litorali del delta del Po e delle foci dell'Adige e del Brenta. *Mem. Sci. Geol. Padova* 36, 1–48

Brambati, A. 1984: Erosione e difesa delle spiagge adriatiche. *Boll. Oceanologia Teorica e Applicata* 2, 91–184

Brambati, A. 1987: *Studio sedimentologico e marittimo-costiero dei litorali del Friuli-Venezia Giulia*. Regione Autonoma Friuli Venezia Giulia, Direz. Regionale Lavori Pubblici, 67 pp.

Brambati, A., Marocco, R., Catani, V., Carobene, L. and Lenardon, G. 1978: Stato delle conoscenze dei litorali dell'alto Adriatico e criteri di intervento per la loro difesa. *Mem. Soc. Geol. Italiana* 19, 389–98

Bruun, P. and Schwartz, M. L. 1985: Analytical prediction of beach profile change in response to a sea level rise. *Zeitschr. Geomorphologie*, Suppl. Bd. 57, 33–55

Butzer, K. W. 1976: *Early Hydraulic Civilization in Egypt, a Study in Cultural Ecology*. Chicago, University of Chicago Press

Carbognin, L., Gatto, P. and Marabini, F. 1985a: Correlations between shoreline variations and subsidence in the Po river delta, Italy. *Third Int. Symp. on Land Subsidence*, Publ. IAHS No. 151, 367–73

Carbognin, L., Gatto, P. and Mozzi, G. 1981: La riduzione altimetrica del territorio veneziano. *Inst. Veneto Sci. Lettere Arti, Rapp. Studi* 8, 55–83

Carbognin, L., Gatto, P. and Mozzi, G. 1985: Case history 9.15. Ravenna, Italy. In, Poland, J. F. (ed.) *Guidebook to Studies of Land Subsidence due to Ground Water Withdrawal*. UNESCO, 291–305

Catani, G. Marocco, R., Brambati, A., Carobene, L. and Lenardon, G. 1978: Indagini sulle cause dell'erosione nel tratto orientale di Valle Vecchia (Caorle, Adriatico Sett.). *Mem. Soc. Geol. Ital.* 19, 399–405

Cavaleri, L., Curiotto, S., Dallaporta, G. and Mezzoldi, A. 1981: Directional wave recording in the northern Adriatic Sea. *Il Nuovo Cimento* 4C, 519–32

Cencini, C., Cuccoli, L., Fabbri, P., Montanari, F., Sembeloni, F., Torresani, S. and Varani, L. 1979: *Le spiagge di Romagna: uno spazio da proteggere*. Consiglio Nazionale delle Ricerche, I, Bologna. Progetto Conservazione del Suolo, Quaderno I

Ciabatti, M. 1967: Ricerche sull'evoluzione del delta Padano. *Giornale di Geologia* sr. 2, 34, 26 pp.

CNR (Consiglio Nazionale delle Recerche) 1985: *Atlante delle spiagge italiane* (Scale 1:100,000, sheets 40, 40A, 51, 52, 53, 65, 77, 89, 100, 101)

Coleman, J. M., Roberts, H. H., Murray, S. P. and Salama, M. 1980: Morphology and dynamic sedimentology of the eastern Nile delta shelf. *Marine Geology* 41, 325–39

Corre, J.-J. 1988: Le littoral du Golfe du Lion face à un rechauffement de l'atmosphère dans les décades à venir. Nairobi, UNEP (OCA), Rept WG, 2/3

Coutellier, V. and Stanley, D. J. 1987: Late Quaternary stratigraphy and paleogeography of the eastern Nile Delta, Egypt. *Marine Geology* 27, 257–75

Dalcin, R. 1983: I litorali del delta del Po e delle foci dell'Adige e del Brenta, caratteri tessiturali e dispersione dei sedimenti, cause dell'arretramento e previsioni sull'evoluzione futura. *Boll. Soc. Geol. Ital.* 102, 9–56

El Askary, M. A. and Frihy, O. E. 1986: Depositional phases of Rosetta and Damietta promontories on the Nile Delta coast. *Journ. Afric. Earth Sciences* 5, 627–33

Emery, K. O. 1980: Relative sea-levels from tide-gauge records. *Nat. Acad. Sciences Proc.* 77, 6968–72

Emery, K. O., Aubrey D. G. and Goldsmith V. 1987: Coastal neo-tectonics of the Mediterranean from tide-gauge records. *Marine Geology* 81, 41–52

Erinc, S. 1953: Cukorovanin aluvyal morfolojisi hakkinda. Istanbul Univ. Cografya Enst. Derg 3–4, 149–59

Erol, O. 1983: Historical changes on the coastline of Turkey. In, Fabbri, P. and Bird, E. C. F. (eds) *Coastal Problems in the Mediterranean Area*. Bologna, IGU Comm. Coastal Environments, 95–107

Evans, G. 1971: The recent sedimentation of Turkey and the adjacent Mediterranean and Black Sea. In, Campbell, A. S. (ed.) *The Geology and History of Turkey*. Annual Field Conference 13, Petrol. Explo. Soc. Libya, 305–406

Fabbri, P. C. 1985: Coastline variations in the Po Delta since 2500 BP. *Zeitschr. Geomorph.*, Suppl. Bd. 57, 155–67

Favero, V. and Serandrei Barbero, R. 1978: La sedimentazione olocenica della piana costiera tra Brenta e Adige. *Mem. Soc. Geol. Ital.* 19, 337–43

Fontes, J. C. H. and Bertolami, G. 1973: Subsidence of the Venice area during the past 40,000 years. *Nature* 244, 339–41

Franco, P., Jeftic, L., Melanotte-Rizzoli, P., Michelato, A. and Orlic, M. 1982: Desriptive model of the northern Adriatic. *Oceanologica Acta* 5(3) 379–89

Frihy, O. E. 1988: Nile delta shoreline changes: aerial photographic study of a 28-year period. *Journal of Coastal Research* 41, 597–606

Frihy. O. E., El Fishawi, N. M. and El Askary, M.A. (in press): Geomorphological features of the Nile delta coastal plain, a review. *Journ. Afric. Earth Sciences*

Gatto, P. 1984: Il cordone litorale della laguna di Venezia e le cause del suo degrado. *Ist. Veneto Sci. Lett. Arti., Rapp. Studi.* 9, 163–93

Gatto, P. and Carbognin, L. 1981: The lagoon of Venice: natural environmental trend and man-induced modification. *Hydrological Sciences Bulletin* 26, 379–91

Gazzi, P., Zuffa, G. C., Gandolfi, G. and Paganelli, L. 1973: Provenienza e dispersione litoranea delle sabbie delle spiagge adriatiche tra le foci dell'Isonzo e del Foglia: inquadramento regionale. *Mem. Soc. Geol. Ital.* 12, 1–37

Georgas, D. and Perissoratis, C. 1989: Implications of future climatic changes for the inner Thermaikos Gulf, Greece. Nairobi, UNEP (AOCA), Rept. WG 2/9

Goby, J. E. 1952: Histoire des nivellements de l'Isthme de Suez. *Bull. Soc. Etudes Hist. Géogr. de l'Isthme de Suez* 5, 23–43

Göney, S. 1976: Adana ovalari. *Istanbul Univ. Cografya Enst. Yay.* 88, 178 pp.

Gornitz, V. and Lebedeff, S. 1987: Global sea-level changes during the past century. In, Nummedal, D. et al. (eds) *Sea Level Fluctuations and Coastal Evolution.* (SEPM Spec. Publ. 41), 3–16.

Hamid, S. 1984: Fourier analysis of Nile flood levels. *Geophysical Research Letters* 11, 843–58

Hassan, F. A. 1981: Historical Nile floods and their implication for climatic change. *Science* 212, 1142–5

Hekstra, G. P. 1986: Will climatic changes flood the Netherlands? Effects on agriculture, land-use and wellbeing. *Ambio* 15, 316–26

Hoffman, J.S. 1984: Estimates of future sea level rise. In, Barth, M. C. and Titus, J. G. (eds) *Greenhouse Effect and Sea-level Rise: A Challenge for This Generation.* New York, van Nostrand Reinhold, 79–103

Inman, D. L. and Jenkins, S. A. 1984: The Nile littoral cell and Man's impact on the coastal littoral zone in the SE Mediterranean. Sydney, *Proc. 17th Int. Coastal Eng. Conf., ASCE*, 1600–17

Khafagy, A. and Manohar, M. 1979: Coastal protection of the Nile delta. *Nature and Resources* 15, 7–13

Khashef, A. A. 1983: Saltwater intrusion in the Nile delta. *Groundwater* 21, 160–7

Lamb, H. H. 1982: *Climate, History and Modern World.* London, Methuen

L'Homer, A., Bazile, E., Thommeret, J. and Thommeret, L. 1981: Principales étapes de l'édification du delta du Rhône de 7000 BP à nos jours; variations du niveau marin. *Oceanis* 7(4), 389–408

Manohar, M. 1981: Coastal process at the Nile Delta coast. *Shore and Beach* 49, 8–15

Marino, M. G. 1988: Climatic change implications for the Ebro Delta. Nairobi, UNEP (OCA), Rept WG 2/3

Meininger, P. L. and Mullie, W. C. 1981: *The significance of Egyptian wetland for wintering waterbirds*. New York, The Holy Land Conservation Fund, 110 pp.

Paskoff, R. 1985a: Tunisia. In, Bird, E. C. F. and Schwartz, M. L. (eds) *The World's Coastline*, New York, van Nostrand Reinhold, 523–8

Paskoff, R. 1985b: Les côtes de l'Albanie. Aspects géomorphologiques. *Bull. Assoc. Géogr. Franç.* 2, 77–83

Peltier, W. R. and Tushingham, A. M. 1989: Global sea level rise and the greenhouse effect. *Science* 244, 806–10

Pirazzoli, P. A. 1983: Flooding ('Acqua Alta') in Venice (Italy): a worsening phenomenon. In, Bird, E. C. F. and Fabbri, P. C. (eds) Coastal problems in the Mediterranean Sea. Proceedings of a Symposium, Venice, 10–14 May 1982. Bologna: International Geographical Union, Commission on the Coastal Environment, 23–31

Pirazzoli, P. A. 1987: Sea level changes in the Mediterranean. In, Tooley, M. J. and Shennan, I. (eds) *Sea-level Changes*. Oxford, Basil Blackwell, 152–81

Raper, S. C. B., Warrick, R. A. and Wigley, T. M. L. (in press): Global sea level rise. In, Milliman, J. D. (ed.) *Rising Sea Level and Subsiding Coastal Areas* (Scope, Bangkok Seminar 1988). Chichester, J. Wiley and Sons

Robin, G. de Q. 1986: Changing the sea level. In, Bolin, B. et al. (eds) *The Greenhouse Effect, Climatic Change and Ecosystems* (Scope 29), Chichester, John Wiley and Sons, 323–59

Rossignol-Strick, M. 1983: African monsoons, an immediate climate response to orbital insolation. *Nature* 304, 46–9

Said, R. 1981: *The River Nile*. Berlin, Springer Verlag

Sestini, A. 1940: Le pianure costiere dell'Albania. *Boll. Reale Societa Geogr. Ital.* sr. 7, 5, 513–27

Sestini, G. 1988: Nile Delta: a review of depositional environments and geological history. In, Whateley, M. K. G. and Pickering, K. T. (eds) *Deltas: Sites and Traps for Fossil Fuels*. (Geol. Soc. London, Spec. Publ. 41), Oxford, Blackwell Scientific Publications, 99–127

Sharaf El Din, S. H. and Moursy, Z. A. 1977: Tide and storm surges on the Egyptian Mediterranean coast. *Rapp. Comm. Int. Mer. Medit.* 24, 33–7

Shuisky, Y. D. 1985: Albania. In, Bird, E. C .F. and Schwartz, M. L. (eds) *The World's Coastline*. New York, van Nostrand Reinhold, 443–4

Sivignon, M. 1987: La mise en valeur du delta de l'Axios. In, Bethemont, J. and Villain-Gandossi, C. (eds) *Les Deltas méditerranéens*. Vienna, Centre Europ. Coord. Rech. Sciences Sociales, 279–99

Smith, S. E. and Abdelkader, A. 1988: Coastal erosion along the Egyptian Nile Delta. *J. Coastal Research* 4(2), 245–55

Sneh, A., Weissbrod, T. and Ehrich, A. 1986: Holocene evolution of the northeastern corner of the Nile Delta. *Quaternary Research* 26, 194–206

Sorribes, J. and Grau, J.-J. 1987: El delta del Ebro: una vision dew conjunto. In, Bethemont, J. and Villain-Gandossi, C. (eds) *Les Deltas méditerranéens*. Vienna, Centre Europ. Coord. Rech. Sciences Sociales, 179–210

Stanley, D. J. 1988: Subsidence in the northeastern Nile Delta: rapid rates, possible causes and consequences. *Science* 240, 497–500

Stournaras, G. and Marcoupoulu-Diacantoni, A. 1985: Les dépôts Plio-Pleistocènes deltaïques du Nestos (Grèce). *Rapp. Comm. Int. Mer. Médit.* 29(2), 175–82

Summerhayes, C. P., Sestini, G., Misdorp, R. and Marks, N. 1978: Nile delta, nature and evolution of continental shelf sediments. *Marine Geology* 27, 43–65

Tousson, O. 1934: Mémoire sur les anciennes branches du Nil. *Mém. de la Soc. Géogr. d'Egypte* (Cairo) 4, 144 pp.

UNDP–UNESCO 1978: *Arab Republic of Egypt. Coastal Protection Studies. Final Technical Report*. Paris, UNESCO, 2, 206–483.

UNEP/WMO/ICSO 1986: *International Assessment of the Role of Carbon Dioxide and Other Greenhouse Gases in Climate Variations and Associated Impacts*. Nairobi, UNEP

United Nations 1982: *Demographic Handbook*. New York

Veggiani, A. 1974: Le variazioni idrografiche del basso corso del Fiume Po negli ultimi 3000 anni. *Padusa* (Rovigo) 1–2, 1–22

Venzo, G. A. and Brambati, A. 1968. Evoluzione e difesa delle coste dell'alto Adriatico da Venezia a Trieste. *Riv. Italiana Geotecnica* 3, 1–18

Wigley, T. M. L. 1988: Future climate of the Mediterranean basin, with particular emphasis on changes in precipitation. Nairobi, UNEP (OCA), Rept WG 2/6

Zunica, M. 1971: *Le spiagge del Veneto*. CNR, Centro di Studi di Geografia Fisica, Univ. Padova, 144 pp.

Zunica, M. 1976: Coastal changes in Italy during the past century. In, Italian Contributions to the *23rd International Geographical Congress*, Moscow, USSR. Rome, Consiglio Nazionale delle Ricerche, 275–81

Zunica, M. 1978: Il delta del Po: elementi per un approccio ecologico. *Boll. Mus. Civico Storia Naturale* (Venice) 24, 19–30

10

The Evolution of the Coastal Lowlands of Huelva and Cadiz (South-west Spain) during the Holocene

C. Zazo, C. J. Dabrio and J. L. Goy

10.1 INTRODUCTION

The Gulf of Cadiz opens towards the south-west limiting the flat lowlands and hilly countryside of Andalusia, made up of glacis and alluvial fan sediments, from the Atlantic Ocean. The morphology of the coasts of south Europe and north-western Africa greatly reduces the effective fetch of Atlantic gales and only wave fronts and surges moving towards the north-east are able to reach the shore (figure 10.1). Daily and prevailing winds blowing from the sea also generate wave fronts which progress roughly in the same direction. As most of the wave fronts approach the coast obliquely, active longshore transport takes place triggering a prominent littoral drift which is directed towards the east and south-east on the Spanish side of the Gulf of Cadiz. Besides direct observation and measurements well-developed spit bars along the coast bear witness to the direction of longshore drift (figure 10.1).

The coast of the Gulf of Cadiz can be described as a semidiurnal, mesotidal coast with mean tidal ranges of 2.1 m and a variation up to 0.45 m between successive flood or ebb tides (Borrego and Pendon 1989). Wave energy is medium because 75 per cent of waves do not exceed 0.5 m in

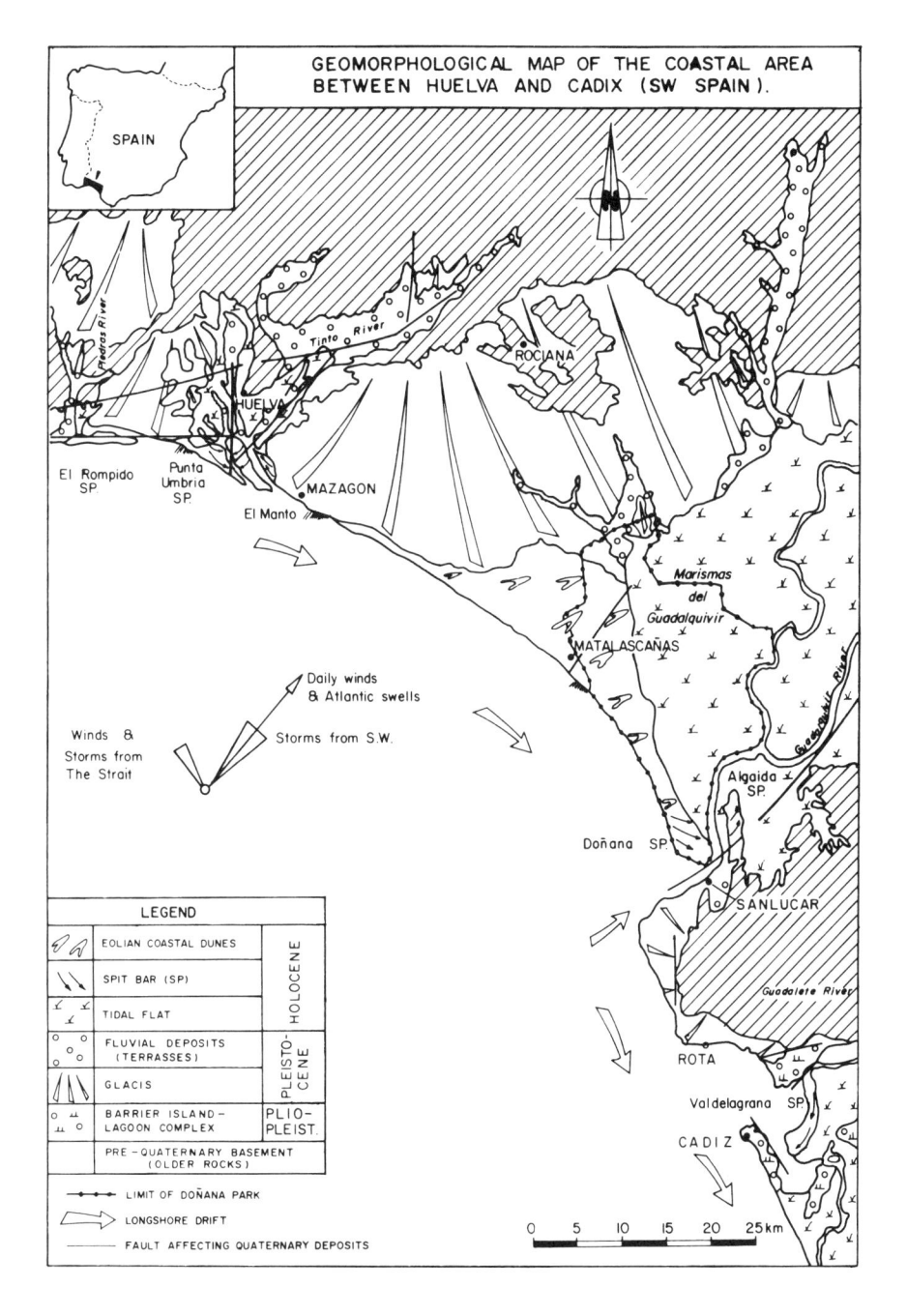

Figure 10.1 Geomorphological map of the coastal area between Huelva and Cadiz (south-west Spain)

height. These conditions favour the development of broad littoral lowlands, usually sheltered by fast-growing spit bars, where fresh water marshes and/or tidal flats extend several kilometres inland (figure 10.1). The most impressive of these marsh-tidal flat complexes is located in the lower reaches of the River Guadalquivir (Menanteau 1982; Vanney et al. 1985) and consequently named Marismas del Guadalquivir. The marshes occupy 1400 km^2, and have been much altered since the 1930s by human activities (Villalon 1987). A large part of these *marismas* (the Spanish name for tidal flats and marshes) are officially protected as the Doñana National Park (boundary shown on figures 10.1 and 10.5) because it is considered a wildlife sanctuary of vital importance for north European birds to nest during the winter.

On the other hand, much economic profit is obtained also from salt pans installed in some tidal flats of the Gulf of Cadiz, from fishing, cattle farming, summer resorts and active harbours. In other words, as in many other countries, there are many reasons of an ecological and environmental nature for being interested in the coastal lowlands.

The aim of this chapter is to present some of the major changes along the coast, considering their causes and the processes involved, to indicate some areas of coastal progradation and retreat (figure 10.2), and to outline the effects of a rise of sea-level of 1 m above the geographic datum used in the Spanish maps (mean level of the Mediterranean Sea in Alicante).

10.2 MORPHOSEDIMENTARY CONTROLS

The complex pattern of sedimentary environments and ecosystems recognized along the coast results from the intricate interaction of coastal dynamics, availability of sediment (there are no definite data about sediment supply by rivers but it is likely that most of the sediment transported is presently retained by dams), sea-level changes, recent tectonism and the multiple, but often subtle, direct and indirect human impacts.

Changes of sea level during the Quaternary caused major palaeogeographic changes due to the dramatic displacement of the shore either towards the land or the sea. However, as far as we know, no written information concerning minor fluctuations of relative sea-level in recent times along the Gulf of Cadiz has been published, and we also lack the detailed maps necessary for comparison. Evidence of changes of coastal morphology in this particular area is common but only the largest ones have been recorded in maps, as they would create problems for navigation. As expected, the

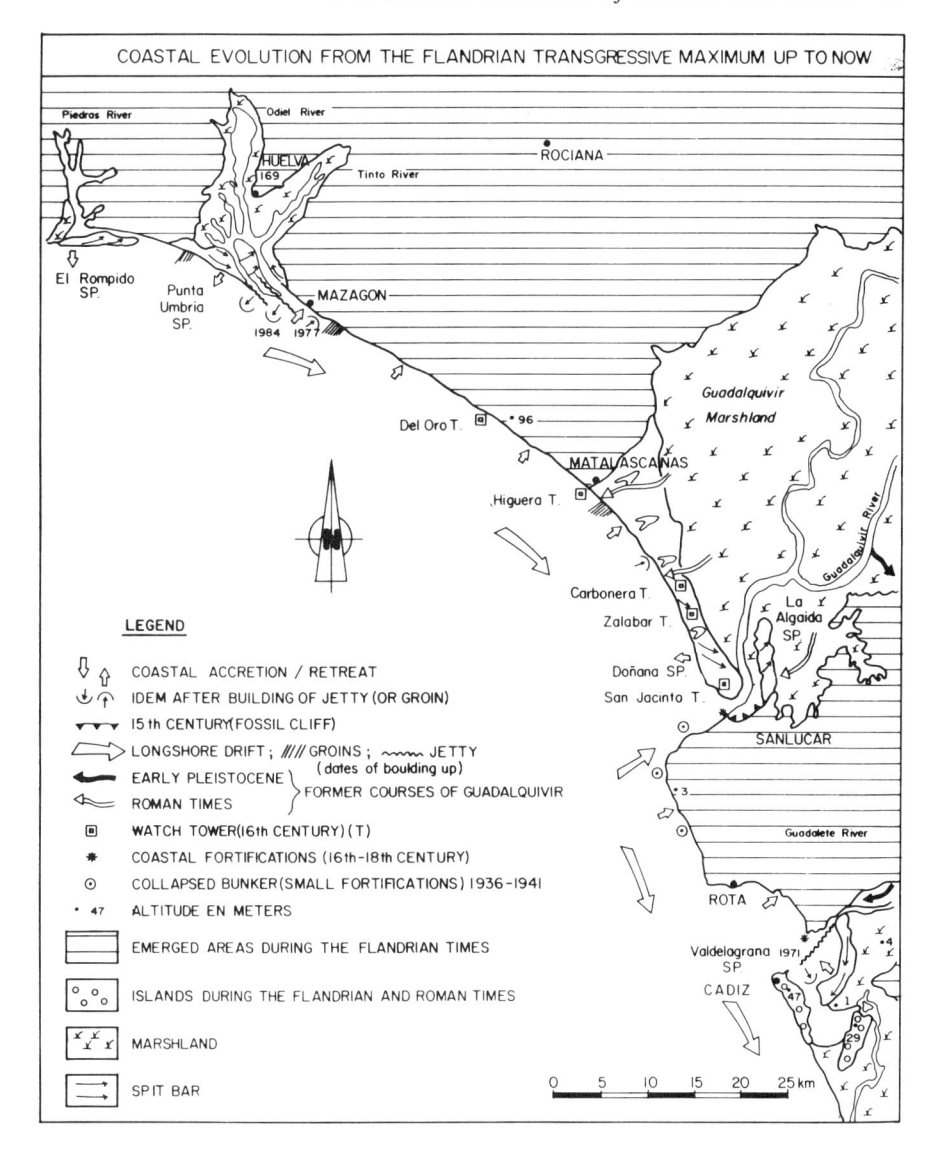

Figure 10.2 Coastal evolution from the Flandrian transgression maximum up to the present, showing the pattern of coastal advance or retreat

most dramatic changes occurred in spits and barrier islands sheltering tidal flats from the sea between the River Guadiana and the Odiel-Tinto estuary (figure 10.3).

The morphosedimentary evolution of the area of the Gulf of Cadiz (Zazo et al. 1986) since the Middle Holocene has followed three main steps:

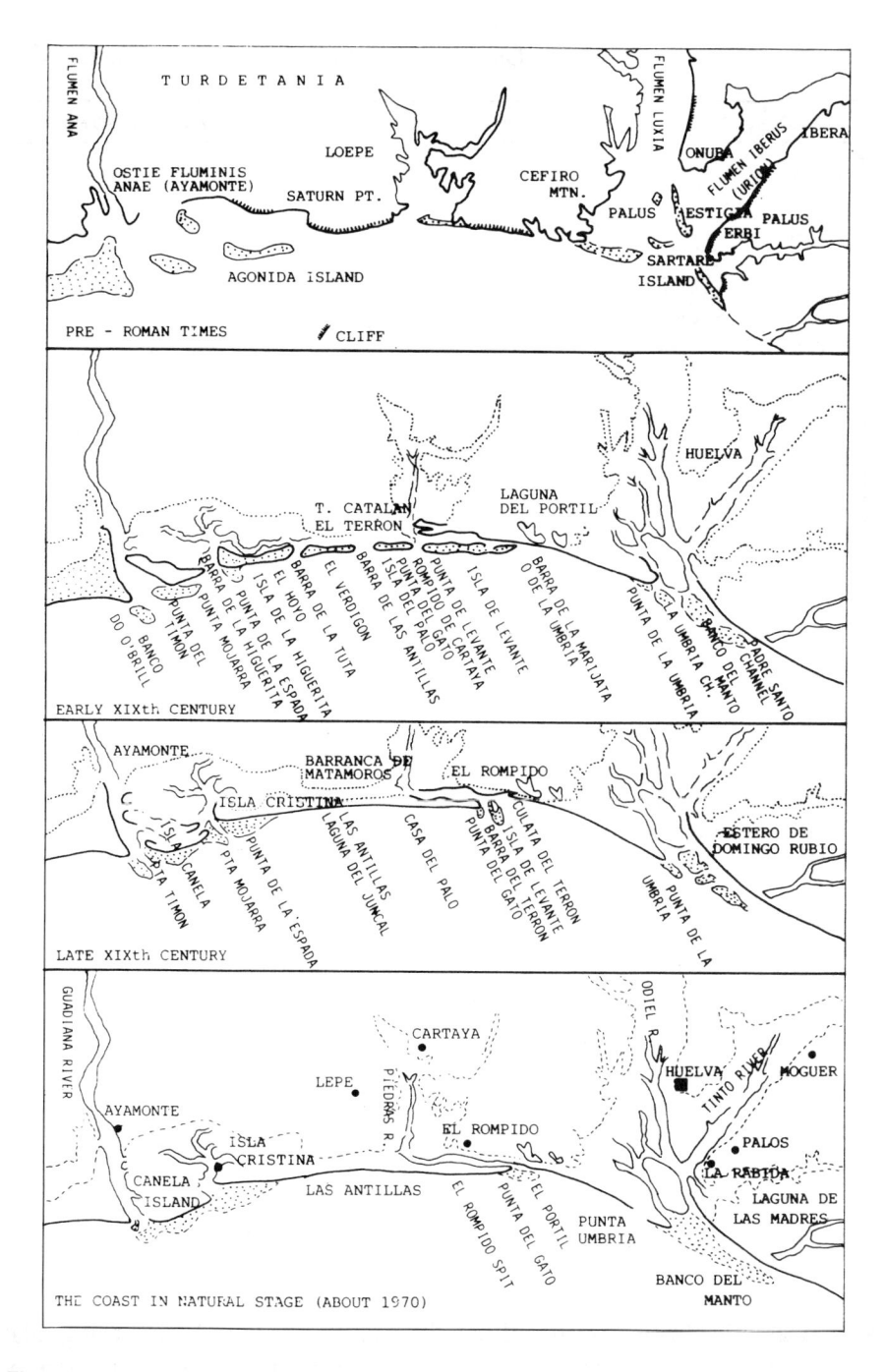

Figure 10.3 Changes of coastal morphology between the Guadiana river and the estuary of Huelva since Early Holocene and pre-Roman times to the early 1970s before the massive coastal works.
Source: (Dabrio and Polo 1987)

1   After the post-glacial sea-level rise (the maximum in Spain occurred about 7500 BP) the lower reaches of the rivers (Guadiana, Piedras, Odiel-Tinto, Guadalquivir and Guadalete) became estuaries that were partly blocked by barrier islands and filled with sediments forming tidal flats (Dabrio and Polo 1987). Although there is some subsidence in the lower Guadalquivir River area (Menanteau 1982) no delta was formed;

2   Around 5000 BP the fall of sea-level favoured the progradation of the shoreline as suggested by carbon-14 dating in Punta Umbria and Guadalquivir spits (Gabinete ANOP 1971). In pre-Roman times, Himilco, a Carthaginian who sailed along the coast about 2500 BP, described large estuaries with some islands at the mouth of the rivers but a great part of the shore was occupied by cliffs (figure 10.3). It is likely that progressive colonization and deforestation of southern Spain produced an enormous input of sediment which favoured the development and the progradation of sandy barrier islands and beaches attached to the former coastal cliffs that turned into fossil features (figure 10.3). Such changes took place very rapidly, in particular after the intensive deforestation of the Iberian peninsula in the Middle Ages and subsequently, as shown for the Mediterranean estuaries by Hoffman and Schulz (1987; an example from the Ebro delta is given in chapter 6). As said before, no data are available to reveal if minor sea-level fluctuations were involved in these changes.

Archaeological and historical data reveal the rapid progradation, but also the erosion, of this coastal segment (figure 10.2): Roman harbours filled with sand, coastal watch towers built by the end of the sixteenth century that have been separated from the sea (170 m to the so-called Torre de Zalabar and 600 m to the Torre de San Jacinto) or lie ruined on the foreshore (Torre del Oro, Torre de la Higuera), artillery bunkers dating from the late 1930s presently overturned in the sea after erosion of the coastal cliffs upon which they were built;

3   In the last century prominent changes of coastal dynamics and morphology caused by human impacts have occurred. They can be illustrated for instance in three basic study cases that may serve as models of many other situations:

   1   the 'free', natural growth of El Rompido spit and related tidal flats;
   2   the modified growth of Punta Umbria spit and the erosion it caused downdrift;
   3   the aborted spit of the Guadiana River.

El Rompido spit and related tidal flats

Longshore drift induced the deposition of a spit bar attached to the western margin of the Rio Piedras estuary, forcing the channel to turn eastward (downdrift) for some kilometres before entering the sea. In this way a sound parallel to the coastline was formed. This study case is considered a suitable model of the natural processes involved in the development and growth of such coastal features under largely natural (not directly man-induced) conditions.

The emerged spit grew parallel to the mainland under the combined effects of waves and tides (figure 10.4) and it is preceded by a shoal complex corresponding to the spit platform (Meistrell 1972). The oblique incidence of wave fronts induced longshore transport involving migration of ridge and runnel systems towards the upper part of the shoreface until they form berms that lie parallel to the shore. In the distal end of the spit, berms bend inland under the combined action of refracted waves and (flood) tidal currents. Tidal currents tend to produce morphologic features oriented normal to the shoreline (figure 10.4(1)). The largest part of the flood current flow is restricted to channels because the water level is too low for them to surpass the shoals that partly block the entrance to the sound. Ebb currents flow during high-water stages when the shoals are basically submerged: consequently they are free to sweep the top surface of the shoals until the water level drops below the mean elevation of the shoals. Reflected waves and also small waves piled up by winds blowing from the land (north) generate small, steep berms in the inland side of the spit closing in the process the swales sheltered behind the curved beach ridges. Strong tidal currents along the channel-shaped sound cut a rather steep channel profile.

Waves accumulate swash bars at the seaward side of the shoals. These bars migrate towards the land until they merge into the shoals placed closer to the spit's end sheltering the formerly active swales (figure 10.4(2)). The new hydrodynamic pattern greatly modifies the original morphology of the shoal: the sheltered swales are subsequently filled up with sediment brought by overwash and/or are colonized by sea grass, clams and fiddler crabs until they develop into tidal flats (figure 10.4(3)).

Morphosedimentological analysis of the systems of beach ridges, swales and shoals along the whole spit revealed three distinct configurations that have been interpreted as the result of differences in the availability of sediment for waves and tides through time (Dabrio et al. 1986). The western sector was formed around 1860–75 under apparent tidal supremacy interpreted as the result of the sudden increase of volume of water forced

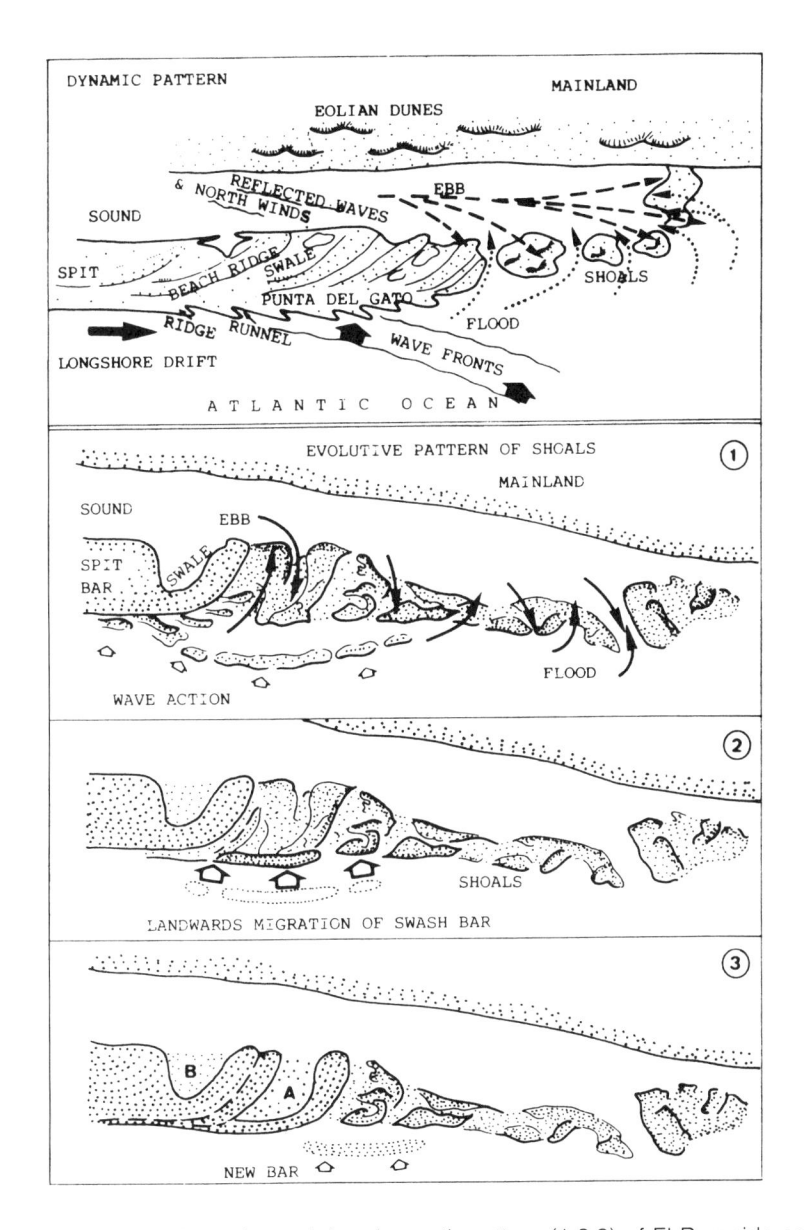

Figure 10.4 Hydrodynamic model and growth pattern (1,2,3) of El Rompido spit bar (see text for explanation): (A) swale filled up with sediment brought by overwash; (B) swale being converted into a tidal flat.
Source: (After Dabrio, Boersma and Fernandez 1986)

through the easternmost inlet of the sound after the closing of (at least) two tidal inlets located up-drift. The central sector grew rapidly (between the 1870s and 1960s) under apparent control of wave action interpreted as the result of the erosion and mobilization alongshore of the sediment previously retained in the ebb tidal deltas off the two tidal inlets that had recently been closed. The distal, eastern, sector (formed after 1960) displays again features suggesting tidal dominance; it is thought to be related to the easy mobilization by tidal currents of the sediment set into suspension by the screws of fishing ships which remove the top of the spit platform during high tide periods. Consequently, the eastern sector experiences the effects of human actions in recent times (after the early 1970s) which are not always noted but are recorded in the morphology of the spit and also in its internal structure.

A preliminary calculation of the volume of sand trapped in the spit bar during its progradation in the last century yields a figure of about 4×10^6 m^3, that is to say, an average of some 4×10^4 m^3 y^{-1}, but the amount of sand removed along the spit must necessarily be higher.

The modified growth of the Punta Umbria spit bar and the resulting downdrift erosion

The large spit of Punta Umbria extends in a south-easterly direction almost closing the estuary of the rivers Odiel and Tinto (figures 10.1 and 10.3). It is well known nowadays that inlets and estuaries trap sediment in tidal deltas and bars reducing, or completely cutting, transport along the coast. However a large part of the sediment slowly bypasses the obstacle and longshore transport is progressively reassumed. Probably that was the case of the spit bar (the so-called Punta Umbria spit) attached to the western margin of the Odiel-Tinto estuary.

In this case, the growth of the spit in its natural stages lasted until the 1860s, when systematic dredging of the harbour of Huelva (placed between the two rivers forming the estuary) started. Unfortunately most of the sand extracted was dumped roughly on top of (or some distance off) the shoals at the end of the spit (the so-called El Manto shoal). Subsequent storm surges and tidal currents again introduced the sand into the estuary (as they did also in the case of El Rompido). Dumping of sediment in the wrong place solved only in part the problem of maintenance of draught, but apparently it also broke the natural bypass of sediment and increased the erosion eastwards (downdrift) of the estuary (figure 10.2). About the late 1960s, the increasing size of ships demanded deeper waters in the channels and also required largely improved safety against swells. As a result a huge jetty, some 25 km long, was built in 1977 extending parallel to the main coast far beyond the outer distal margin of the spit platform. The jetty blocked the longshore drift trapping large amounts of sand. The deficit of sand was clearly visible in the flat beach profiles with poorly developed berms and small (or absent) ridge and runnel systems. The lack of sand magnified the erosional effects of the oblique wave fronts for about 45 km towards the south-east (figure 10.2) and a series of groynes was erected to prevent further degradation. A few kilometres south-east from Matalascañas the longshore drift regains its normal values producing beach progradation at the distal parts of the Doñana Spit.

The highly polluted waters of the Odiel-Tinto estuary form a plume that flows parallel to the shoreline enhancing the degradation of the coastal zones. The plume moves under the driving force of the ebb currents reaching in its way several active summer resorts (which already suffer from coastal erosion) and, still more downcurrent, the coastal side of Doñana National Park.

The aborted spit of the River Guadiana

In its natural stages, a spit bar tended to grow attached to the western side of the estuary of the River Guadiana (the one separating Spain and Portugal) in a way similar to the preceding cases. However, such a development is not desirable from a political point of view owing to the need of keeping stable borders between the two countries and well-defined offshore home waters. In the case described, political reasons prevailed and a jetty was built to prevent any spit growth on the Portuguese side of the river mouth. Later on other jetties and groynes were laid down to guarantee navigability at the entrances of fishing marinas. The combined effect of all these coastal works has been to trap much sediment, reducing the supply of sediment to the regional coastal drift necessary for the development of healthy beach profiles. An appreciable reduction of the sand supply is felt in the El Rompido Spit placed some kilometres towards the east.

Other examples

Examples of erosion and progradation are also known south of the Guadalquivir River mouth. They are visible mostly in damaged or ruined castles and bunkers. The most recent example of erosion, well inside the bay of Cadiz, affects the Vadelagrana spit through the drastic reduction of sand supply caused by new jetties built at the mouth of the River Guadalete (figure 10.2). Further degradation of the spit will follow the massive building of tourist resorts at present.

10.3 POSSIBLE EFFECTS OF A RISE OF SEA-LEVEL

A relative rise of sea level of 0.5 m to 1.0 m by the end of the next century, as specified at the Noordwijkerhout Workshop (see chapter 11 and Kwadijk and de Boois 1989), may generate a varied series of consequences:

1. Increased risk of inundation of lowlands. This means the loss of a large part of Doñana National Park (figure 10.5) much of which lies below 2 m, as well as direct and indirect damage to two cities (Huelva and Cadiz), some ten villages (for instance Rota) and numerous industrial plants and factories, salt pans, rice farms and summer resorts that lie below the +5 metre contour line;
2. Damage or loss of bridges and structures built for coastal and shore protection;
3. Loss of beaches and increase of erosion. A rapid rise of sea-level would magnify the negative impacts whereas a fairly gentle rise would permit better adaptation and perhaps give time to take advantage of the changing conditions by means of well-planned coastal engineering;
4. Modifications of the drainage patterns and probable migration upriver of the loci of sedimentation. In our opinion, the effect is to create problems for navigation in the lower reaches of the rivers (mainly the River Guadalquivir) because of sedimentation of sand bars and shoals requiring dredging;
5. Migration towards the mainland of the sea water/fresh water interface and degradation of coastal aquifers. The consequences are to be felt mostly in lowland areas which depend heavily on fresh water to survive: Doñana National Park and fresh water lakes (for example, El Portil, Laguna Madre). It would also affect the rich agriculture systems of the Marismas del Guadalquivir and human water supply (although no precise data of economic value are available);
6. Disruption and probable loss of coastal ecosystems and large areas of wildlife reserves, mainly those based on fresh water lakes.

The possible strategies against such major events are as simple as they are expensive:

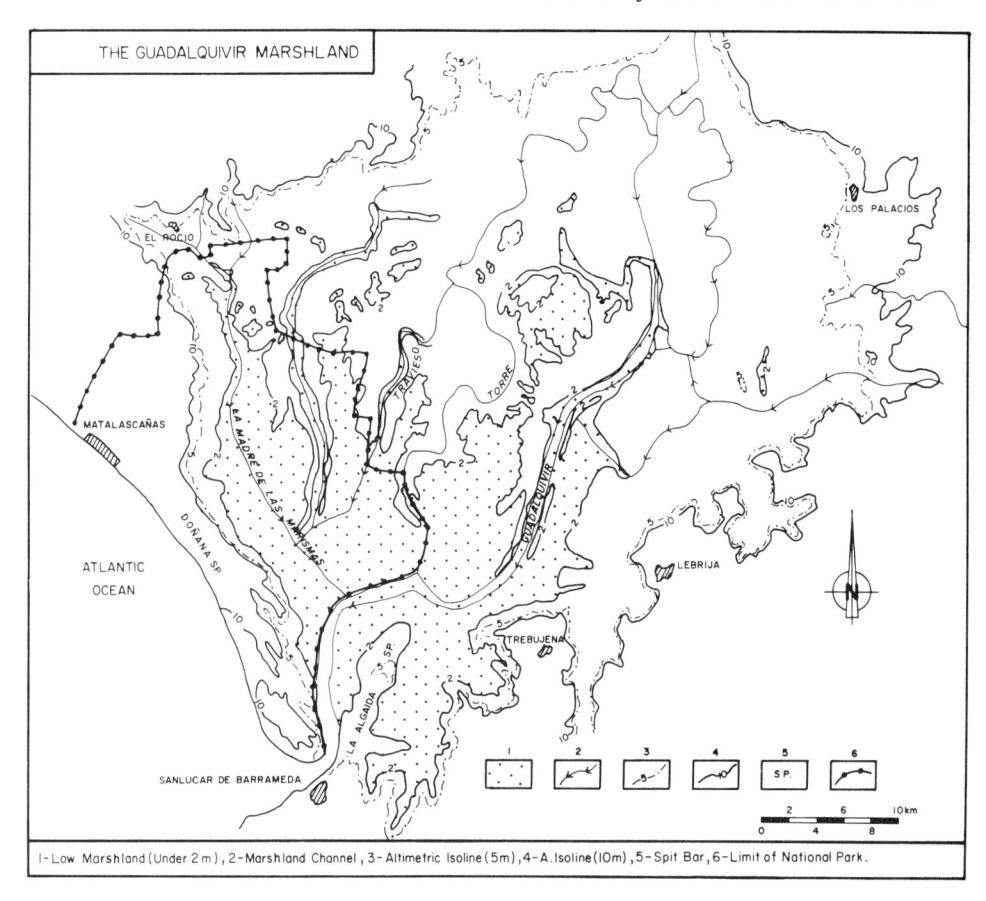

Figure 10.5 The Guadalquivir marshland

1 To begin a losing battle directed to lowland protection by means of complex coastal works. This is not always possible because of negative environmental impacts and economic constraints.
2 To move human settlements and industrial concentrations some distance inland, taking into account the huge economic and social consequences.

In any case, it is obvious that more detailed studies of coastal dynamics and morphosedimentology coupled with a precise dating of events are required to get a clear idea of the precise evolution of coastal systems and the main controls to which they are subjected. The authors feel that now is the proper time to start planning future investments of money and social effort in the littoral lowlands of the Gulf of Cadiz.

ACKNOWLEDGEMENTS

This paper is part of a wider research programme with financial support from the Spanish CICYT Project PR 83–2460 and DGICYT Project PB 88–0125. C.J.D. enjoyed financial support from CSIC Programme 630/070.

REFERENCES

Borrego, J. and Pendon, J. G. 1989: Caracterización del ciclo mareal en la desembocadura del Rio Piedras (Huelva). *Comunicaciones XII Congreso Español de Sedimentología*, 97–100

Dabrio, C. J., Boersma, J. R. and Fernandez, J. 1986: Evolución sedimentaria de la flecha del Rompido. *Actas IX Congreso Nacional de Sedimentología*, 1, 329–41

Dabrio, C. J. and Polo, M. D. 1987: Holocene sea-level changes, coastal dynamics and human impacts in southern Iberian Peninsula. In, Zazo, C. (ed.) *Late Quaternary Sea-level Changes in Spain*. Museo Nal. Ciencias Naturales, CSIC, 227–47

Gabinete de Aplicaciones Nucleares a las Obras Publicas, 1971: Determinación de la eded, mediante Carbono 14 en muestras de conchas procedentes de los litorales de Huelva y Valencia (unpublished internal report, Madrid)

Hoffman, G. and Schulz, H.D. 1987: Holocene stratigraphy and changing coastlines at the Mediterranean coasts of Andalucia (SE-Spain). In, Zazo, C. (ed.) *Late Quaternary Sea-level Changes in Spain*. Museo Nal. Ciencias Naturales, CSIC, 153–8

Kwadijk, J. and Boois, H. de (eds) 1989: *Final Report. European Workshop on Interrelated Bioclimatic and Land-use Changes*, 17–21 October 1987. Bilthoven, National Institute of Public Health and Environmental Protection

Meistrell, F. J. 1972: The spit-platform concept: laboratory observation of spit development. In, Schwartz, M. I. (ed.) *Spits and Bars*. Stroundsburg, Penn., Dowden, Hutchinson and Ross, 225–83

Menanteau, L. 1982: Les marismas du Guadalquivir. Example de transformation d'un paysage alluvial au cours du Quaternaire récent (Thèse 3ème cycle, Université de Paris-Sorbonne)

Vanney, B. R., Menanteau, L., Zazo, C. and Goy, J. L. 1985: Mapa fisiográfico del litoral atlántico de Andalucia, E: 1/50,000, M. F. O2 – Punta Umbría–Matalascañas; M. F. O3 – Matalascañas–Chipiona. Seville, Edit. Serv. Pub. y BOJA

Villalon, J. C. 1987: Intervenciones humanas en el estuario del Guadalquivir. In, Bethemont, J. and Villain-Gandossi, C. (eds) *Les Deltas méditerranéens*. Vienna,

Centre Européen de Coordination de Recherche et de Documentation en Sciences Sociales, 59–106

Zazo, C., Goy, J. L. and Dabrio, C.J. 1986: Later Quaternary and recent evolution of coastal morphology of the Gulf of Cadiz (Huelva–Cadiz), southwestern Spain. *First Int. Symp. on Harbours, Port Cities and Coastal Topography*, Haifa, Israel

11

The Future of the European Coastal Lowlands

M. J. Tooley and S. Jelgersma

11.1 INTRODUCTION

There was a consensus amongst those attending and giving papers at the session on 'Impact of a future rise in the sea-level on the European coastal lowlands' at Noordwijkerhout in 1987 that these areas were already at risk and that any rise in sea-level due to the enhanced greenhouse effect would exacerbate an already serious situation. The fact that these areas were already at risk arose in part from the geology and in part from man's activities. Geological history has endowed the coastal lowlands with their distinctive sediments, landforms and resources. Man has utilized these resources since the mesolithic period, and progressively more intense and enlarged use has been made of them since Roman times. Through reclamation, agriculture, ground water and river management, settlement and industrialization, irreversible changes have been wrought in the coastal lowlands. These comprehend the lowering of ground and water levels, the advance of the salt water front and salinization especially in Mediterranean Europe, a massive sediment supply to beaches by way of river catchments followed by interruption as water resources were managed, the destruction of natural coastal protection such as salt marshes and lagoons and capital-intensive developments and investments in industry, infrastructure, residential property and recreational facilities.

The sediments and landforms of the coastal lowlands have been revealed as the result of the economic and social activities of man. These sediments and landforms are a natural archive of environmental change, preserving a record of climatic and sea-level changes, as well as registering changes in the drainage basins that feed them. They contain a record, waiting to be deciphered, of great storms and the impacts of surges. They have registered episodic, catastrophic events to which the shoreline and lowland have adjusted: what Fairbridge (1989) refers to as 'crescendo events in sea-level changes' clearly need investigation, and these, together with the sea-level, temperature and carbon dioxide signals from the past 140,000 years provide a context for present land-/sea-level relationships, rates of land and sea-level changes at different spatial scales and the prognostications that have seized the scientific and political imagination. The step from the world of imagination to the world of decision-making, the management of change and development control in coastal lowlands at risk is a big one, that has been confounded by the lack of adequate environmental databases, covering a sufficiently long period at a high level of resolution. The data are uneven spatially and temporally, so that conclusions obtained in one coastal lowland are not generally applicable to another without modification. The tide gauge record displays an unevenness of cover and a concentration in developed countries in the northern hemisphere, and yet, even here, there are marked differences in density, length of record and its quality. Extreme water levels are rarely recorded accurately, because of the failure of the gauge. It is well over one hundred years ago that HMS *Challenger* set sail with scientific orders requiring, *inter alia*, the establishment of tidal poles to determine once and for all the mean level of the sea and establish permanent benchmarks in relation to mean sea-level at the tidal pole sites (Anon. 1873). If these poles and benchmarks had been maintained on the islands and mainlands abutting the world's oceans then the scientific community would now be in possession of the empirical data to arrive at an unequivocal answer to questions that are being asked about the rates and directions of sea-level changes over the last 100 years and the nature of the evidence. Equivocal answers and ambiguous evidence invite criticism, and are a prescription for delaying action, so effectively promulgated by the George Marshall Institute (1989) and others.

In the opposing camp are those who argue that earth systems have been so transformed by man, particularly during the last 10,000 years (the Holocene) that change is irrevocable and irreversible, and planning for change is a pressing need. Changes in the past, it is argued, can no longer provide the analogue for future changes. In support of this, it has to be admitted that the present-day level of carbon dioxide and its equivalents in the atmosphere is higher than that of any time in the past 140,000 years, and certain consequences, not hitherto experienced on earth must arise from this simple fact. Whether it is higher global temperatures or higher sea-levels, both unevenly distributed, is a matter of conjecture and speculation. However, the natural rates of change of temperature and of sea-level have been extraordinarily rapid in the past and the impact of such rates of change on plant and animal communities and on sediment budgets and sedimentation cycles is only now beginning to be explored.

What is not conjecture and speculation are the risks to which coasts and coastal lowlands are already exposed. In this chapter, some of the deleterious processes to which coastal lowlands in Europe are being exposed, as described in the foregoing chapters, will be summarized and the likely impacts of sea-level changes in the future will be considered. The response of human communities in the past to the impact of sea-level changes will be outlined, and the implications for coastal lowlands at risk in developing countries will be explored. Finally the conclusions reached at Noordwijkerhout in 1987 will be reiterated and the coastal systems most at risk from sea-level rise will be described as a basis for the management of European coastal lowlands.

11.2 THE PRESENT STATUS OF EUROPEAN COASTAL LOWLANDS

The present-day condition of European coastal lowlands has arisen as the result of an interaction of natural environmental factors and human factors operating at different scales and over different time periods. The distribution of coastal wetlands, saltmarshes and sand dunes (figures 11.1 and 11.2) is explained by variations in global sea-level, working on available and abundant clastic sediments at a regional scale and affected fundamentally by a transient wind climate. The plants and animals adapted to the stressful environments that typify these coastal systems form distinct and diverse communities, the richness and beauty of which have been recognized as sites worthy of conservation and protection (van der Meulen et al. 1989: Dijkema et al. 1984: Doody 1991: Davidson et al. 1991). Yet these communities are most vulnerable to the impacts of sea-level and climatic change, which during the past 4,000 years at least have disrupted and damaged them. From the damaged remnants new communities have reassembled: it is a moot point whether or not with sea-level rise and climatic change in the future they will retain the capacity to re-form in the areas they occupy at present. Alternatively new areas may be created in

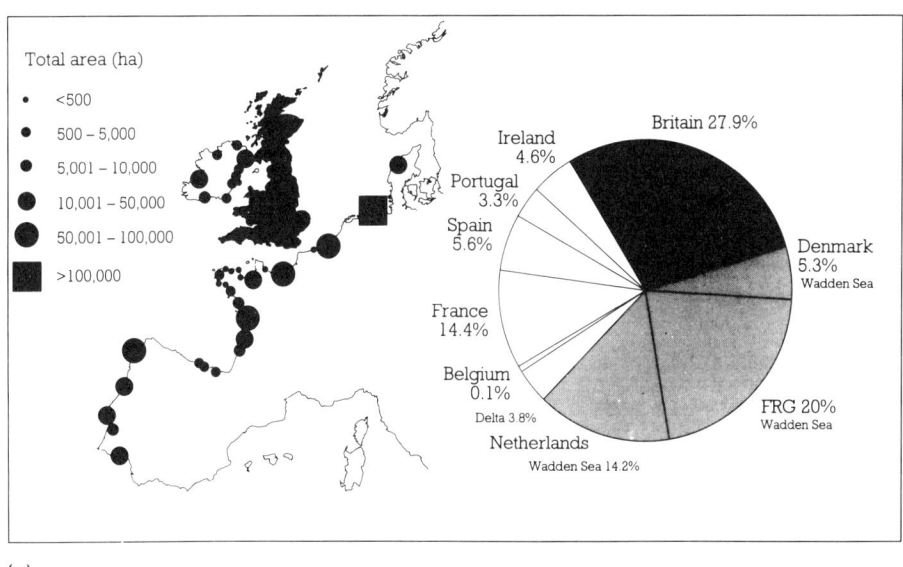

The Future of the European Coastal Lowlands 221

(a)

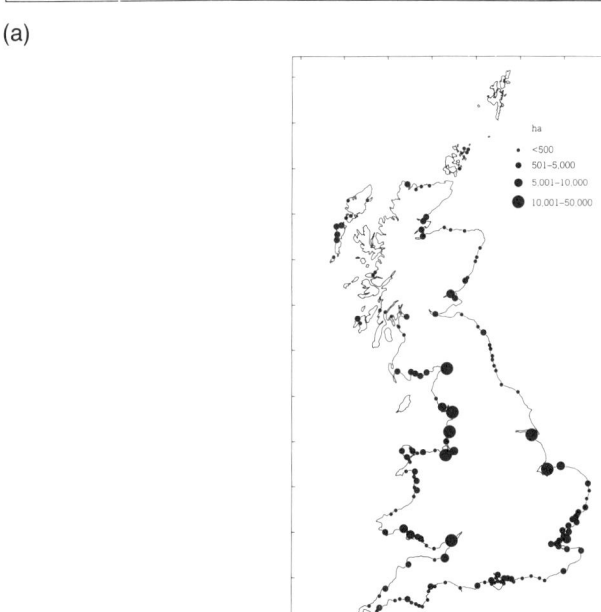

(b)

Figure 11.1a The distribution and extent of estuarine areas, excluding Britain. The pie diagram shows the percentage of intertidal and subtidal habitats out of a total of 1.8m ha for the North Sea and Atlantic seaboard.

Figure 11.1b Estuarine areas, including saltmarshes, in Britain (after Davidson et al. 1991). The distribution, type and extent of saltmarshes for the rest of Europe, including the Mediterranean, can be found in Dijkema et al. 1984). Some reference to saltmarshes in the Baltic Sea area can be found in Bird and Scwartz (1985).

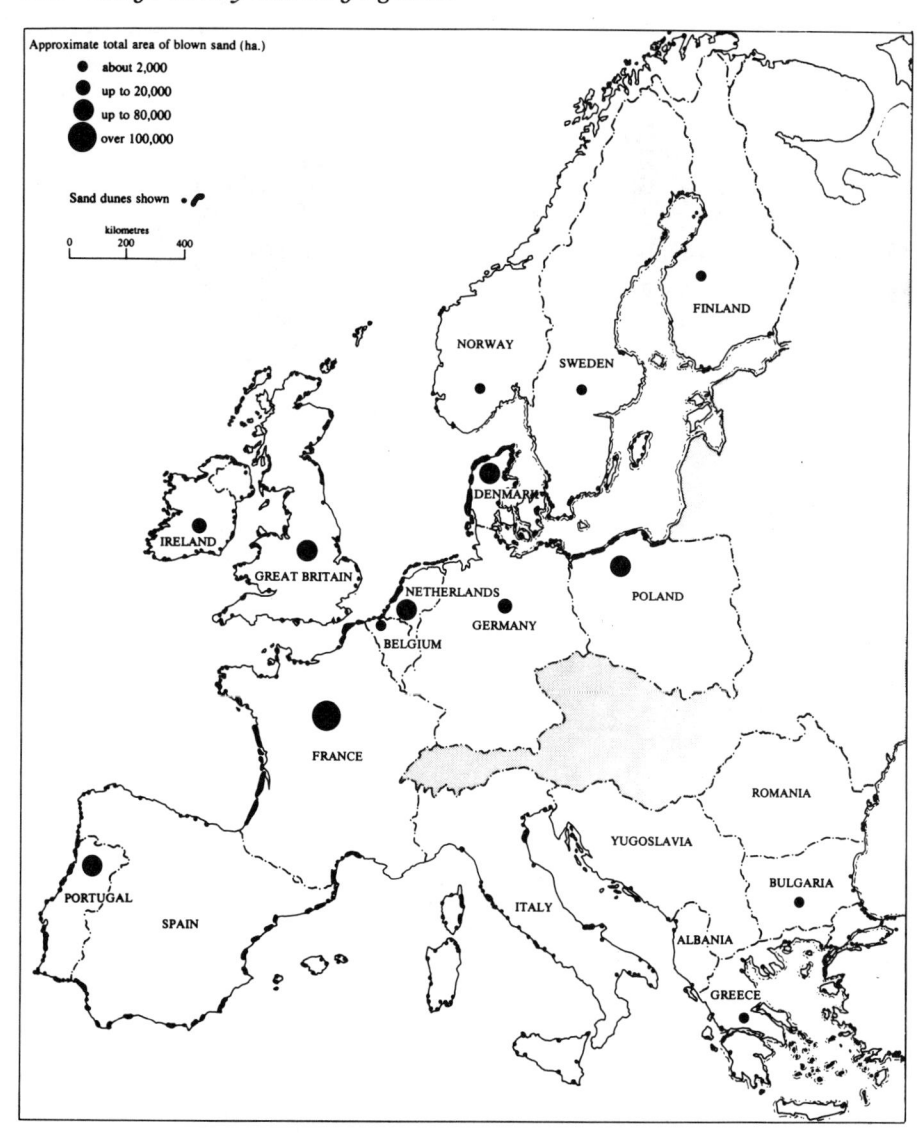

Figure 11.2 Map of Europe showing the distribution and approximate area of sand dunes for most European countries based on surveys by members of the European Union for Coastal Conservation (after Doody 1991). The map supersedes earlier versions, such as the one in Klijn (1990). For Great Britain, a detailed sand dune inventory has been drawn up by Doody (1989, 1991).

inland sites as a retreat scenario and one of the adaptive responses to sea-level rise (IPCC 1990a). In this case, wetlands, saltmarshes and dunes can be allowed to develop, perhaps where the historic and geological record points to precursors.

History of the coastal lowlands

In the first chapter, the similar sedimentary record of the coastal lowlands was stressed. The thick marine clastic sediments interleaved with autochthonous, brackish water, fresh water and terrestrial organic sediments are as much a characteristic of the Baltic (Gudelis and Konigsson 1980) as they are of the North Sea (Oele et al. 1979 and chapters 3 to 5), the Irish Sea (Kidson and Tooley 1977), the Atlantic seaboard (Ters 1973 and chapter 10) and the Mediterranean (Pirazzoli 1987a and chapters 6 to 9). The classic area for such sedimentary sequences is the Dutch coastal lowlands which are described in chapter 5; corroborating data come from The Wash and Fenland coastal lowlands of eastern England (chapter 4) and the Belgian coastal lowlands (chapter 3). In the Mediterranean, it is mainly the deltaic areas that furnish the evidence considered here: the Ebro delta in chapter 6, the Rhône delta in chapter 7 and the Po and Nile deltas in chapter 9. But there are also tidal flat and lagoonal coastal plains of which one example is given here from the Gulf of Lions. Other examples would be the Pontine Marshes in Italy, the coastal lagoons and lowlands of Albania (see figure 9.2), the Marathon Plain of Greece (Baeteman 1985), and the Kara Menderes Cayi west of Troy in Turkey (Kraft et al. 1980). These and other coastal lowlands and deltaic areas are shown in figures 1.3 and 9.1 and are described briefly in chapters 1 and 9. Further descriptions can be found in Bethemont and Villain-Gandossi (1987). The alternating organic and inorganic sediments bear witness to extensive marine transgressions and regressions over an area that Skertchly (1877) described for the Fenland and Wash of eastern England, but equally applicable to all the coastal lowlands of the world, as that 'debatable ground between land and sea.' Variations in thickness of the clastic facies indicate a local response to subsidence or sediment input, or a regional response to isostatic uplift. Correlations between uplifted and subsiding areas can be made on the basis that a transgressive overlap in an isostatically rising area is regionally significant, whereas a regressive overlap in a subsiding area is regionally significant (Shennan et al. 1983). In this way a supra-regional sea-level signal can be identified. During the Holocene all the coastal lowlands became infilled with clastic and organic deposits, which contain a record of sea-level changes and of shoreline advances and retreats.

Microtopography (microlevees along tidal creeks) was significant not only for the pattern of plant and animal communities (for example, sea purslane, *Halimione portulacoides* prefers such sites where sediments are coarser and better drained, and immersion times are less, and Shelduck, *Tadorna tadorna*, nest there on north-west European saltmarshes, such as those in the Ribble estuary of Lancashire, north-west England), but also for seasonal and semi-permanent settlement in prehistoric and historic times. The habitation mounds of the Danish, German and Dutch coastal lowlands are well known as a response to high tides, storm surges and rising sea-levels in a high risk coastal location (for example, see Van Giffen 1940; Rohde 1978; Louwe Kooijmans 1974; Godwin 1975; Behre et al. 1979). Their successful adaptation to increasingly high storm surges is manifest by the successive house platforms as shown on figure 11.3. But reclamation, enclosures, drainage and resource exploitation, particularly the cutting of peat for salt and fuel, controlled by turbary rights, and the construction of evaporating pans for salt making, began to have a profound effect on the ground altitudes of the coastal lowlands, where hitherto the influx of clastic sediments and the upward growth of ombrogenous peats had for centuries on end kept ground surfaces above the rising sea-level and all but the most extreme storm surges.

Ground altitudes

The surface altitudes of the coastal lowlands are an expression of natural processes and man-induced changes, which can have both a positive and a negative effect.

The natural processes comprise sediment influx up to an altitude controlled by sea-level (especially the High Water Mark of spring tides) and

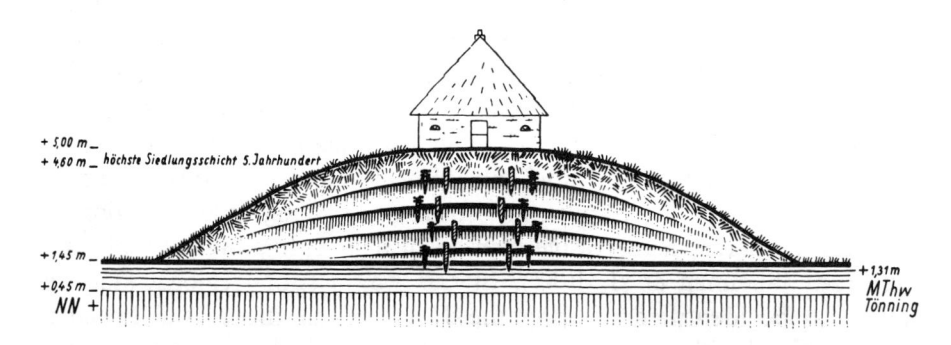

Figure 11.3 Cross-section through a 'warf' or 'wurt', which is a mound constructed of silts, sands and clays from the tidal flats and saltmarshes, and stable manure, located at the top of the unreclaimed marshes. They were raised continually from AD 500 onwards as extreme storm water levels rose. They are found along the German and Dutch coasts (after Rohde 1978). NN = Normal Null.
Source: (Rohde 1978)

the extreme water levels associated with storm surges, storm waves and river floods. In addition, peat-forming communities, particularly after the climatic deterioration of the sub-Atlantic period 2,500 years ago in the cooler and moister parts of Europe, dominated by species of *Sphagnum* and other bryophytes and nourished by nutrient-poor rainfall, raised ground surfaces over 5 m above the surrounding marshland, restricting the penetration of sea floods. In the British Isles, the great raised bogs of the Dovey estuary (Borth), Firth of Forth (Flanders Moss), Solway Firth (Wedholme Flow), Roundstone, Galway (Godwin 1975; Walker 1966; Huddart et al. 1977; Osvald 1949), in their stratigraphy bear witness to their ability to prevent inundation, though elevated ground water tables may have resulted in swamping of parts of the bog surfaces with nutrient-rich waters, leading to eutrophic conditions (see discussion in Tooley 1979).

Other natural processes were subsidence and uplift following deglaciation, tectonic subsidence within the great sedimentary basins of north-west Europe, and subsidence due to sediment and water loading in the deltaic areas of the Mediterranean Sea and southern North Sea. In addition, seismic activity is not uncommon in Europe and earth movements associated with earthquakes particularly in the Mediterranean area have dislocated beaches by several metres. In areas of ice-loading and subsequent release during deglaciation, such as Fenno-Scandinavia and Scotland, such activity may also have had an effect on the altitude of Holocene beaches.

In the Netherlands (see chapter 5), subsidence as the result of tectonic downwarping at the edge of Tertiary sedimentary North Sea basin has been calculated to 1.5 cm/100 years or 0.15 mm/year. Locally, as the result of subsidence of thick Pleistocene layers and neotectonics, ground lowering of more than 0.6 mm/year has been recorded (Brouwer et al. 1989). In the Rhône delta, subsidence is a feature of the area from Saintes-Maries to Fos and as far inland as the north shore of Lake Vaccarès: subsidence ranges from 1.5 to 4 m (chapter 7). The rate of subsidence in the Venice lowlands due to both tectonic factors and sediment consolidation ranged from 1.3 to 4.5 mm/year. In the Po delta, between 1900 and 1950, it was 0.5 mm/year (chapter 9). Pirazzoli (1987b) has assigned the increase in flooding levels of 40 cm in Venice since 1872 to several factors: 14 cm are attributed to pumping of ground water for industrial purposes in the port of Marghera (ended in 1975); 3–7 cm to geological subsidence; 14 cm to hydrodynamic factors, including man-induced tidal changes due to alterations in the inlets where exchanges between the lagoon and the Adriatic Sea took place. For the Nile delta, Milliman et al. (1989) give a general value of 3.5 mm/year, but this conceals a range of values within the delta from 1.2 to 4.8 mm/year (chapter 9). Subsidence rates on the north-east margin of the delta have been calculated to 5 mm/year over the past 7,500 years and have led to the diversion of at least four of the Nile distributaries (Stanley 1988).

Rates of ground lowering, due to the drainage of organic soils, the dewatering of the organic sediments, sediment consolidation and oxidation of the surface layers, are greater then those associated with quiet tectonic subsidence. In chapters 1 and 5 examples are given of ground lowering due to these processes of 30 mm/year, with extreme values of 220 mm/year for a few years immediately following drainage. In the Netherlands, at an experimental site at Zegweld west of Utrecht, 2 m below NAP, drainage between 1969 and 1979 led to surface subsidence of up to 155 mm or 14 mm/year (Schothorst 1982). Greatest subsidence occurred during the dry summers of 1975 and 1976. In the Po delta pumping of methane-bearing water has resulted in subsidence rates of 300 mm/year, and in Venice ground water withdrawal resulted in subsidence of 7 mm/year between 1952 and 1969 (chapter 9).

These rates, as Milliman et al. (1989) have already noted for deltaic regions exceed by several orders of magnitude the rates of sea-level rise over the last 100 years, thereby exposing not only deltaic lowlands but also coastal lowlands to an increasing risk of flooding given present rates of sea-level rise and extreme water levels, without invoking the operation of an enhanced greenhouse effect on global temperatures and sea-level.

Ground altitudes in coastal lowlands and tidal flats can also be raised as a result of reclamation and pumping sediments from the sea floor of continental shelf seas on to reclaimed land. Under these conditions, ground altitudes can be raised more than 5 m above natural ground levels. In prehistoric times, salt workings on the coast yielded considerable volumes of silt and fine sand that was discarded, and now forms chaotic landforms. In the Lincolnshire marshes of eastern England silt hills 4.8 to 6.1 m high and each one occupying c 1.2 ha bear witness to medieval salt making (Rudkin 1975). For this area, Pattison and Williamson (1986) have calculated that the spoil from salt making (salterns) contain 23 million m^3 of silt and occupy 8.5 km^2 (see discussion in Tooley 1990b).

Shoreline protection

The protection of the coastal lowlands of Europe from aperiodic flooding by the sea and by rivers and long-period inundation is haphazard, uncoordinated both within and between countries, and comprises natural features and man-made defences.

The natural features are sand dunes (figure 11.2) and saltmarshes (figure 11.1) many of which are being actively eroded and are inadequately managed in many parts of Europe. In chapter 3, attention has been drawn to the narrowness and low altitude of the dunes on the central Belgian coast, which protect the Belgian coastal lowlands to the south. Similarly, the sand dunes on the east Lincolnshire coast of the United Kingdom between Mablethorpe and Donna Nook (figure 4.6) are narrow and vulnerable to erosion. In the Gulf of Lions (chapter 8) dune heights and widths vary, so that some stretches of coast and low-lying areas to landward are extremely vulnerable. Many of the littoral lowlands of Huelva and Cadiz in south-west Spain are protected by extensive dune systems, which become attenuated and vulnerable near river mouths (chapter 10).

Man-made defences include earth embankments, revetments and seawalls, often with groyne fields to intercept longshore drift of sediments. The effects of artificial structures on coasts was one of the projects pursued by members of the Commission on the Coastal Environment (International Geographical Union), and this led, in part, to the data on the world's coastlines assembled by Bird and Schwartz (1985) and to a book by Walker (1988) on artificial structures and shorelines, containing examples from European coastal lowlands.

Land use and land-use changes

The natural ecosystems of the coastal lowlands – sand dunes, saltmarshes, tidal flats, lakes, reedswamps, fresh water marsh and fen and bogs – have been completely transformed at the hand of man. Reclamation has occurred on an extensive scale for agriculture, and raw materials (peat, sand, gravel, methane, petroleum) have been extracted. Wetlands have become reduced in area at the expense of wildlife: in the Mediterranean (European side) there were over 46,500 ha of protected coastal wetlands in 1987 (UNEP 1987) of which 1,500 ha were in the Ebro delta (chapter 6) and 13,117 ha in the Camargue, Rhône delta (chapters 7 and 8). In the western Algarve in 1987–8, there was a total wetland area of 2,240 ha, of which less than half was natural: the balance had been drained or reclaimed and used as irrigated rice fields, or for grazing or for salt works or as waste disposal dumps (Pullan 1988).

Many of these wetlands and what is defined as 'useful coast' (i.e. flat, coastal land suitable for development) in the Mediterranean Action Plan (UNEP 1987) are products of man's activities. Forest clearance in drainage basins and along coastal uplands has resulted in the liberation of sediments, that have been added to the coastal sediment budgets and coastal progradation has occurred. In the case of the Ebro delta (chapter 6) some 20,000 tonnes/year of sediment (now in part intercepted by the Mequinenza dam) were added and the delta grew from the tenth century 26 km seaward and now occupies 350 km^2 (figure 6.5). The Rhône delta has built seawards in a series of stages summarized by L'Homer et al. (1981): from the Middle Ages onwards, significant additions were made in the east as the Vieux Rhône extended. In the nineteenth century, the two distributaries of the Rhône carried over 49×10^6 tonnes/year of sediment, and rates of advance reached 20–25 m/year. Since the management of the Rhône sediment loads have decreased six-fold, and 62 per cent of the coastline has receded (chapter 8). The development of the Po delta began in Etruscan and Roman times, advancing at a rate of 4 m/year: the lobate delta began to grow from the sixteenth century onwards at 100 m/year. The consequences of reafforestation in the catchment and construction of dams have resulted in a reduction in sediment from 16.9 million to 10.5 million tons/year, and together with subsidence in the delta have caused coastal erosion and shoreline retreat (chapter 9). In southern England, the great cuspate foreland of Dungeness with the reclaimed tidal flats of Romney Marsh and Walland Marsh occupy 27,000 ha which began accumulating 5,000 years ago. But great influxes of fine-grained sediments – the products of clearance in the High Weald – infilled the shingle swales from 3,000 years ago, and the harbours and havens that served them became choked with sediment from the tenth century onwards (Tooley and Switsur 1988; Tooley 1990b). On the Durham coast in eastern England, colliery waste comprising washery waste (including 2 per cent coal) and coal measure shales and sandstone have been dumped into the sea: in the 1960s and 1970s from four disposal points some 2.4 million tons/year were dumped. In places, the high water mark retreated seawards some 122 m between 1858 and 1970 (Hydraulics Research Station 1970).

Much land in the coastal lowlands became available for reclamation as the consequence of man's activities, clearing forest and freeing sediments in catchments. The subsequent management of water resources in these catchments has resulted in a diminution of sediment supply to the coast and initiated erosion and coastline retreat.

However, the consequences of reclamation and settlement from Roman times onwards for agriculture and resource exploitation have been land use changes, largely because flat coastal land can be more economically developed and is close to ports.

From the 1950s to the early 1980s during a period of state intervention and regional planning in Europe, planned industrial areas were located in coastal areas and focused on a port. These areas of industrial port development became known as Maritime Industrial Development Areas (MIDAS) (Takel 1974); in France, as *Zones Industrielles Portuair* (ZIP) (Vigarie 1981). In the 1920s, there were prototypes in the Mediterranean: for example, at Livorno in 1924 where an industrial zone of 250 ha attached to the port on flat reclaimed marshland of the Arno delta was developed. In addition, the first group of port refineries was established in the 1920s and 1930s: Etang de Berre, Marseille and La Spezia in 1929, Naples in 1934, Genoa in 1935 and Leghorn in 1936 (Verlaque 1981).

However, the main development of the MIDAS in Europe was along the southern shore of the North Sea and at the western end of the Mediterranean (figure 11.4). The Botlek Scheme in Rotterdam initiated the Europort development, which involved the allocation of over 10,000 ha of reclaimed land devoted to oil, petrochemical and shipbuilding industries. A parallel development occurred at Antwerp, and the Rhine–Meuse delta MIDAS model was adopted in Germany (Wilheimshaven and Hamburg) and in France. In France, there were three major growth poles – at Dunkerque, Le Havre and Fos-Marseille. By 1970, some 100,000 ha of coastal land had been developed in western coastal land between the Seine and Elbe estuaries (Vigarie 1981).

In the United Kingdom, possible MIDAS were identified in 1970 (G. Hallett and P. Randall in Takel 1974): these were, in Scotland, the Cromarty Firth, the Firth of Tay, the Upper Firth of Forth, the Clyde estuary; in Wales, Cardiff and Newport; and in England, the Tees estuary, Humberside, The Wash, Thames/Medway and Weston-super-Mare/Clevedon. Of these, Humberside, Thames/Medway and Cardiff/Newport were regarded as the best sites for port industrial development.

Teesside, in north-east England, where ICI, British Steel, British Titan and Phillips Petroleum have developed on flat reclaimed land around a deep water port was regarded in 1966 by the National Ports Council as one of the four sites in the country left where industrial port development could take place (Takel 1974); but boom and slump have been a characteristic, deindustrialization and attempts at industrial regeneration a feature of an area that in the 1960s and 1970s had excellent prospects from involvement in large capital-intensive process plants, placing it among Europe's maritime industrial complexes (Benyon et al. 1989; Champion and Townsend 1990). However, notwithstanding these setbacks, some £2,426 million has been invested in fixed capital between 1974 and 1984 (Cleveland County Council 1984), and this excludes the investment in electricity generating plant – the Seaton Carew Nuclear power station. Much of this investment is located on reclaimed ground below +5m (Ordnance Datum) which is less than a metre above the highest water level recorded here during the 1953

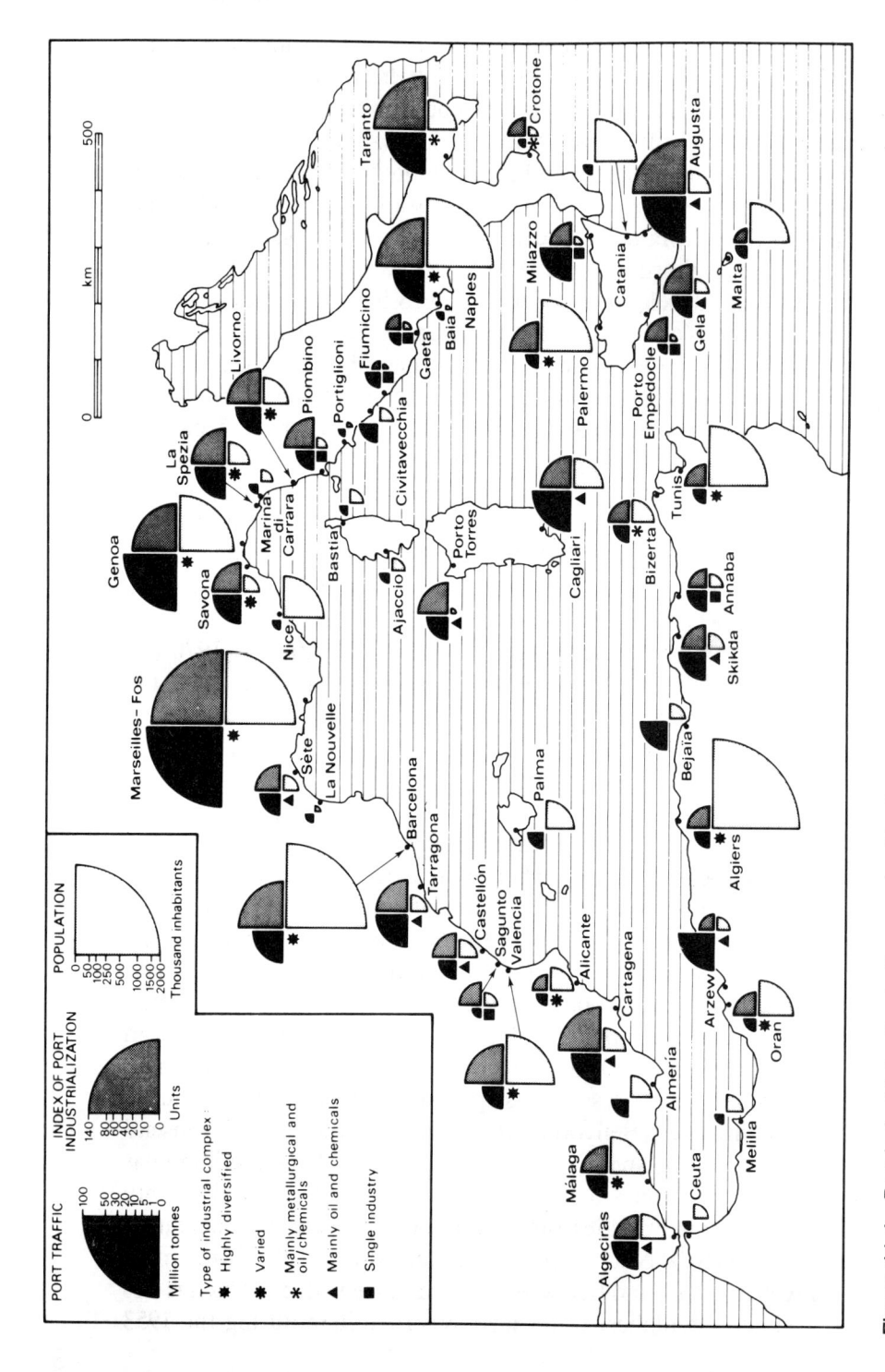

Figure 11.4 Port industrialization in the western Mediterranean in 1977. The index of port industrialization is explained in Verlaque (1981).

storm surge (Shennan and Sproxton 1990). Cleveland County in north-east England has a population of 552,400 (1986) and a workforce of *c.* 260,000 of which *c.* 19 per cent are unemployed. Heavy industry is located on either side of the River Tees estuary, and ten major industrial complexes lie within or adjacent to the 35 km^2 of reclaimed tidal flats and marshland lying below +5m (OD) (figure 11.5). Within this area, 1.2 km^2 are occupied by residential properties, two A-class roads bisect it and there are at least thirty landfill sites, some of which are known to contain heavy metals and other toxins (Shennan and Sproxton 1990). This area is clearly at risk from future sea-level rise and the extreme water levels of storm surges associated with it (figure 11.6).

The coastal lowlands of Europe have experienced spectacular changes in land use during the past 100 years, and, particularly during the past forty years. Natural areas – lagoons, wetlands, tidal flats and marshes – have been reclaimed, protected from the sea and converted initially to product-

Figure 11.5 The Tees coastal lowlands in north-east England showing the area occupied by industry (diagonal and black shading). The black shaded areas are below +5 m OD. One industrial plant lies at +3 m OD and has a capital value of £100 million and an annual production value of £75million.
Source: (Shennan and Sproxton 1990)

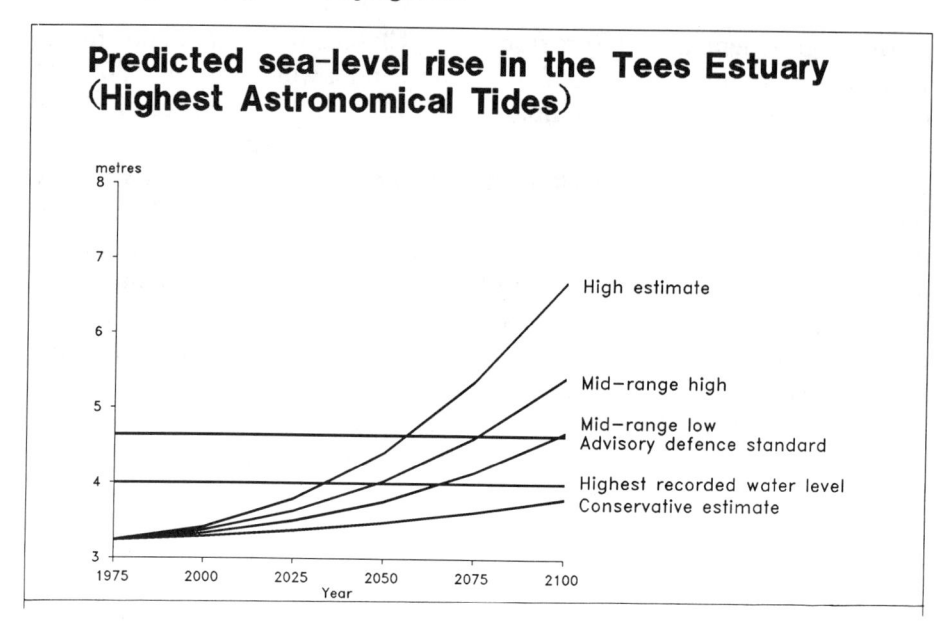

Figure 11.6 Predicted sea-level rise in the Tees estuary based on the sea-level rise scenarios of Hoffman et al. (1983), the altitudinal origin of which is +3.25 m OD – the Highest Astronomical Tide.

ive agricultural land and thence either to industrial land focusing on ports or to residential land with a significant development of tourism. Some 8,800 km or 22 per cent of the north coast of the Mediterranean Sea is being used or is potentially usable for tourism (Henry 1977). Associated with these changes in land use has been an increase in population in the coastal strip of Europe. For the Iberian peninsula, Cendrero and Charlier (1989) have noted that the population of the interior is declining and of the coast is increasing: of the 4 million population of the Asturias, Cantabria and the Basque country no less than 3 million live in a coastal strip 15 km wide. In the Mediterranean basin, Vallega (1988) has identified the period 1950–75 as a phase of continuous change, during which time economic growth in the West and the expansion of maritime trade led to littoral industrialization, urbanization and international tourism. This pattern is to a greater or lesser extent universal in the coastal lowlands of the European Community. Hence the impacts of future sea-level rise will be greater, more disruptive and more costly to society than the earlier stages of development from natural ecosystems to agriculture.

11.3 IMPLICATIONS FOR DEVELOPING COUNTRIES

The coastal lowlands of Europe share many characteristics with those of the rest of the world (figure 11.7) differing only in the degree of integrity of use and stage of development. The European perspective, therefore, could provide insights into responses in developing countries with coastal lowlands, rising sea-levels from man-induced and natural subsidence (relative sea-level rise) and the consequences of the enhanced greenhouse effect. The technological and resource transfer from developed European countries to developing countries, however, needs to be balanced by an awareness of the accumulated experience that exists there in relation to the adaption and adjustment of populations in coastal lowlands at risk from aperiodic and periodic marine inundation. The solution does not necessarily reside in applying capital-intensive, high-technological solutions to developing countries and ignoring traditional adjustments.

The juxtaposition of intensely developed coastal lowlands in Europe to coastal lowlands in developing countries in North Africa permits a ready comparison. Sestini (chapter 9) has traced the stages of development, both natural and man-dominated, in the heavily industrialized Po delta with 1.8 million people below +5 metres and the agriculturally dominated Nile delta

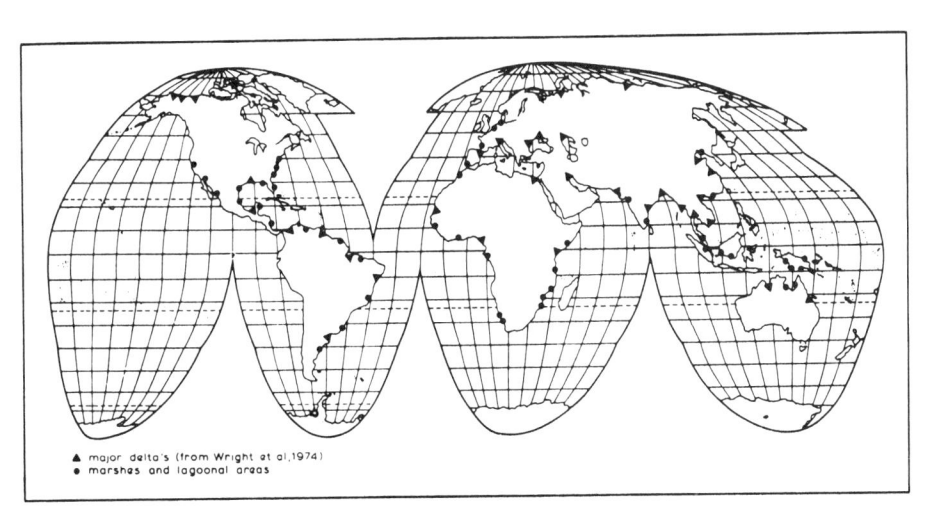

Figure 11.7 Map of the world to show the distribution of the major deltas (from Wright et al. 1974) and marshes and lagoonal areas.
Source: (From S. Jelgersma in Vreugdenhil and Wind 1987)

with 10 million people living below +3 metres. It has been estimated that a 1 m rise of sea-level could impact on Egypt by inundating 12–15 per cent of Egypt's arable land in the delta (IPCC 1990b).

A comparable situation exists in Bangladesh where rural population densities are high (locally in excess of $1000/km^2$), ground altitudes are low (less than +5 m for much of the southern half of the country) and subsidence rates range from 10 to 25 mm/year and may locally be greater (Milliman et al. 1989). Up to 35 per cent of the total land area of Bangladesh (144,000 km^2) is flooded annually during the monsoon and there are high river discharges, which damage a significant proportion of the rice and jute crop. Water management in the catchment of the River Ganges in India, and, hence, outside the jurisdiction of the government of Bangladesh, has reduced the amount of sediment delivered to the Ganges delta, the rate of sedimentation in the delta and the sediment budget of the deltaic coast. Natural protection from inundation is afforded by the 400,000 ha belt of mangrove forest in the Sunderbans, but storm surges in the Bay of Bengal are irresistible. During the 12 November 1970 cyclone, regarded as the deadliest tropical cyclone in history (Frank and Husain 1971), water levels gave maximum water depths on land of up to 6 m, highest on the coast and declining landwards. The impacts were 300,000 people drowned, 4.7 million people directly affected and crop losses amounting to $63 million. The response to flooding in Bangladesh is a five-year Flood Action Plan (1990–5), with funding coordinated by the World Bank, in which a key element is the construction of coastal embankments at the delta edge and main embankments along the distributaries of the Ganges (FAP 1990). Unfortunately, no account appears to have been taken of projected sea-level rise enhancing the relative sea-level rise already resulting from regional subsidence. It has been estimated (Holdgate 1989) that a 1 m rise of sea-level in Bangladesh by the middle of next century, resulting from 90 cm eustatic sea-level rise and 10 cm subsidence would have the following impacts: 2000 km^2 or 16 per cent of the total land area would be inundated and 14 per cent of the net cropped area destroyed; 10 per cent of the current population of 100 million would be displaced; 1.9 million homes would be destroyed; loss in output equivalent to 13 per cent of GDP and loss of assets of $c.$ 450 billion taka. Some consideration should be given to locating villages on artificially raised ground, which would afford some protection for people and livestock from extreme water levels during storm surges and river floods, as it did in the coastal lowlands of north-west Europe.

The coastal lowlands of Brazil will experience similar problems associated with sea-level rise as the populous and industrialized coastal lowlands of Europe. The geological record and sea-level changes have been described for many parts of the coast (for example, Suguio et al. 1980; Ireland 1987) and the stratigraphic record from the lagoons displays the alternation of marine clastic and organic sediments that appears to be characteristic of most of the coastal lowlands of the world. From these sediments, rates and directions of sea-level movements can be calculated. There is evidence from Lagoa do Itaipú of an interglacial or interstadial sea-level attaining an altitude of $c.$ -4.0 m (Imbituba datum). Whereas the maximum altitude attained by sea at a regressive overlap dated to 2270 ± 50 at Lagoa do Padre was $+1.4$ m, which is supported by the vermetid evidence of Delibrias and Laborel (1971). No evidence has been found for a high sea-level $+3.5$ to $+4.7$ m above present mean sea-level about 5,000 years ago, which is a time of marked negative sea-level tendency on the coast of Rio de Janeiro state (Ireland 1987). The stratigraphic data do not support the conclusion of a migrating geoid (Martin et al. 1985). In the Lagoa do Padre, there are data to calculate rates of sea-level change: between 6,800 and 5,200 years ago, rates of rise were close to 1.5 mm/year – similar to values calculated for the past 100 years.

The impacts of sea-level rise on the coast of Brazil have been summarized by Muehe and Neves (1989), who describe the contemporary geomorphological conditions of the coast. Of the coastal states of Brazil, seven have over 1 million people each living in the coastal zone, and of these Rio de Janeiro state has 7.7 million (1980 census) or 68 per cent of the state population living in the coastal zone. Most of this population is concentrated in Rio de Janeiro City which has been developed on barrier beaches enclosing lagoons such as Leblon/Ipanema barrier enclosing the Lagoa Rodrigo de Freitas, and the Restinga de Jacarepaguá enclosing the Lagoa da Jijuca and the Lagoa de Jacarepaguá, and on narrow, reclaimed coastal plains. Muehe and Neves (1989) have noted erosion along Leblon beach and flooding during storms.

A final example can be taken from China, where much of the population is nourished from food production in the great coastal lowlands and deltas of the Pearl River and Yellow River. The impacts of sea-level rise on the Zhujiang (Pearl River) delta, have been investigated by Englefield (1990). More than 15 million people live in the delta, occupied in agriculture or fishing. Some 81 per cent of the delta lies below $+0.9$ m (Yellow Sea Datum) and 43 per cent below $+0.3$ m YSD. Mean High Water ranges from $+2.2$ to $+2.5$ m YSD and the highest recorded water level was in 1937 and reached $+3.3$ m YSD. Sea embankments are formed of silt and clay, and continual extension seaward by embanking fish ponds ensures that the protective belt of *Phragmites* or mangroves is either narrow and ineffective in reducing wave energy or has been destroyed by the reclamation process. The honeycomb of fishponds, rather like the natural lagoons in unreclaimed coastal lowlands affords some protection from rising water levels of an aperiodic or periodic nature, but no attention has been given to safeguarding populations by locating villages on ground above extreme water levels that have been recorded or are predicted. The expedient of plugging breaches in sea embankments with successive phalanxes of soldiers from the Red Army is perhaps not the most satisfactory solution, and a solution incapable of export to other states with coastal lowlands in the world.

The annual costs of preventing marine inundation and coastal erosion arising from a 1 m rise of sea-level along the coasts of fifty developing countries have been reported (IPCC 1990b) as a percentage of Gross National Product. They range from 34 per cent for the Maldives, to 2.4 per cent for Mozambique, 0.8 per cent for Sri Lanka and 0.5 per cent for Madagascar.

11.4 THE FUTURE OF EUROPEAN COASTAL LOWLANDS

The prognostications for the coast and coastal lowlands in the Mediterranean indicate an increased intensity of all activities from population density to tourism, from agricultural land use to industrial land use and from transport to energy. This intensification of land use in coastal lowlands will occur throughout Europe, although there will be regional variations.

In the Mediterranean, the coastline extends for 46,000 km, of which 19,500 km (42 per cent) are identified as 'useful coast', that is where coastal lowlands back the coastline and are flat and easily developed. Land use conflicts exist along 6500 km of 'useful coast', and industrial and energy-generating facilities will be increasingly located on the coast. The maps (figures 11.8 and 11.9) show the distributions of operating and planned capacity of petroleum refineries and thermal power stations on the coast. In summary, there are at present fifty-eight main oil loading and unloading terminals, fifty refineries with eleven more planned, and sixty-two thermal power stations with thirty-two more planned. What is not indicated in the Blue Plan Scenarios is the number of nuclear power stations in the Mediterranean. There are seven on the coast – in Egypt, Greece and Italy – and in catchments feeding into the Mediterranean, there are four on the River Po, three on the River Rhône, four on the River Ebro and one on the River Jucar (Couper 1983). Altogether there are 27 reactors in operation in the catchment, excluding the Black Sea catchment (Varley 1991).

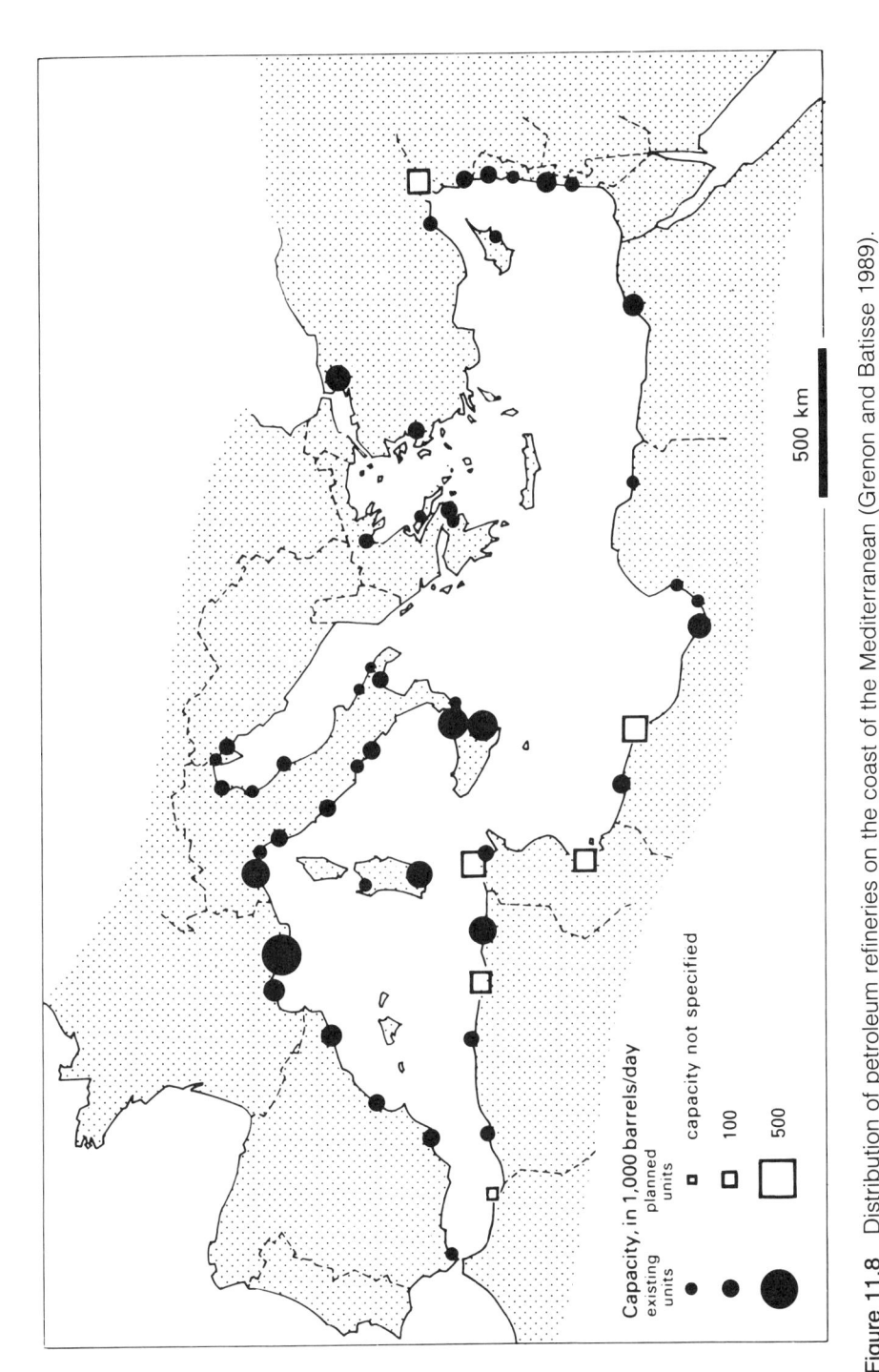

Figure 11.8 Distribution of petroleum refineries on the coast of the Mediterranean (Grenon and Batisse 1989). *Source:* (Reproduced by permission of the Director, Mediterranean Blue Plan Regional Activity Centre)

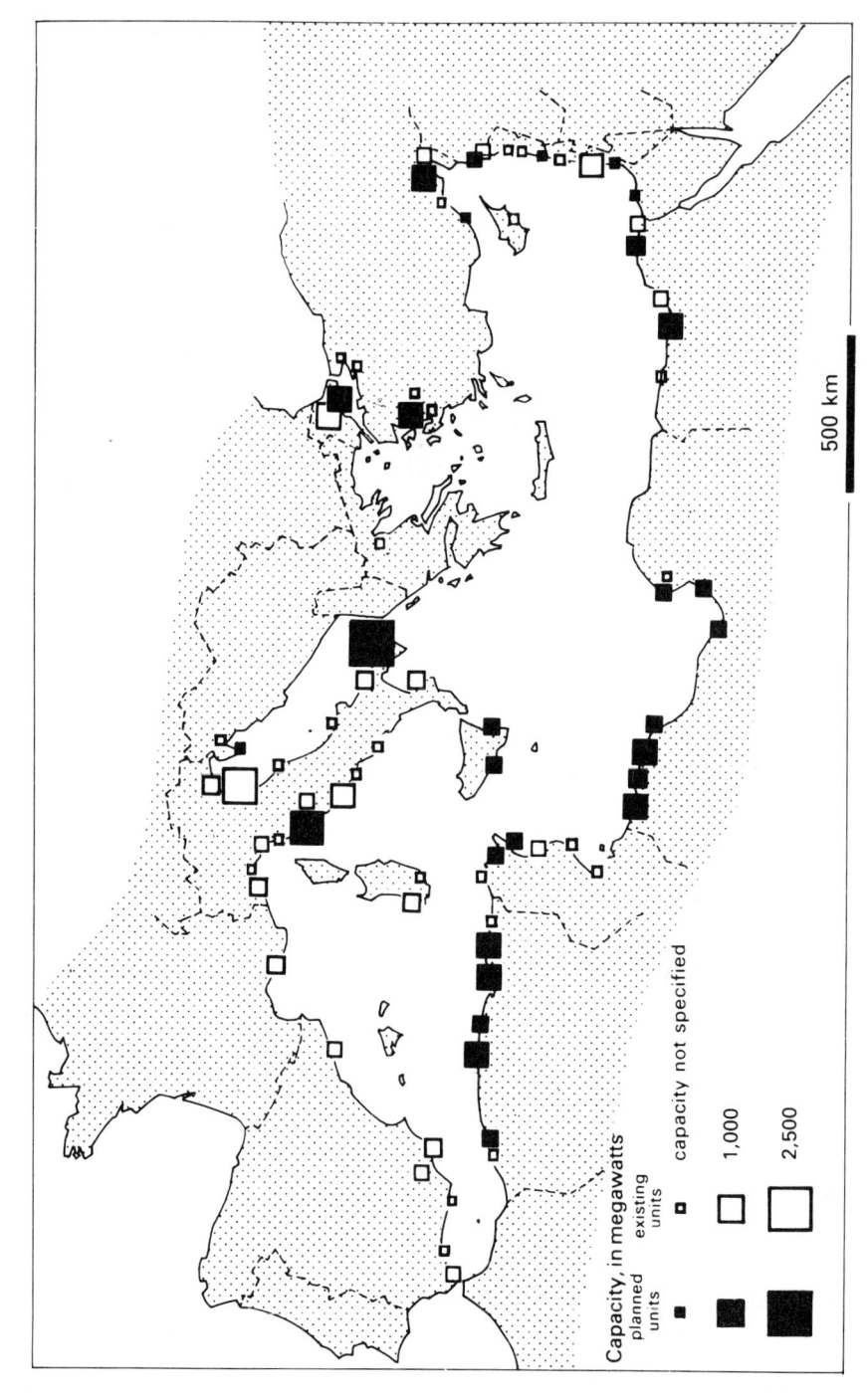

Figure 11.9 Distribution of thermal power stations on the coast of the Mediterranean (Grenon and Batisse 1989).
Source: (Reproduced by permission of the Director, Mediterranean Blue Plan Regional Activity Centre)

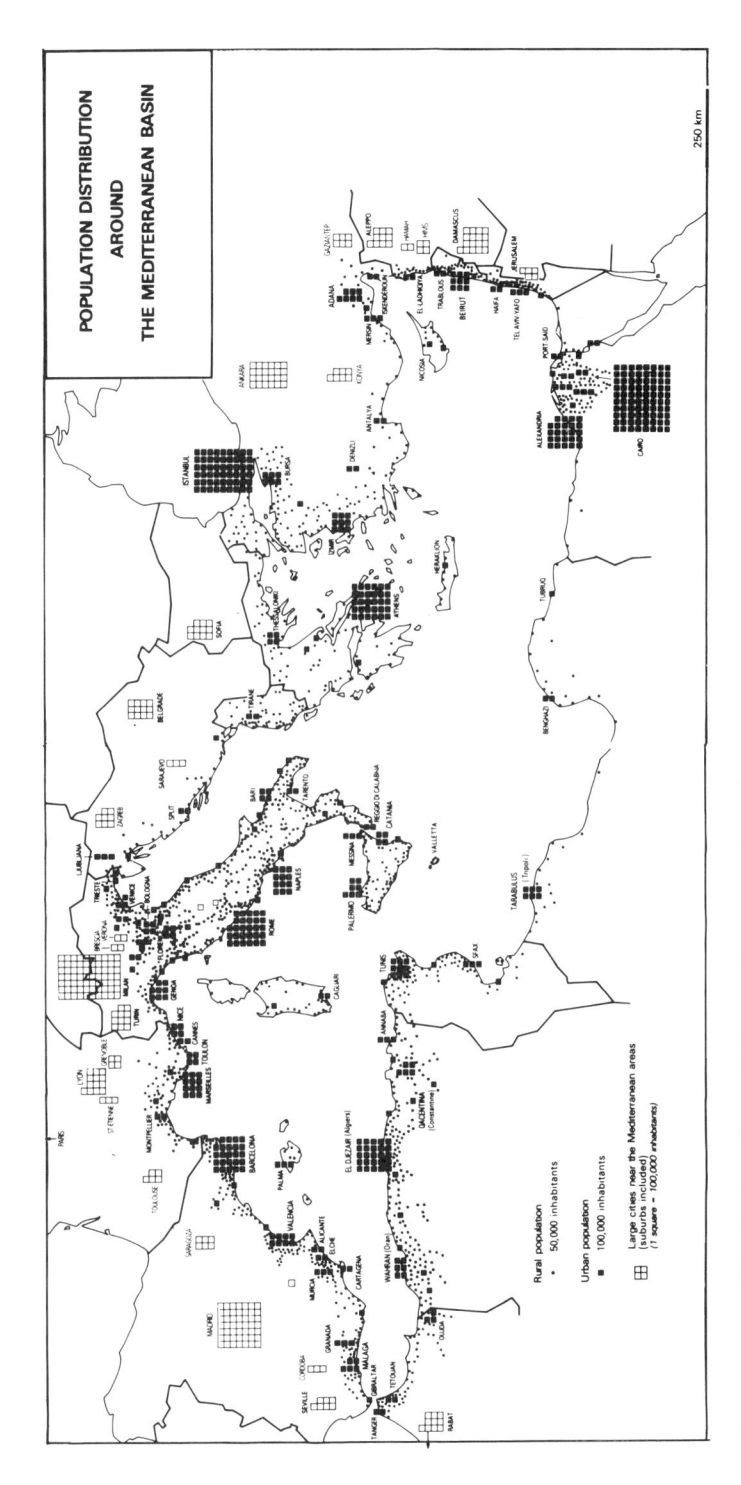

Figure 11.10 Distribution of population on the coast of the Mediterranean. (Grenon and Batisse 1989).
Source: (Reproduced by permission of the Director, Mediterranean Blue Plan Regional Activity Centre)

In 1985, the total population of the coastal regions of the Mediterranean (figure 11.10) was 133 million, and this is expected to rise to between 200 and 220 million by AD 2025 with the greatest increases in the east and southern parts of the 'useful coast' and marked urbanization. Tourism (Tangi 1977) is expected to increase, as is industrialization. Salt, fresh and brackish water aquaculture for fish species such as mullet, perch and bream is expected to require 1 million ha of coastal lowland by AD 2025. Hence, land use conflicts will be severe and will increase greatly beyond the present 6,500 km of 'useful coast,' impinging on natural and wild areas that occupy space.

Such intensification of the land use of coastal lowlands implies a considerable increase in the damage from rising sea-level, an increase in the incidence of aperiodic storms and long-term inundation.

At the workshop on Coastal Zone Management in Miami in December 1989 (IPCC 1989), a series of adaptive options was proposed and the working group urged that the nations should assess the implications of sea-level rise and develop site-specific strategies. Three strategies were proposed:

Technical engineering and structural options Coastal engineering methods are well established and effective and include the construction of seawalls, breakwaters, dikes, levees, floodgates and bulkheads, the use of beach nourishment, and the raising of ground levels of coastal lowlands by fill material. However, these are capital intensive solutions, requiring continuous maintenance and adaptation, and often have severe environmental impacts.

Natural, biological and ecological options Impacts of sea-level rise can be mitigated by replacing lost resources or developing alternative habitats. Wetlands can be created, dunes stabilized and mangroves planted.

Non-structural options The simplest approach is to allow coastal resources and land uses to respond naturally to changing conditions. Structures can be removed and populations resettled away from vulnerable areas.

Such options, particularly the retreat from coastal lowlands threatened by the impacts of sea-level rise, require a coordinated response. The workshop on coastal zone management proposed 'new institutional arrangements to coordinate various levels of government decision making.'

Attempts have been made in the past, with little success, to establish institutional frameworks to oversee coastal development and establish guidelines for the management of the coast. In the United Kingdom, Steers (1944) concluded that what was required was a 'national authority as a co-ordinator and judge, in the last resort, of all forms of planning for the use and enjoyment of the coast, whether scientific, economic, or popular'. In the 1980s, the EC made what Cendrero and Charlier (1989) regarded as a 'timid move' by publishing the European Coastal Charter and outlining an action programme, which has come to nothing. For the coast of the Mediterranean, reference in the Mediterranean Action Plan (UNEP) has been made to the 'narrow and disputed strip of (coast) land' where 'lucid choices will have to be made, development policies will need to be clearest and control of their impacts on the environment firmest . . . protection policies will have to be implemented most vigorously and severely. Special institutions for conservation of the coast, drainage, protection of natural areas, must be conceived and set up.'

In the context of Louisiana State in the USA Meo (1988) has explored the institutional response at local, state and federal levels to coastal land loss due to sea-level rise, and has concluded that an intergovernmental effort is required. Even with the inexorable rise of sea-level and the range and variability of associated impacts, Meo is not sanguine about the development of a comprehensive mitigating strategy. In 1990, UNEP's report on *The State of the Marine Environment* stressed the need for coastal planning: 'planning the development of the coastline as a whole . . . should be undertaken more widely. International guidelines, including criteria and standards would provide valuable assistance in such planning in different geographical areas, but awareness, resources and political will are needed if the health of the resident and the transient populations, the survival of marine wildlife and the functional integrity of the vital land–sea interface are to be maintained.'

Also in 1990, the Coastal Zone Management Subgroup of IPCC Working Group 3 (IPCC 1990b) recommended the establishment of comprehensive coastal zone management plans to be implemented by AD 2001. National coordination and international cooperation were stressed, and response support to meet the implementation deadline date was costed to $US 10 million for the period 1992–7. Commendably, stress has been laid on the need for data collection and survey studies of key physical, social and economic parameters. Typical data comprise topographic information, tidal and wave data, land use data and population statistics. It is gratifying to note that these data have already been in the course of assemblage since 1987 for the United Kingdom in a geographical information system as part of the European Programme on Climatic Hazards (EPOCH) I and II (Shennan and Tooley 1987; Shennan and Sproxton 1990; chapter 4, this volume).

It is to be hoped that these and other ideas for special institutions to oversee and guide coastal development, integrating conflicting demands in the face of rising sea-levels and increased storminess in the coming decades, will be considered seriously and with urgency by decision-makers as a basis for implementing a planning strategy for European coastal lowlands. In 1987, recommendations had already been made for action in relation to the coastal lowlands of Europe, and to reach a larger constituency they are reiterated here.

11.5 THE NORDWIJKERHOUT RECOMMENDATIONS OF 1987 ON EUROPEAN COASTAL LOWLANDS

Statement of the Problem

The European coastal lowlands comprise deltas and plains that include extensive zones of intense economic and agricultural activity, natural areas and wetland sites, and are the home of dense populations. Around the southern North Sea basin alone the coastal plains are the home of more than 20 million people. These people live close to present sea-level, and are already at risk from inundation as a consequence of coastal erosion and storm surge, and the rise of sea-level that has occurred during the past 100 years. Sea-level has risen by 10 to 15 cm in the past century as the result of global warming, due to the increase of atmospheric CO_2 and other radiatively active gases and other factors, and will continue to rise in the next 100 years by amounts estimated between 0.5 and 3.5 m. Unless remedial actions are taken now, the social and economic impacts will be profound and widespread. During the session these problems were addressed in twelve papers grouped by geographical areas – the southern North Sea Basin, the Atlantic seaboard of Europe and the Mediterranean. The following conclusions were agreed:

Conclusions

1 The unequivocal signal from the session was that *all* the European coastal lowlands and their shorelines were already experiencing damage from erosion, inundation during storm surges, storm waves, subsidence and salt water intrusion as the consequences of sea-level rise, increased incidence of storminess and man's activities. Man's economic and social activities are exacerbating the situation and comprise:

- sand extraction from beaches and offshore for reclamations and for the construction industry;
- the destruction of natural shoreline defences, such as sand-dunes, for the provision of hotel accommodation and amenities for the tourist industry;
- the interruption and diversion of longshore sediment transport by groynes, jetties and harbours;
- the reduction of the sediment load of rivers by water management in drainage basins and the construction of dams and reservoirs, cutting off sediment supply to nourish beaches and deltas;
- the canalization of rivers for navigational purposes;
- the reclamation of coastal lowlands for agriculture, industrialization and residential development;
- the utilization of ground water for drinking water and irrigation which has led to subsidence, and the penetration of salt water.

2 The maximum rise of sea-level during the recent geological past did not exceed 2.2 m/100 years, and during the past 4,000 years, when the rate of rise was considerably less, sedimentation has kept up with and locally exceeded sea-level rise. In natural areas, where man's activities are not pre-eminent, such as the Dutch, German and Danish Wadden Sea, sedimentation has kept up with the present rise of sea-level of 10–15 cm/100 years. Investigations of the sediments of the coastal lowlands indicate that, given a natural sediment budget, these areas responded and adjusted to a range of rates of sea-level change and climatic change in the past.

3 Current databases are uneven spatially and temporally, and are inadequate for the purposes of prediction. Environmental, social and economic data are required from existing sources and new investigations in compatible and accessible forms. A notable deficiency in the environmental database is analysed information on the sediment budget of the shorelines of coastal lowlands and adjacent shorelines, waves, tides, the strength and direction of tidal currents and the range of variation of these parameters in time. Geographic information systems should be employed to integrate data collection, analysis, management and display.

4 The risks from sea-level rise and the impacts on coastal lowlands can be evaluated by the application of cost–benefit analysis. Land use conflicts are a feature of coastal lowlands, and decisions will need to be made on which activities and functions need protection and which can be lost without cost to the human community. The concept of sacrificial areas is accepted.

Research proposals

1 Monitoring the movement of sea-level employing the tide gauge data maintained and updated by the Permanent Service for Mean Sea level, Institute of Oceanographic Sciences (now Proudman Oceanographic Laboratory), Bidston, UK, to determine the increased rate of sea-level rise, and local variations.

2 Monitoring the impact of sea-level rise, neotectonic movements (earthquakes) and man's economic and social activities on European shorelines and coastal lowlands using satellite imagery analysis.

3 Investigations on the coastal dynamics of European shorelines to determine changes in the sediment budget consequent upon sea-level rise.

4 Investigations on the recent geological history of coastal lowlands and the determination for each lowland of rates of sea-level change and the responses in terms of sediment type and distribution, of land forms and of palaeogeography. These investigations will serve as an analogue for future sea-level changes. Data from Unesco/IGCP projects 61 and 200 on sea-level change can be utilized and enhanced by new data, contributed by the International Union for Quaternary Research (INQUA) Commission on Quaternary Shorelines and the International Geographical Union (IGU) Commission on the Coastal Environment.

5 The development and application of objective criteria to be applied to land use planning in the coastal lowlands of Europe, and the production of maps showing high, intermediate and low risk zones in relation to the Environmental Protection Agency (EPA) sea-level scenarios to AD 2100.

6 The updating of the Council of Europe's map of saltmarshes of Europe, and an extended inventory to include coastal wetlands and natural areas. The impact of sea-level rise, using different sea-level scenarios should be undertaken for selected areas of international significance for the world conservation strategy as identified by the United Nations (UN) and the International Union for the Conservation of Nature (IUCN) and within the framework of the European Community (EC), United Nations Environment Programme (UNEP) and the INQUA Man and Biosphere programme.

7 An investigation into the history of the dunes of the Atlantic seaboard of Europe. The investigation will establish the cyclicity of sand blowing and provide a link with climatic changes and sea-level changes. A calculation for

The Future of the European Coastal Lowlands

each dune system of the volume of sand mobilized and stabilized will provide data on the sand volume added to or abstracted from the coastal sand budget, and how much sand will be liberated as sea-level rises and the climate changes.

Recommendations

Action now

- Control coastal development, to minimize risks to human life from sea-level rise.
- Control land reclamation, to reduce the growing area of coastal lowlands susceptible to inundation.
- Control ground water exploitation to reduce subsidence and salt water intrusion.
- Zone land in lowlands into high, medium and low risk categories. Strategic industries, such as electricity generation, should be located in future away from high risk zoned land.
- Suspend all dumping/storage of toxic and radioactive wastes in high risk zones in coastal lowlands susceptible to long-term inundation from sea-level rise.

Action over the next five years

- Sponsor research programmes to generate new data from coastal lowlands.
- New and existing environmental data and social and economic data will be integrated in Geographical Information Systems (GIS), developed on PCs to permit interchange of information.
- Models of sea-level change will be tested rigorously against empirical data collected from the coastal lowlands at different spatial scales from the local scale (individual estuary) to the regional scale (southern North Sea basin, Mediterranean basin).
- An evaluation of the cost-effectiveness of shoreline protection of coastal lowlands will be undertaken and an inventory of European coasts made to determine which segments of coasts need protection and which segments can be sacrificed.

REFERENCES

Anon. 1873: The scientific orders of the Challenger. *Nature* 7, 191–3

Baeteman, C. 1985: Late Holocene geology of the Marathon Plain (Greece). *Journal of Coastal Research* 1(2), 173–85

Behre, K.-E., Menke, B. and Streif, H. 1979: The Quaternary geological development of the German part of the North Sea. In, Oele, E. et al. (eds), *op. cit.*, 85–113

Bethemont, J. and Villain-Gandossi, C. (eds) 1987: *Les Deltas méditerranéens*. Vienna Centre Européen de Coordination de Recherche et de Documentation en Sciences Sociales

Benyon, H., Hudson, R., Lewis, J., Sadler, D. and Townsend, A. 1989: 'It's all falling apart here': coming to terms with the future in Teesside. In, Cooke, P. (ed.) *Localities: The Changing Face of Urban Britain*. London, Unwin Hyman, 267–95

Bird, E. C. F. and Schwartz, M. L. (eds) 1985: *The World's Coastline*. New York, van Nostrand Reinhold

Brouwer, F. J. J., Murre, L. M. and Noomen, P. 1989: Recent vertical crustal movements and sea level rise (in the Netherlands). Paper presented at the annual meeting of the EC, EPOCH project 'Sea-level', Cork, Ireland, 6pp.

Cendrero, A. and Charlier, R. M. 1989: Resources, land use and management of the coastal fringe. *Geolis* 3(1–2), 40–60

Champion, A. G. and Townsend, A. R. 1990: *Contemporary Britain: A Geographical Perspective*. London, Edward Arnold

Cleveland County Council 1984: *Information Note 277*. Middlesbrough, Cleveland County Council, Research and Intelligence Unit

Couper, A. (ed.) 1983: *The Times Atlas of the Oceans*. London, Times Books

Davidson, N. C., Laffoley, D.d'A., Doody, J. P., Way, L. S., Gordon, J., Key, R., Drake, C. M., Pienkowski, M. W., Mitchell, R., and Duff, K. L. 1991: *Nature Conservation and estuaries in Great Britain*. Peterborough, Nature Conservancy Council

Delibrias, C. and Laborel, J. 1971: Recent variations of the sea level along the Brazilian coast. *Quaternaria* 14, 45–9

Dijkema, K. S. (ed.), Beeftink, W. G., Doody, J. P., Géhu, J. M., Heydeman, B. and Rivas Martinez, S. 1984: *Salt Marshes in Europe* (Nature and Environment Series No. 30). Strasbourg, Council of Europe

Doody, P. 1989: Conservation and development of the coastal dunes in Great Britain. In, F. van der Meulen et al. (eds), *op. cit.*, 53–67

Doody, J. P. 1991: *Sand Dune inventory of Europe*. Peterborough, Join Nature Conservation Committee, and Strasbourg, European Committee for the Conservation of Natural Resources.

Englefield, G. J. H. 1990: The use of a geographical information system to assess the impact of marine flooding on coastal lowlands in Southern China: applications to southeast Asia. In, *Conference on Geography in Asian Region*. Vol. 2: *Geography Education/Environment and Resources/Socio-economic/Techniques*. Brunei Darussalam, University of Brunei Darussalam, 464–80

Fairbridge, R. W. 1989: Crescendo events in sea level changes. *Journal of Coastal Research* 5(1), ii–vi

Flood Action Plan 1990: *Review February 1990*. Dhaka, Bangladesh, Flood Plan Coordination Organization and World Bank Action Plan Coordinator, 11pp.

Frank, N. L. and Husain, S. A. 1971: The deadliest tropical cyclone in history. *Bull. Am. Met. Soc.* 52(6), 438–44

The George C. Marshall Institute 1989: *Scientific Perspectives on the Greenhouse Problem*, Washington, D.C., George C. Marshall Institute

Giffen, A. E. van 1940: Die Wurtenforschung im Holland. In, Haarnagel, W. (ed.) *Probleme der Küstenforschung im südlichen Nordseegebiet*. Hildesheim, August Lax

Godwin, H. 1975: *The History of the British Flora: A Factual Basis for Phytogeography*. Cambridge, Cambridge University Press

Grenon, M. and Batisse M. (eds) 1989: *Futures for the Mediterranean Basin: The Blue Plan*. Oxford, Oxford University Press

Gudelis, V. and Königsson, L.-K. (eds) 1980: *The Quaternary History of the Baltic*, Uppsala, Acta Univ. Ups. Symp. Univ. Ups. Annum Quingentesimum Celebrantis 1

Henry, P.-M. 1977: The Mediterranean: A Threatened Microcosm. *Ambio* 6(6), 300–7

Hoffman, J. S., Keyes, D. and Titus, J. G. 1983: *Projecting Future Sea-level Rise: Methodology, Estimates to the Year 2100 and Research Needs*. Washington, D.C., Environmental Protection Agency

Holdgate, M. W. (Chairman) 1989: *Climate Change: Meeting the Challenge* (Report by a Commonwealth Group of Experts). London, Commonwealth Secretariat

Huddart, D., Tooley, M. J. and Carter, P. A. 1977: The coasts of north west England. In, Kidson, C. and Tooley, M. J. (eds), *op. cit.*, 119–54

Hydraulics Research Station 1970: *Colliery Waste on the Durham Coast: A Study of the Effect of Tipping Colliery Waste on the Coastal Processes* (Report No. Ex 521). Wallingford, Hydraulics Research Station

Intergovernmental Panel on Climate Change (IPCC) 1989: *Adaptive Options and Policy. Implications of Sea Level Rise and Other Coastal Impacts of Global Climate Change* (Workshop Report of the Coastal Zone Management Subgroup, 27 November–1 December 1989)

Intergovernmental Panel on Climate Change (IPCC) 1990a: *Adaptive Responses and Policy. Implications of Sea Level Rise and Other Coastal Impacts of Global Climate Change* (Summary report of the Coastal Zone Management Subgroup of Working Group 3, 1 May 1990)

Intergovermental Panel on Climate Change (IPCC) 1990b: *World Oceans and Coastal Zones* (Report of the Ocean and Sea level Subgroup of Working Group 2, Assessment of Environmental and Socio-economic Impacts of Climate Change)

Ireland, S. 1987: The Holocene sedimentary history of the coastal lagoons of Rio de Janeiro State, Brazil. In, Tooley, M. J. and Shennan, I. (eds) *Sea-level Changes*. Oxford, Basil Blackwell, 25–66

Kidson, C. and Tooley, M. J. (eds) 1977: *The Quaternary History of the Irish Sea (Geological Journal* Special Issue No. 7). Liverpool, Seel House Press

Klijn, J. A., 1990: Dune forming factors in a geographical context. In, Bakker, Th. W., Jungerius, P. D. and Klijn, J. A. (eds) *Dunes of the European Coasts; Geomorphology – Hydrology – Soils. Catena*, Supplement 18, 1–13

Kraft, J. C., Kayan, I. and Erol, O. 1980: Geomorphic reconstructions in the environs of ancient Troy. *Science* 209, 776–82

L'Homer, A., Bazile, F., Thommeret, J. and Thommeret, Y. 1981: Principales étapes de l'édification du delta du Rhône de 7000 BP à nos jours; variations du niveau marin. *Oceanis* 7(4), 389–408

Louwe Kooijmans, L. P. 1974: *The Rhine/Meuse Delta: Four Studies on its Prehistoric Occupation and Holocene Geology*. Leiden, E. J. Brill

Martin, L., Flexor, J.-M., Blitzkow, W. D. and Suguio, K. 1985: Geoid change indications along the Brazilian coast during the last 7000 years. *Proc. Vth Intl. Coral Reef Congress*, Tahiti, 3, 85–90

Meo, M. 1988: Institutional response to sea-level rise: the case of Louisiana. In, Glantz, M. H. (ed.) *Societal Responses to Regional Climatic Change: Forecasting by Analogy*. Boulder and London, Westview Press, 215–42

Meulen, F., van der Jungerius, P. D. and Visser, J. H. (eds) 1989: *Perspectives in Coastal Dune Management*. The Hague, SPB Academic Publishing

Milliman, J. D., Broadus, J. M. and Gable, F. 1989: Environmental and economic implications of rising sea level and subsiding deltas: the Nile and Bengal examples. *Ambio* 18, 340–5

Muehe, D. and Neves, C. F. 1989: Potential impacts of sea level rise on the coast of Brazil. In, Titus, J. G. (ed.) *Changing Climate and the Coast* 2, 311–39 (Report of the IPCC Working Group 3, Miami 1989, 27 November–1 December)

Oele, E., Schüttenhelm, R. T. E. and Wiggers, A. J. 1979: *The Quaternary History of the North Sea*, Uppsala, Acta Univ. Ups. Symp. Univ. Ups. Annum Quingentesimum Celebrantis 2

Osvald, H. 1949: Notes on the Vegetation of British and Irish mosses. *Acta Phytogeographica Suecica* 26, 1–62

Pattison, J. and Williamson, I. T. 1986: The saltern mounds of north-east Lincolnshire. *Proc. Yorks. Geol. Soc.* 46, 77–9

Pirazzoli, P. A. 1987a: Sea level changes in the Mediterranean. In, Tooley, M. J. and Shennan, I. (eds) *Sea-level changes*. Oxford: Basil Blackwell, 152–81

Pirazzoli, P. A. 1987b: Recent sea-level changes and related engineering problems in the Lagoon of Venice (Italy). *Prog. Oceanog.* 18, 323–46

Pullan, R. A. 1988: *A Survey of the Past and Present Wetlands of the Western Algarve*. (Liverpool papers in Geography, 2). Liverpool, University of Liverpool, Department of Geography, iii + 100 pp.

Rohde, H. 1978: The history of the German coastal area. *Die Küste: Archiv für forschung und technik an der Nord- und Ostsee* 32, 6–29

Rudkin, E. H. 1975: Medieval salt making in Lincolnshire. In, Brisay, K. W. de and Evans, K. A. (eds) *Salt: the Study of an Ancient Industry*. Colchester, Colchester Archaeology Group, 37–40

Schothorst, C. J. 1982: Drainage and behaviour of peat soils. In, Bakker, H. de and Berg, M. W. van den (eds) *Proceedings of the Symposium on Peat Lands below Sea Level*. Wageningen, International Institute for Land Reclamation and Improvement (ILRI) Publication 30, 130–63

Shennan, I. and Tooley, M. J. 1987: Conspectus of fundamental and strategic research on sea-level changes, in Tooley, M. J. and Shennan, I. (eds) *Sea-level Changes*. Oxford, Basil Blackwell, 371–90

Shennan, I., Tooley, M. J., Davis, M. J. and Haggart, B. A. 1983: Analysis and interpretation of Holocene sea level data. *Nature* 302, 404–6

Shennan, I. and Sproxton, I. 1990: Possible impacts of sea level rise: a case study from the Tees estuary, Cleveland County. In, Doornkamp, J. C. (ed.) *The Greenhouse Effect and Rising Sea Levels in the U.K.*, Nottingham, MI Press, 109–33

Skertchly, S. B. J. 1877: *The Geology of the Fenland*, London, Memoirs of the Geological Survey, England and Wales

Stanley, D. J. 1988: Subsidence in the northeastern Nile delta: rapid rates, possible causes, and consequences. *Science* 240, 497–500

Steers, J. A. 1944: Coastal preservation and planning. *Geogr. Jl.* 104, 7–18

Suguio, K., Martin, L. and Flexor, J.-M. 1980: Sea-level fluctuations during the past 6000 years along the coast of the State of São Paulo, Brazil. In Mörner, N.-A. (ed.) *Earth Rheology, Isostasy and Eustasy*. Chichester, John Wiley and Sons, 471–86

Takel, R. E. 1974: *Industrial Port Development, with Case Studies from South Wales and Elsewhere*. Bristol, Scientechnica (Publishers)

Tangi, M. 1977: Tourism and the environment. *Ambio* 6(6), 336–41

Ters, M. 1973: Les variations du niveau marin depuis 10,000 ans, le long du littoral atlantique français. In, *Le Quaternaire, géodynamique, stratigraphie et environnement*. Paris, Centre National de la Recherche Scientifique, 114–35

Tooley, M. J. 1979: Sea-level changes during the Flandrian Stage and the implications for coastal development. In, Suguio, K., Fairchild, T. R., Martin, L. and Flexor, J.-M. (eds) *Proceedings of the 1978 International Symposium on Coastal Evolution in the Quaternary*. São Paulo, Universidade de São Paulo, 502–33

Tooley, M. J. 1990a: The chronology of coastal dune development in the United Kingdom. In, Bakker, Th. W., Jungerius, P. D. and Klijn, J. A. (eds) *Dunes of the European Coasts: Geomorphology – Hydrology – Soils. Catena*, Supplement 18, 81–8

Tooley, M. J. 1990b: Sea level and coastline changes during the last 5000 years. In,

McGrail, S. (ed.) *Maritime Celts, Frisians and Saxons*. Council for British Archaeology, Research Report 71, xi + 1–16

Tooley, M. J. and Jelgersma, S. 1989: INQUA commission on Quaternary Shorelines. *Journal of Coastal Research* 5(1), 166–7

Tooley, M. J. and Switsur, V. R. 1988: Water level changes and sedimentation during the Flandrian Age in the Romney Marsh area. In, Eddison, J. and Green, C. (eds) *Romney Marsh: Evolution, Occupation, Reclamation*. Oxford University Committee for Archaeology Monograph No. 24, 53–71

UNEP (United Nations Environment Programme) 1987: *Mediterranean Action Plan. Preliminary Report of Blue Plan Scenarios*. Eighth Meeting of National Focal Points for the Blue Plan, 20–2 July 1987, Sophia Antipolis. UNEP/WG 171/3

UNEP (United Nations Environment Programme) 1990: *The State of the Marine Environment* (UNEP Regional Seas Reports and Studies, No. 115)

Vallega, A. 1988: A human geographical approach to semienclosed seas: the Mediterranean case. *Ocean Yearbook* 7, 372–93

Varley, J. (ed.) 1991: *World Nuclear Handbook*. Sutton, Nuclear Engineering International.

Verlaque, C. 1981: Pattern and levels of port industrialization in the western Mediterranean. In, Hoyle, B. S. and Pinder, D. A. (eds) *Cityport Industrialization and Regional Development: Spatial Analysis and Planning Strategies*. Oxford, Pergamon Press, 69–85

Vigarié, A. 1981: Maritime Industrial Development Areas: structural evolution and implications for regional development. In, Hoyle, B. S. and Pinder, D. A. (eds) *Cityport Industrialization and Regional Development: Spatial Analysis and Planning Strategies*. Oxford, Pergamon Press, 23–36

Vreugdehil, C. B. and Wind, H. G. 1987: Framework of analysis and recommendations. In, Wind, H. G. (ed.) *Impact of Sea Level Rise on Society*. Rotterdam, A. A. Balkema, 1–20

Walker, D. 1966: The Late Quaternary history of the Cumberland Lowland. *Phil. Trans. R. Soc.* B 251, 1–210

Walker, H.J. (ed.) 1988: *Artificial Structures and Shorelines*. Dordrecht, Kluwer Academic Publishers

Wright, L. D., Coleman, J. M. and Erickson, M. W. 1974: Analysis of major river systems and their deltas: morphologic and process comparisons (*Technical Report* No. 156. Baton Rouge, Coastal Studies Institute, Louisiana State University, 114 pp.

The Contributors

Baeteman, C.   Belgian Geological Survey, Jennerstraat 13, 1040 Brussels, Belgium, and Earth Technology Institute, Vrije Universiteit Brussel, Pleinlaan 2, 1050 Brussels, Belgium

Cauwenberghe, C. van   Dienst der Kust-Hydrografie, Administratief Centrum, Vrijhauenstraat 3, 8400 Ostend, Belgium

Corre, J.–J.   Institut de Botanique, 163 rue Auguste Broussonet, 34000 Montpellier, France

Dabrio, C. J.   Departmento de Estratigrafía, Faculdad de Ciencias Geológicas, Universidad Complutense, 28040 Madrid, Spain

Goy, J.L.   Departmento de Estratigrafía, Faculdad de Ciencias Geológicas, Universidad Complutense, 28040 Madrid, Spain

Jelgersma, S.   Rijks Geologische Dienst, PO Box 157, Richard Holkade 10, 2033 PZ Haarlem, Netherlands

Lannoy, W. de   Vrije Universiteit Brussel, Pleinlaan 2, 1050 Brussels, Belgium

L'Homer, A.   12 rue de L'Ecole Normale, 45000 Orléans, France. Lately, Bureau de Recherches Géologiques et Minières (BRGM), BP 6009, 45060 Orléans, Cedex 2, France

Malde, J. van   Rijkswaterstraat, Dienst Getijdewateren, PO Box 20907, 2500 EX, The Hague, Netherlands.

Marino, M. G.   Escuela Nacional de Sanidad, Ciudad Universtaria, 28040 Madrid, Spain

Paepe, R.   Belgian Geological Survey, Jennerstraat 13, 1040 Brussels, Belgium, and Earth Techno-

	logy Institute, Vrije Universiteit Brussel, Pleinlaan 2, 1050 Brussels, Belgium
Sestini, G.	Via della Robbia 28, 50 132 Florence, Italy
Shennan, I.	Department of Geography and Environmental Research Centre, Science Laboratories, University of Durham, Durham DH1 3LE, Great Britain
Tooley, M. J.	Department of Geography and Environmental Research Centre, Science Laboratories, University of Durham, Durham DH1 3LE, Great Britain
Zazo, C.	Departmento de Estratigrafía, Faculdad de Ciencias Geológicas, Universidad Complutense, 28040 Madrid, Spain

Author Index

Abdelkader, A. 187
Alestato, J. 12
Andrade, C. F. de 8
Astier, A. 138

Baeteman, C. 14, 62, 63, 223
Bakker, Th.W. 6
Barth, M. 73, 74, 75, 174
Batisse, M. 237, 238, 239
Baudière, A. 158
Bazile, F. 156
Behre, K.-E. 7, 224
Beltagy, A. I. 191
Ben Menachem, A. 187
Bennema, J. 100
Benyon, H. 229
Bertolami, G. 179
Bethemont, J. 223
Bevilacqua, E. 181
Bindschadler, R. A. 73
Bird, E. C. F. 5, 28, 165, 221, 227
Blanc, J. J. 138, 141, 165
Boerjan, P. 68
Bolin, B. 2, 124, 174
Bondesan, M. 179
Boois, H. de xi, 7, 214
Borowkwa, R. K. 12
Borrego, J. 204
Brambati, A. 176, 181
Britton, R. H. 161
Brouwer, F. J. J. 225

Brückner, H. 14
Bruun, P. 113, 115, 190
Burton, R. G. O. 81
Butzer, K. W. 174

Caillaud, A. 145
Carbognin, L. 179, 181, 190
Carter, R. W. G. 8, 27
Catani, G. 190
Cauwenberghe, C. van 60
Cavaleri, L. 179
Cencini, C. 179, 181
Cendrero, A. 16, 24, 232
Ceunynck, R. de 63, 65
Chamley, H. 138
Champion, A. G. 229
Charlier, R. M. 16, 24, 232
Ciabatti, M. 179
Clark, J. A. 4, 75
Coleman, J. M. 184
Colenbrander, H. J. 21, 108, 109
Consiglio Nazionale delle Recerche 180, 181
Corre, J.-J. 13, 166, 172
Couper, A. 236
Coutellier, V. 184, 186
Crane, A. 124
Currie, R. G. 50, 52–3

Dabrio, C. J. 208, 209, 210, 211
Dalcin, R. 181
Davidson, N. C. 221

Author Index

Davies, J. L. 10
Delibrias, C. 235
Depuydt, F. 59, 63
Dijkma, K. S. 220, 221
Donner, J. 12
Donoghue, D. N. M. 85
Doody, P. 222
Doornkamp, J. C. 26
Duboul-Ravazet, C. 156

Eisma, D. 13
El Askary, M. A. 184
Emberger, L. 158
Emery, K. O. 175, 184
Englefield, G. J. H. 10, 235
Erinc, S. 172
Erol, O. 190
Eronen, M. 12
Estienne, P. 158
Evans, G. 8, 172

Fairbridge, R. W. 84, 219
Flemming, N. C. 13
Fontes, J. C. H. 179
Fowler, G. 7
Franco, P. 179
François, F. 145
Frank, N. L. 234
Frihy, O. E. 182, 184, 187

Géhu, J. M. 222
Georgas, D. 172
Giffen, A. E. van 102
Glopper, R. J. de 101
Goby, J. E. 184
Godard, A. 158
Godwin, H. 7, 224, 225
Goemans, T. 22, 26, 118
Göney, S. 172
Gornitz, V. 4, 74, 175
Got, H. 139
Gottschalk, M. K. E. 10, 36
Graff, J. 10
Grau, J.-J. 128, 170
Grenon, M. 237, 238, 239
Greslou, M. 157, 159
Grupiner, A. 11
Gudelis, V. 12, 223

Hageman, B. P. 7
Hamid, S. 174
Hansen, J. 2
Harris, R. 73
Hassan, F. A. 174
Hekstra, G. P. 124, 175
Henrionet, J. 60
Henry, P.-M. 232
Heurteaux, P. 137
Hobbs, A. J. 85
Hoeven, P. C. T. van der 41
Hoffman, J. S. 2, 3, 175, 209, 232
Hofstede, J. 10
Holdgate, M. W. 234
Houghton, J. T. xi, 2, 73
Huddart, D. 225
Husain, S. A. 234
Hutchinson, J. 5, 81
Hydraulics Research Station, (UK) 228

Inman, D. L. 184
Ireland, S. 235

Jelgersma, S. 6, 7, 22, 97, 98, 102, 233
Jenkins, S. A. 184
Jensen, J. 46
Jongman, R. H. G. 21

Kakkuri, J. 12
Karl, T. R. 73
Kerr, R. A. 84
Keyes, D. 73, 84
Khafagy, A. 187
Khashef, A. A. 191
Kidson, C. 8, 10, 223
Königsson, L.-K. 12, 223
Kraft, J. C. 14, 223
Kramer, J. 10
Kraus, N. C. 60, 66
Kruit, C. 137, 139, 140
Kwadijk, J. xi, 7, 214

Laborel, J. 235
Lamb, H. H. 10, 84, 174
Langendoen, E. J. 48
Leatherman, S. P. 113
Lebedeff, S. 74, 175
L'Homer, A. 142, 157, 156, 186, 228

Linden, H. van der 100, 103
Linke, G. 84
Lisitzin, E. 11
Liss, P. 124
Lohrberg, W. 46
Long, D. 11
Louwe Kooijmans, L. P. 7, 224

Malde, J. van 101
Maldonado, A. 128, 129, 131
Manohar, M. 184, 188
Marcoupoulu-Diacontini 190
Marino, M. G. 170
Martin, L. 235; R. 156
Mattana, O. 181
Maymó, J. 130
McIntyre, A. 73
Meininger, P. L. 172
Meistrell, F. J. 210
Menanteau, L. 209
Mercer, J. H. 73
Meulen, F. van der 220
Milliman, J. D. 225, 226, 234
Misdorp, R. 25
Monaco, A. 154
Moor, G. de 58, 60
Mörner, N.-A. 12
Mostaert, F. 63
Moursy, Z. A. 186
Muehe, D. 235
Mullie, W. C. 172

Neftel, A. 73
Neilson, G. 11
Neves, C. F. 235
Newman, W. S. 84
Nijpels, E. H. T. M. ix
Normann, J. O. 12

O'Connor, W. P. 52
Oele, E. 8, 223
Oerlemans, J. 2, 4, 74
Oomkens, E. 139
Osvald, H. 225

Palomäki, M. 12, 28
Paskoff, R. 172, 190
Passant, F. 22

Pattison, J. 226
Pauc, H. 139
Peltier, W. R. 175
Pendon, J. G. 204
Perissoratis, C. 172
Picon, B. 160
Pilkey, O. H. 60
Pirazzoli, P. A. 4, 13, 74, 159, 171, 175, 179, 223, 225
Podlejski, V. D. 161
Polo, M. D. 208, 209
Pons, A. 157
Postma, H. 117
Prescott, V. 28
Primus, J. A. 4, 75
Pullan, R. A. 227

Raper, S. C. B. 2
Raukas, A. 12
Regy, P. 159
Reid, H. F. 11
Richez, G. 137
Robin, G. de Q. 2–4, 111, 175
Roebart, A. J. 26
Rohde, H. 7, 10, 224
Romariz, C. 8
Ronde, J. de 41
Ritsema, A. R. 11
Rossignol-Strick, M. 187
Rossiter, J. R. 21, 84
Ruddiman, W. F. 73
Rudkin, E. H. 226
Rueda, F. 161, 165
Rycroft, M. J. 73

Said, R. 187
Schneider, S. H. 2, 28, 85
Schröder, P. C. 27
Schoorl, H. 113
Schothorst, C. J. 100, 226
Schulz, H. D. 209
Schwartz, M. L. 113, 190, 221, 227
Seale, R. S. 81
Seidel, S. 73, 84
Seró, R. 130
Sestini, A. 172; G. 186, 190, 233
Sharaf El Din, S. H. 186

Author Index

Shennan, I. 8, 10–11, 15, 24, 77, 78, 79, 81, 82, 85, 87, 223, 231, 241
Shuisky, Y. B. 190
Simeoni, C. 179
Sissons, J. B. 11
Sivignon, M. 172
Skertchly, S. B. J. 223
Smith, D. E. xi; S. E. 187
Sneh, A. 186
Sorribes, J. 128, 170
Sproxton, I. W. 24, 87, 231, 241
Stanley, D. J. 184, 186, 226
Steers, J. A. 241
Stewart, R. W. 74
Stournaras, G. 190
Straaten, L. M. J. U. van 96, 97, 138, 165
Suguio, K. 235
Summerhayes, C. P. 188
Surrel, E. 165
Streif, H. 7–8, 10
Switsur, V. R. 228

Takel, R. E. 229
Ters, M. 223
Titus, J. 61, 73, 74, 75, 84, 111
Tooley, M. J. 4, 6, 8, 23, 24, 74, 87, 222, 223, 225, 226, 228, 241
Tousson, O. 186, 187
Townsend, A. R. 229

Tushingham, A. M. 175
Tziavos, C. 14

Vallega, A. 232
Vanhove, N. 68
Varela, J. 128
Varley, J. 236
Veen, C. J. van der 111
Vellinga, P. vii
Verlaque, C. 229, 230
Vernier, E. 138
Vigarie, A. 229
Villain-Gandossi, C. 223
Vreugdenil, C. B. 233

Waalewijn, A. 51
Walker, D. 225; H. J. 227
Warrick, R. A. 2, 4, 74
Weele, P. I. van der 36
Wigley, T. M. L. 2, 174
Wijn-Nielson, A. C. 73
Williamson, I. T. 226
Wind, H. G. 233
Wood, F. B. 73
Woodworth, P. L. 5, 85
Wright, H. L. 60
Wunsch, C. 52

Zagwijn, W. H. 7, 98
Zunica, M. 181

Subject and Place Index

Page numbers in italic indicate a reference to a figure or plate

Aberdeen 80
acque alte 180
Adana 172
Adige 179
Adriatic Sea 13, 25, 172, 225; coast 176–82, *177*, *178*, 188; economic activity, *194*; population density *195*
Aegean Sea 14; coast 172
aerial photographs 161
agriculture 128, 132, 137, 149, 174, 192, 193
Aigues-Mortes 137, 145, Plate 7
Albania 172, *173*, 188, 223
Albères 154, 156, 158, 161, 162
Alfaques 132
Algarve 8, 227
Alicante 206
Alps 174, 176
amphidromic point 48
Amsterdam 36, 37, 48, 51, 95, 103
Antalya 172
Antarctica 4, 75, 175
Antwerp 48, 69–70, 229
Apennines 176
ARC-INFO 85
Arles 138, 167
Atlantic Wall, Belgium 59, Plate 6
Aude 161

Baltic Sea 11–13, 223
Bangladesh 234
barrier beaches 176; islands 97
Bas-Rhône 160
Bays, Abuqir 184, 186, 190; Arta 172; Bengal 234; Morecambe 10
beaches, erosion 58; Holocene 13; nourishment 118
beachrock 13
Beauduc 140, 141, 149, 163
benchmark 51, 60
Bergen an Zee Plate 3
Béziers 167
Biosphere Reserve 161
birds 97; Camargue 161; Ebro delta 128, *130*; Lake Ichkeul 172, *173*; migrating 117, 125, 128, 172; overwintering 172; Shelduck 223–4
Blue Plan 236, *237*, *238*, *239*
bora 179
Brazil 234–5
Bristol Channel 10
Brugges 56, 67
Bruun Rule 113, *115*
Butterwick 80

Cabo Tortosa 129, 132, 134
Cadiz 214

Subject and Place Index

Callantsoog Plate 5
Carmague 137, 158, 159, 160, 166; Basse 157, 160; Haute 157; Moyenne 157; Petite 141, 143, 145, 148, 156, 157: afforestation 148; National Reserve 161, 166; rainfall 158; Regional Park 148, 160; Seawall 148
Canet 154
CFCS viii, 1, 72
CO_2 1, 72, 73, 124, 174, 175, 219, 220; predictions 2, 73, 124; rates of increase 73; scenarios 175
Carnon 149, 156, 163
Caspian Sea 28
Catalonia 132
Cervia 179
Chandler Effect 52
China 235–6
Cie des Salins du Midi 141, 143
La Clape 156, 162
Cleveland County Council 229, 231
cliffs 12
coastal barriers 94, 139, 176, 184; Mediterranean 139, 147; Netherlands 96, 97; types 139
coastal dynamics 215
coastal engineering 179, 214
coastal hazards 27
Coastal Hydrographic Office (Belgium) 61
coastal lowlands, Adriatic 172, 176–82, *177*; Belgium 57, 223; Cadiz 204–15, 227; Eastern England 77, 223; Europe 6–16; Greece 172; Huelva 204–15, 227; Mediterranean 127, *171*; Netherlands 94–8, 103–5; North Africa 172; Romagna 179, 180; Teesside 24, 229; Veneto 179, 190, 225: distribution 9; future 218–45; landfill sites 231; land-use changes 103–4, 227–32; planning strategies 26–7, 228–32; population increase 105; population trends 191; sediments 6–7, 13–14, 61, 95, *96*, 223
coastal zone management 191, 240, 241
coastline changes 113, 180; advance 79, 113, 161, 180, 181, 187, *207*; monitoring 143; rates 161, *162*, *163*, 164, 180; retreat 28, 113, 161, 180, 187, 188, *207*
coasts, Albania 172, *173*, 188, 223; Atlantic 8–11; Algarve 8, 227; Baltic 11–12; Belgium 10; Catalonia 126; Denmark 10; Durham 228; France 10; Friuli 180; German 8, 10, 224; Greece 172; Irish Sea 8; Mediterranean 12–16, 139–40, 143–5; Netherlands 10, 95–7; North Sea 8; Norway 10, 11; Poland 12; Tunisia 172, *173*; Turkey 172, *173*
colliery waste 228
Comacchio 179, 191
Commission on the Coastal Environment 227
consolidation of sediments 5–6, 81, 98, 100–1
cost-benefit analysis 87, 118–21
La Courbe-à-la-Mer 145, 147
crescendo events 219
Cukorova plain 172
currents 159, 180, 184; ebb 204, 210; flood 204, 210; longshore 186

dams, Aswan High 176; on Ebro 128, 170, 228; impact on sediment supply 170, 190; Mequinenza 128, 228; on Rhône 138
deforestation 181, 209
De Lage Moere 56
Delfzijl 42, *43*, 46, 49, 50
delta, Arno 229; Axios 172; Büyük Menderes 172; Ebro 13, 14, 15, 127, 128–34, 170, 186, 209 238; Europe 9; Evros 172; Goksu 172; Isonzo 176; Kücük Menderes 13; Mediterranean 126, *171*; Nile 13, 176, 182–8, 223, 225, 228, 233; Po 13, 15, 172, 176 *178*, 181, 223, 225, 228, 233; Rhône 13, 15, 127, 136–50, 154, 157, 170, 186, 223, 225, 228: advance 181; lobate 181, 186; shorelines 190; subsidence rates 225
De Moeren 56
Den Helder 38, 41, 45, 46, 49, 50, 51–2
La Dent 147

Subject and Place Index 259

De Panne 59, 63, 68
dikes, earth 25; river 148, 176, 181; polder 103; Hondsbossche 97, 113, 114, 116–17; Sea 25, 61, 86, 94, 102, 111, 236
Doñana National Park 206, 213, 214
drainage 81, 100, 102–3, 226
dredging 181, 182
drinking water 21, 105
dunes 6, 7, 15, 21, 25, 56, 140, 156, 227; Adriatic coast 176, 179, 181; Belgium 56, 58, 60, 63–5, *65*, 66–7, 69, 227; Egypt 182, 184, 187; Gulf of Lions 156, 159, 160, 165, 166; Netherlands 94, 97, 101, 103, 105, 113–16, Plates 4 & 5; Lincolnshire 81, 227; Poland 12; Schouwen *115*: artificial 160; distribution *18*, 140, *222*; drinking water from 105; erosion *114*; formation 140; ground water 105; loss from development 69; management 67, 227; natural sea defence 103; Old Dunes 7, 63, 65, 97; planting 140, 160; prehistoric occupation 101; reconstruction 148; tourism 105; Young Dunes 7, 64, 97, 103, *104*
Dungeness 228, Plate 2

earthquake 11, 187
Easter Scheldt *109*, 117
Ebro delta, fisheries 128; impacts of sea-level rise 131–4; lobe development *129*; migratory birds *130*; population 128; sediments 228
Egypt 172, 174, 176, 191
Emden 42
Emilia-Romagna region 172
La Encanizada 128, 131
English Channel 10, 11, 67
Environmental Protection Agency (EPA, USA) 74
Ephesus 13
Espiguette 147, 156, 157, 163
Estuary, Dovey 225; Europe *9*; Huelva 208; Meuse 96, 97, 101, 111; Odiel-Tinto 207, 212; Rhine 96, 97, 101, 111; Ribble 10, 224; Scheldt 69, 86, 96, 97, 111; Severn 10; Tees 231; Thames 11
Etang de Berre 229
Etang de la Ville Plate 7
European Coastal Charter 241
European Community xi, 232, 241; Commission 23, 192
European Union for Dune Conservation (EUCD) 222
European Workshop on Interrelated Bioclimactic and Land-use changes ix, 23, 218, 242–5
Europoort 105

faults 11
Federal Emergency Management Agency (FEMA: USA) 27
Fenlands 5–6, 7, 8, 72, 223; reclamations *82*
Ferrara 179
Firth of Forth 225; valley 11
fisheries 97, 117, 128, 132, 159, 192, 193, 197, 206; fishponds 236
Flandrian transgression 207
Flood Action Plan 234
flood alleviation 87
Flood Insurance Rate Maps (USA) 27
flooding 81, 94, 160, 174; probability of exceedence 111, *112*
floodmarks 36
forebulge 98
forest fires 125
Fos 137, 141, 149, 154, 160, 164, 172, 229
Fosdyke 87
France 25; Rhône delta 136–50; Gulf of Lions 154–65
French Electricity Board 138, 150
Friesland 116
Freiston Marsh 80

Gard 163
geographic information system 24, 85, 87, *88*, 231, 241
geoid 27
geomorphology 205

George Marshall Institute, Washington, DC 219
Germany 7–8, 10, 12, 22, 25, 105; Ruhr 105; sea embankments 10, 22; subsidence 12
Gibraltar Point 86
La Gracieuse 140, 141, 147
La Grande Motte 149, 156, 161, 163, 164
Grau 159; Grau de la Dent 145, 147; Le Grau-du-Roi 149
Greece 172, 188, 223
Greenland 4, 175
Groningen 116
groundwater 19, 21, 108; brackish 108; extraction 100, 179; methane-bearing 179; polluted 105; recharge 26, 174
groynes 15, 81, 140
Gulf, Bothnia 12, 28; Cadiz 204, 206, 207, 213, 215; Fos 141, 159, 163; Gdansk 12; Leucate 156; Lions 136, 139, 153–67, 170, 223, 227; Narbonne 156; Thermaikos 172
Gulf of Lions, climate 158; development 160–1; geography 154; geology 154; geomorphology 154–7; tides 158; waves 159

Haarlem 21
Haarlemmermeer 103
The Hague 95
Hamburg 22, 25, 229
Harbour, Abuqir 182, 190, 193; Alexandria-Dakheila 182, 190, 193; Antwerp 69–70, 97, 109; Buso 179; Camargue 137, 147; Cervia 179; Cesenatico 179; Chioggia 179, 191, 192; Corsini 179, 192; Dürres 172; Fos 137, 142, 149, 229; Garibaldi 179; Grado 179, 192; Huelva 212; Leghorn 229; Livorno 229; Malamocco 179; IJmuiden 116; Marghera 181, 192, 225; Port-La-Nouvelle 159; Ravenna 172, 179, 181, 192; Rotterdam 105, 111, 121; Romney Marsh 228; Said 186, 190; Sete 159; Venice 192; Vlone 172: impact of sea-level rise 118; industrialization 230, MIDAS 229, 230; Roman 209; yachting 137

Harlingen 38, 43, 44, 46, 49
Hérault 156, 160, 161, 165
heritage sites 192, Plate 7
HMS Challenger 219
Holme Fen 5–6
Holocene 219; sediments 94, 98, 101; transgression 180
Hondsbossche dike 97, 113, 114, 116–17, Plate 4
Hook of Holland 36, 97, 112
Hudde stones 37, 51
Huelva 214
hydroelectricity 138, 181

ice caps, Antarctica 4, 75, 175; British 98; Greenland 4, 175; Scandinavian 98; Scottish 98; West Antarctica 175
IGN 161
IJmuiden 116
IJsselmer 103, 108, 117
Immingham 80
Intergovernmental Panel on Climate Change xi, 2, 221, 234, 236, 240, 241
International Geographical Union 227
International Organizations, EPOCH xi, 241; EC xi, 232, 241; Global Change Programme ix; ICSU viii, 174, 175, 193; IGU 227; INQUA xi; IPCC xi, 2, 222, 234, 236, 240, 241; UNEP viii, 5, 131, 174, 175, 193, 227; WMO viii, 124, 174, 175, 193; World Climate Research Programme ix; Villach Conference viii, 2, 124–5
Irish Sea 8, 10, 223
Isonzo 179
Isla de Buda 131, 132
isostasy 10, 12, 98–9
Israel 186
Izmir 172

Katwijk 36, 37, 38, 48
Kent 87
Koksijde 58

lagoons 147, 154, 172, 176, 187, 223; Adriatic 172, 176, 179; Albania 223; Brazil 235; Berre 137; Caorle 176; Ebro delta 128; Egypt 172; Gado

180; Grando-Marano 176; Ireland 8; Netherlands *96*, 98; Portugal 8; Vaccares 141; Venice 180, 181: salinity 138; world distribution *233*
Landsat Thematic Mapper (TM) 85
land-use, Belgium 67–9; Netherlands 101–9; Tees 231: planning 23
Lake IJssel 21
Languedoc 160
Leucate 156, 161
Leverton Marsh 86
Libeccio 179
Lincolnshire 15, 80, 227; altitudes *89*; coast 84, 227; County Council 81; car ownership 90; GIS map output *89*, *90*; industry *90*; marshes 226; sea-levels *78*, *79*
liquefaction 66
Lithuania 12
Little Ice Age 10, 181
Lowestoft 80

Mablethorpe 81, 227
Madagascar 236
malaria 125
Maldives 236
Marathon Bay 14, 223
marine inundation 25; transgression 7
Maritime Industrial Development Areas (MIDAS) 229, 230
Marseille 137, 141, 170, 229
marshes 206, 214; Adriatic coast 176; Camargue 157; Nile delta 187; Pontine 223; Spain 206, 214; Teesside 231; world distribution *233*
mediaeval warm period 10
Mediterranean Sea, 13–14, 15, 28, 127, 154, 170 *171*, 197–98, 206, 225, 232; catchment *14*, *171*; coastline 236; scenery 125–7; tectonic activity 13; tourism 232, 240; useful coast 236; water depths 13: basin, population *239*; petroleum refineries *237*; power stations *238*
Mediterranean Action Plan 227, 236, *237*, *238*, *239*, 241
MED-POL 131
Meuse Waterway 112

Miami 240
Middelkerke 67
Ministry of Transport and Public Works (Netherlands) 120–1
mistral 137
Monaco 158
Monfalcone 176
Montpellier 158, 167
Morecambe Bay 10
morphosedimentology 206, 207, 215, 223
mosquito 159

Narbonne 156, 167, 170
National Company for the Rhône 150
National Flood Insurance Program 27
National Geographic Institute, France (IGN) 159
Natural Environment Research Council (NERC) 83
Nature Conservancy Council (NCC) 86; reserves 97, 132
NATO Advanced Workshop 26
neolithic farmers 157
neotectonics 225
Netherlands 6, 7, 21, 22, 25, 36–54, 94–121, 126, 226; area below sea-level *106*–*7*; dunes 7, 25, 97; economic value of lowlands 120; geology *95*, *96*; grand lowering 226; risk hazards *119*; tectonic subsidence 225
NGF 157
Nieuwport 57, 58, 61, 67
Nieuwe Waterweg 97, 112
Nile delta 182–8, 193
Nimes 167
Noordwijkerhout ix, 23, 109, 214, 218, 220, 242; Recommendations 242–5
Normaal Amsterdams Peil (NAP) 36–7, 39, 50–1, 57, 106, 112; connection to NN 51
Normal Null (NN) 50, 224
North Sea 10, 11, 16, *18*, *19*, *20*, 21, 22, 48, 67, 95, 96, 97, 111, 117, 170, 225; coasts 84; industry *17*, *19*; nuclear power stations *19*; oil and gas fields *20*; population *17*; Quaternary history 8, 223; tectonic downwarping 98, 225; Tertiary *99*, 225

North Shields 80
Northumberland 87
North Wales 10

oil and gas fields *20*, 105, *110*
Oostduinkerke 59
Ordnance Datum (OD) 5, 80, 229
Ostend 60-1, 67

peat, basal 63; ombrogenous 6, 224; oxidation 98; raised 6, 225; wastage 81
perimarine zone 102
Permanent Service for Mean Sea Level 4, 80
Perpignan 167
Petten 36
planning, coastal zone 26-7, 191, 228-32
Pleistocene 61, 94, 98, 101
Po delta 190, 192, 193, 226; distributaries 180, 181; Pila 180, 181, Plate 8; Primaro 180; Reno 180; Volano 180: evolution 180-1; rates of growth 228; sediment load 228; subsidence rates 226
Po river 138, 180, 181, 236
polders 103, 105
pollution 139, 193, 213
population, in Mediterranean coastal lowlands 172; in north-west European coastal lowlands *17*, *19*; Albania 172; Belgium 67; Camargue 137; Ebro delta 128; north-east Adriatic coast *195*; Nile delta *196*
Portugal 25
ports, see harbours
Potsdam 50
power stations, nuclear 16, *19*, 22, 139, 229; decommissioning 22; Mediterranean 236, *238*; radionuclides from 139; thermal 236; waste 139
prehistoric settlement 101
Proudman Oceanographic Laboratory 4
Punta Umbria 209, 212-13
Pyrenees 161, 174

rainfall 174, 180-1
Ravenna 172, 179, 181

reclamation 80, 81; Netherlands 100; Tees 231; Wash 81, *82*
recreation, Belgian coast 67-9; Dutch coast 108-9; Mediterranean 126
Rhône de St. Ferréol 141, 142
Rhône delta, Bouches-du-Rhône 163; climate 137-8; distributaries 228; impacts of sea-level rise 147-50; map evidence 143; rates of erosion and accretion 143, *144*, 145; recession 228; sediment load 228; subsidence 142-3; water management 166
Ribble estuary 10, 224
rice cultivation 128, 137, 160, 192, 227
Riège forest 161
Riga 12
Rimini 176, 181
Rio de Janeiro 235
risk, factors 23-7; hazards map 119; zones 111
River Achelaos 172; Adige 176, 180; Brenta 176, 180; Buna 172; Ceyan 172; Drin 172; Durrance 138; Ebro 128, 170, 236; Fiumi Uniti 176, 181; Ganges 234; Guadalete 209, 213; Guadalquivir 206, 209, 213, 214; Guadiana 207, 208, 209; IJssel 98, 103; Isonzo 176, 180; Jucar 236; Livenza 181; Matia 172; Medjerda 172; Meuse 21, 25, 41, 94, 98, 103, 109, 111; Nile 138, 172; Odiel 207, 209, 212; Pearl 235; Piave 176, 180, 181; Piedras 209, 210; Po 138, 180, 181, 236; Reno 176, 181; Rhine 21, 25, 41, 94, 98, 103, 105, 109, 111: Old Rhine 96, 116; Rhône 136, 138, 170, 236: Grand Rhône 136, 138, 142, 143, 147, 149; Petit Rhône 137, 138, 145, 147, 156, 157; Ribble 10; Scheldt 94, 98, 109, 111; Sile 176, 181; Semeni 172; Seyhan 172; Skumbini 172; Tagliamento 176; Tech 161; Tees 231; Têt 161; Tinto 207, 209, 212; Vistula 12; Vojussa 172; Welland 86; Witham 86; Yellow 235; Yser 56, 62, 65; Zhujiang 235
river, embankments 148, 176, 181; floods 180; discharge *108*, 109, 117, 138,

Subject and Place Index   263

180, 181, 187; sediment 128, 138, 163–4, 170
Romagna 179, 181, 190
Romney Marsh 228
Rotterdam 21, 95, 97, 105, 112, 229; Botlek Scheme 229; oil refineries 105; port 105, 111, 121; waterway 21, 109

Saintes-Maries-de-la-Mere 137, 141, 143, 147, 156, 225
Salin de Giraud 137, 141, 149
salinization 125, 193
Salins d'Aigues-Mortes 145, Plate
salt extraction 137; making 159, 224, 226, 227
salt marshes 12, 18, 20; distribution *18, 221*; creeks 101; Belgium 58; Leverton 86; Poland 12; Rhône 149; Wash 80, 83
salt pans 138, 147, 149, 172, 206; Aigues-Mortes 145, Plate
saltwater intrusion 105, 111, 117, 121, 125, 149
sand, budget 181; dredging 182; transport 212
Scandinavia 12
Scheveningen 36
Schistosomiasis 125
Schouwen 113, *115*
scirocco 179
Sea, Adriatic 13, 25, 172, 225; Aegean 13; Baltic 11–12, 223; Caspian 28; Irish Sea 223; Mediterranean 12–14, 15, 28, 125–7, 206, 225, 232; North 8, 10, 21, 22, 36, 75, 84, 103, 231
sea defence 81, 121, 190; costs 121, 190; design height 111
sea embankments 25, 86, 236; Bangladesh 234; Belgium 59; Camargue 148, 160; Netherlands 103, 111; Wash *80*, 81: damage to 25; earth 25; inadequate 103
sea level, changes Brazil 235; the Netherlands 98, 102; Wash *78, 79*
sea level, mean 15, 22, 36–55
sea level, rise, accelerations 85, 174, 188; Eemian 7; Holocene 78–9: environmental response 21–2, 76–81;

legal implications 28; man's response 81, 84–5; management 26; monitoring 85–6; predictions 2–3, 4, 109, 124–5, 175; rates 3, 4, 7–8, 76, 79, 80, 83, 101, 174–5; scenarios 74–6, 83; Wash 75, EPA 74–5; shoreline retreat 66; thermal expansion 75
sea-level rise, impacts on, cities 191, 214; developing countries 233–6; Dutch coast 94–121; economy 118–21; ecosystems 117, 125; Gulf of Cadiz 214; Gulf of Lions coast 166; East Mediterranean 170–98; West Mediterranean 125–6, 170–2
SeaRISE 73
seawater 52
sediment, supply 128, 138–9, 163–5, 170, 181, 187; compaction 179, 226
seiches 111, 180
seismicity 11
Sète 156, 159, 172
Sheerness 80
shingle 228
shoreline, *see also* coastline Belgium 58; Netherlands 95–7: change 143; erosion 113, 177; forecast 113, 140; prograding 139, 140, 143, 177; rates of change 143; retreat 66, 139–40, 143
Sicily 174
Silifke 172
Skegness 81
snowline 174
soil, erosion 174; salinity 125, 174, 193
solar radiation 74
Southend 80, 81
Spain 25, 170, 174, 227; zero datum 206
spits 187, Beauduc 140, 141, 147; Doñana 213; L'Espiguette 145; La Gracieuse 140, 141, 147; Guadalquivir 209; El Rompido 209, 210–12, *211*, 213; Punta Umbria 209, 212–13; Vadelagrana 213
SPOT 85
Sri Lanka 236
storm penetration maps 191
storm surges 10, 23, 186, 224; Baltic Sea 11–12; North Sea 8, 10, 21, 22, 36, 75, 84, 103, 231: in 1953, 36, 97, 103,

111, 229–30; in 1978, 75; in 1981, 39; in 1983, 75; in 1990 Plate 3 Mediterranean 141, 145, 167: in 1982, 145 167; in 1985, 141; on Teesside 231, 232; on Dutch coast 37, 39, 96, 112; increasing frequency 111–12; increasing altitude 112; response 224; barrier 112, 117, 121
Storrega slide 11
Strait of Dover 11
subsidence 25, 175, 225; Adriatic 179, 190; Germany 11; Nile delta 184, 187; Mediterranean 13; North Sea 98, 100; Netherlands 98; rates 12, 76, 98, 142–3, 179, 225; Rhône delta 142–3, 149; tectonic 25, 98, 99, 179
swash bars 210
Sweden 12

Tab's Head 86
La Tancada 128, 131
TWA (Belgian zero datum) 56, 58, 69
tectonic, activity 13; subsidence 25, 98, 99, 179
Teesside 229, coastal lowlands *231*, industry 229, population 231, tank farm Plate 1; MIDAS 229; nuclear power station Plate 1; storm surges 231
temperature, Gulf of Lions 158; biogeographical consequences 167; oscillations 174; predictions 2, 73, 124–5, 174, 219; rise, effects on pathogenic organisms 125; sea 158
tendencies of sea-level movement 78
Texel 96; Texelstroom 25
Thames barrage 25; estuary 11
Thau Lake 156
thermal expansion 4, 25, 75
Thessaloniki 172
tidal, current 97, 210; creek 223; curve 117; cycle 52; flats 63, 96, 116–17, 206, 209, 210, 211, 223, 224, 231; inlet 97; oscillations 41; range 10, 141
tide, changes in 111, 225: variations 11: Amsterdam *48*; Baltic 11; Bristol Channel 10; English Channel 10; Mediterranean 12–13, 141;

Netherlands 37–54, 96, 111; ebb 204, 210; flood 204, 210; High Astronomical 75, 232; polar 52
tide gauges 4–6, 16, 37, 80–1, 100: longest records 5; Belgium 60–1; Delfzijl 42, *43*, 46, 49; Den Helder *45*, 46, 49; Eastern England 80–1; Germany 53; Gulf of Gabes 180; Gulf of Lions 158; Harlingen *44*, 46, 49; Marseille 141; Netherlands *40*; Nile delta 184, 186; Spain 204
Torre de Zalabar 209
tourism 116, 181, 197; Adriatic coast 172, 181, 192; Camargue 137; Gulf of Lions 159, 160; Mediterranean 125, 232; Rhône Delta 137
tourist development 129; resorts 128
Trabucador isthmus 129, 132
Trieste 176, 180
tsunamis 11
Tunisia 172, 188
turbary 224
Turkey 172, *173*, 174, 188, 223

UNEP viii, 5, 131, 174, 175, 193, 227, 241
USA 241; National Research Council 2; Environmental Protection Agency 74
uplift 11, 225; Bothnian Gulf 12; Norway 11; Scotland 11: coastal effects 12; rates 12
Utrecht 226

Vaccares 141, 142; Lake 157, 225
vermetids 13, 235
Vendres 156, 163; Lake 161
Veneto region 172, 179
Venice 25, 172, 176, 179, 192, 226; lagoon 176, 180, 193; lowlands 225
Veurne 67
Villach conference viii, 2, 124–5, 175
volcanic aerosols 74

Wadden Sea 15, 25, 45, 108–9, *108*, 111; fisheries 117; islands 96, 97, 116–17; migrating birds 117
Walland Marsh 228

Subject and Place Index 265

The Wash 6, 8, 15, 23, 72–91, 223; flood defence levels 75; reclamations *82*; sea-level rise scenarios 75; storm surges 75:
water levels, extreme, Adriatic 180; Belgium 58–9; Netherlands *39*, 111; Nile delta 186; Tees estuary 231, 232
water management 117, 160, 166
water quality 125
waves 10, 159, 166, 186, 210; Adriatic 179; Gulf of Lions 159; Spain 204: action 96; energy 165, 179, 204; fetch 184; height 159, 165, 204; increase in height 165; probability 159; refraction 179, 210; storm 10, 61, 184; swells 184
West Antarctic Ice Sheet 73, 175
Wester Scheldt 48, 69, *109*, 111

wetlands 15, 109, 125, 161, 172, 174, 227, 231; Algarve 227; Camargue 160–1, 227; Ebro delta 227; Egypt 182; Guadalquivir 215; Mediterranean, extent 227; Rhône delta 227: reserves 109, 172, 214
windbreaks 140
winds 158; Adriatic 178; Gulf of Lions 158: bora 179; mistral 137; scirocco 179
World Bank 234
World Meteorological Organisation viii, 124
world population 3
würten 102, 224

Zeebrugge 61, 67
Zuider Zee 36, 48, 101, 117; closure 116

Related Titles: List of IBG Special Publications

1. Land Use and Resource: Studies in Applied Geography (A Memorial to Dudley Stamp)
2. A Geomorphological Study of Post-Glacial Uplift
   John T. Andrews
3. Slopes: Form and Process
   D. Brunsden
4. Polar Geomorphology
   R. J. Price and D. E. Sugden
5. Social Patterns in Cities
   B. D. Clark and M. B. Gleave
6. Fluvial Processes in Instrument Watersheds: Studies of Small Watersheds in the British Isles
   K. J. Gregory and D. E. Walling
7. Progress in Geomorphology: Papers in Honour of David L. Linton
   Edited by E. H. Brown and R. S. Waters
8. Inter-regional Migration in Tropical Africa
   I. Masser and W. T. S. Gould with the assistance of A. D. Goddard
9. Agrarian Landscape Terms: A Glossary for Historial Geography
   I. H. Adams
10. Change in the Countryside: Essays on Rural England 1500–1900
    Edited by H. S. A. Fox and R. A. Butlin
11. The Shaping of Southern England
    Edited by David K. C. Jones
12. Social Interaction and Ethnic Segregation
    Edited by Peter Jackson and Susan J. Smith
13. The Urban Landscape: Historical Development and Management (Papers by M. R. G. Conzen)
    Edited by J. W. R. Whitehead
14. The Future for the City Centre
    Edited by R. L. Davies and A. G. Champion
15. Redundant Spaces in Cities and Regions? Studies in Industrial Decline and Social Change
    Edited by J. Anderson, S. Duncan and R. Hudson

16 Shorelines and Isostasy
    Edited by D. E. Smith and A. G. Dawson
17 Residential Segregation, the State and Constitutional Confict in American Urban Areas
    R. J. Johnston
18 River Channels: Environment and Process
    Edited by Keith Richards
19 Technical Change and Industrial Policy
    Edited by Keith Chapman and Graham Humphrys
20 Sea-level Changes
    Edited by Michael J. Tooley and Ian Shennan
21 The Changing Face of Cities: A Study of Development Cycles and Urban Form
    J. W. R. Whitehand
22 Population and Disaster
    Edited by John I. Clarke, Peter Curson, S. L. Kayasha and Prithvish Nag
23 Rural Change in Tropical Africa: From Colonies to Nation-States
    David Siddle and Kenneth Swindell
24 Teaching Geography in Higher Education
    John R. Gold, Alan Jenkins, Roger Lee, Janice Monk, Judith Riley, Ifan Shepherd and David Unwin
25 Wetlands: A Threatened Landscape
    Edited by Michael Williams
26 The Making of the Urban Landscape
    J. W. R. Whitehand
27 Impacts of Sea-level Rise on European Coastal Lowlands
    Edited by M. J. Tooley and S. Jelgersma

Also published by Blackwell Publishers for the IBG

Atlas of Drought in Britain
Edited by J. C. Doornkamp and K. J. Gregory